普通高等教育电子信息类专业“十三五”规划教材

基于FPGA的电子系统设计

殷卫真 编著

内容简介

本书讨论采用 VHDL(超高速硬件描述语言)与 FPGA(现场可编程门阵列)平台设计电子系统。期望从关注电子设计多姿多彩的需求与形式,到归化式的基于平台的设计验证,再到针对需求的优化实现,最终从满足需求的角度来阐释电子系统设计。书中设立系列主题,带入典型的设计,给出了综合案例,及多种方法的设计、仿真与测试。与此同时,通过注释、点评与说明介绍知识技巧,引导读者了解基于 FPGA 的电子系统设计理念、风格、标准、规范。教材试图向 CPU 及其接口与通信系统设计方面延伸,扩展设计基础。本书介绍的 VHDL 语言的应用超越了通常可综合的 VHDL 编写,丰富了设计描述,采用第三方仿真软件 Modelsim 方式进行 Testbench 等的仿真,可以提高测试共享,提高移植性、仿真效率和仿真复杂度。介绍的在线工具测试验证等手段,可提高设计、验证、测试的功效,降低开发成本。本书采用不同风格和层次介绍设计特点的方式,主要面向电子信息工程专业与计算机专业本科生编写,也可作为研究生的设计参考书,以及电子设计专业人员的设计参考书。

图书在版编目(CIP)数据

基于 FPGA 的电子系统设计/殷卫真编著. —西安:西安交通大学出版社,2016.8(2019.10 重印)

ISBN 978-7-5605-8867-4

Ⅰ.基… Ⅱ.①殷… Ⅲ.①电子系统-系统设计 Ⅳ.①TN02

中国版本图书馆 CIP 数据核字(2016)第 186770 号

书　　名 基于 FPGA 的电子系统设计
编　　著 殷卫真
责任编辑 毛　帆

出版发行 西安交通大学出版社
(西安市兴庆南路 1 号　邮政编码 710049)
网　　址 http://www.xjtupress.com
电　　话 (029)82668357　82667874(发行中心)
传　　真 (029)82668315(总编办)
(029)82668280
印　　刷 西安日报社印务中心

开　　本 787 mm×1092 mm　1/16　**印　　张** 25　**字　　数** 617 千字
版次印次 2016 年 12 月第 1 版　2019 年 10 月第 2 次印刷
书　　号 ISBN 978 - 7 - 5605 - 8867 - 4
定　　价 59.00 元

如发现印装质量问题,请与本社发行中心联系、调换。
订购热线:(029)82665248　(029)82665249
投稿热线:(029)82668818　QQ:354528639
电子信箱:lg_book@163.com

序

应用型本科教学“规划教材”编写的活动，促使我们思考如何突破传统的模式来突出“应用”这两个字。对应用型本人的理解是，引导学生在应用过程中学习，并通过学习掌握知识的精髓进而具有应用能力。因而在本教材的编写过程中，本人尝试突出思想与方法，淡化知识的方式，试图将讲解知识变为知识路径的导引，提高效率与受益面。这种方式需要以下 3 个前提：

1. 教师要积极参与网络多媒体等资源的建设与应用

随着网络的普及和发展，很多书籍的知识在网络上可以迅速地搜索到，借助网络和各种多媒体等工具（如电子课件、电子微课、视频等）学习很有效率和效果。本人尝试编写书籍时将知识淡化，使学生在整个学习过程中依靠网络、电子手段与应用更紧密结合，所以教师要不断地跟踪新技术，积极参与网络多媒体等资源的建设与应用，为这种应用学习提供有力支撑。

2. 教学方式方法的改进

我们在卓越工程师计划的研究探索中建立的新型的学习方式，也突出了加强在过程与应用中学习。为了提高效率，我们建立“翻转”课堂，将教学实验室基于平台的设计与学生采用的“口袋实验室”相结合，把一些知识的学习甚至实验放在课下，课上增加交流讨论、检查答辩和点评的内容，所以本书有一些手册的特点，不仅适合学习，同时非常适合在应用中查阅。

3. 书籍适合不同程度的学生学习

书中每一种设计都追求不同的设计风格和设计层次的表述，对于行为级算法级的描述，可以针对没有专业基础的低年级同学。对于 RTL（寄存器传输级）追求设计优化，学生至少有一定的程序设计基础和数字电路基础。对于复杂电子系统设计，学生最好具有计算机组成原理、微机接口原理、数字信号处理、通信系统原理、ASIC 设计等专业知识。

前　言

1. 再论 FPGA 设计学习

现代电子系统设计的发展与当前大信息大工程的趋势密切相关，被喻为数字系统设计三大基石的核心技术 FPGA（Field Programmable Gate Array，现场可编程门阵列）、DSP（Digital Signal Processor，数字信号处理器）和 CPU（Central Processing Unit，中央处理器单元）中，FPGA 的发展尤其引人瞩目。

FPGA 从最初的传统数字逻辑设计的硬件平台，已经发展成可以集 DSP、CPU、计算机接口与通信系统等跨领域跨行业设计应用的载体，其更大规模、更高性能器件的不断推出，加上它灵活和可顺序又特有的并行方式，使它不仅可能在更低层级实现一些通用 DSP 与 CPU 及其嵌入系统的功能与性能，满足高层次的需求，而且可以表现出非常独特的特点。FPGA 甚至是一些通用 DSP、通用 CPU 与 ASIC（Applications Specific Integrated Circuit，专用集成电路）先期开发与验证的平台，可以有更多的知识产权。FPGA 的集成化反复可编程、可扩展性、可维护性，是保证产品生存周期竞争力的有力武器。利用 EDA 等设计工具，可以自顶向下设计，建立协同设计，提高设计环节的反馈与测试验证的能力，不断仿真后再反复编程设计实现，使设计更生态，使产品在快速面世的同时有好的质量。

一方面，FPGA 依赖工具、工艺、平台支持的部分更具特色，是 FPGA 厂家提供的设计成果，享用这些资源带给设计独特的发展。另一方面，利用与工艺、平台无关的相对独立、灵活、标准的硬件描述语言（HDL）开发，使设计更标准化、可移植、可复用等，有更大的适应与发展的可能。

在教学上，数字电路设计、计算机组成原理、微机原理与接口、计算机体系结构、通信系统原理和数字信号处理等原本相对独立的课程都可进一步向以 FPGA 为核心的电子系统设计渗透联系与融合。FPGA 作为可以实现贯通教育的平台，在教学甚至科研等方面很有意义，因为更低层级的实现方式伸展出的高端设计意味着学生可以对本质有更深刻的理解，可以有更宽广的适应能力和发展能力。

FPGA 编程对可综合的追求，使得优秀的 FPGA 设计者偏爱简单的语言描述，初学者会误认为 FPGA 开发简单，实际学好 FPGA 是比较难的。一般计算机语言都是顺序语句，而描述 FPGA 设计的 HDL 既有顺序语句，又有并行语句，与其他硬件开发不同的还有，比如单片机侧重软件编程，而 FPGA 侧重软硬件可编程并优化，所以 FPGA 开发是很有力度和难度的，值得花时间研究。

现在人人都在讲设计理念，故然，没有理念的设计味同嚼蜡，无灵魂、无生命，但是没有支撑的设计理念则如镜中月水中花。FPGA 可以开发出真实的生机勃勃的有着先进理念面对应用和发展的电子系统，换言之，学习开发 FPGA 可以从技术到原则到理念。当你有一个精妙的理念，通过 FPGA 设计实现了，学习 FPGA 的目标就达到了。

2. 平台和语言选择

FPGA 最常见的设计平台对应着 Xilinx 和 Altera 两个厂家。采用 HDL 进行开发是

FPGA研究的主要手段，VHDL（Very-High-Speed Integrated Circuit Hardware Description Language，非常高速的硬件描述语言）与 Verilog 作为 Ieee 标准，均有大量的应用者，拥有许多资源。系统 C 目前是比较新的描述语言，但是相应的设计资源比 VHDL 与 Verilog 少。对于许多初学者倍感纠结的是采用哪一个厂家，采用何种 HDL，面对两种语言、两个厂家众说纷纭，各执一词。最初国内高校采用 VHDL、Altera 的学校较多，近些年高校采用 Verilog 和 Xilinx 开发均有增加之势。对于 FPGA 开发的专业工作者，无疑都应该掌握，否则就失去了一片资源，一半信息世界，一种根据需求与成本等因素对综合实现平台的选择。

自顶向下的设计从开发资源丰富、利用方便的平台开始，不断求精；就设计而言，我们始于思想，继而方法论，思想不断进步，方法不断突破，实践不断创新；就教学而言，我们主张举一反三。Xilinx 和 Altera 类似平台触类旁通，VHDL 与 Verilog 等，类似语言可混合设计。本书举的“一”是采用 Altera、VHDL。

3. FPGA 设计说明与要求

3.1 基于 FPGA 的电子系统设计的要求

总体要求：依据设计理念、理论、原则，甚至审美直觉审视、阐释设计，实现可优化的工程设计与系统设计；设计功能明确，性能优越、创意新颖，方法技术创新、有应用价值和前景。

(1)工程层面要求：

①提供“产品”原型，有 DFX 的考虑；

②要从需求规律与产品进化角度提出多种方案、多种方法、多种实现；

③充分利用实验室现有的资源和厂家提供的样片、评估板，关心成本；

④撰写完善的系统规格说明书等设计文档；

⑤有“产品”技术规范，数据手册，“产品”演示视频等。

(2)技术层面部分：

①具有方便、实用、美观的人机界面与交互；

②性能好、功能丰富，闭环控制；

③实现与 PC 通信；

④自主设计通信协议；

⑤嵌入微处理器核与任意类型的操作系统；

⑥深入探讨原理和算法的应用；

⑦深入探讨设计实现及其应用；

⑧自由发挥。

3.2 设计与设计约束

设计要考虑的设计约束，主要涉及两个方面，一是设计资源，二是设计优化。

本书以 HDL 为主，设计与平台无关，但是当探讨具体的综合与测试验证就不可能不涉及到设计平台。书里提及不同的设计资源带给我们不同的设计考虑和实现。

本书硬件综合与测试采用的 Altera 平台，有出自厂家的设计库，其在性能等方面高于出自 HDL 的电路，所以在书中介绍如何利用厂家的 Altera 库资源。

Open IP Core（IP：Intellectual Property，开放的知识产权核）是设计需要关注的重要资

源。本书介绍 Open IP Core 微处理器 8051 核的使用。单片机的研究早于 FPGA,其研究资源很丰富,这无形中增加了设计资源,也为学习 CPU 设计提供了资料。

设计优化先要设计多样化,从工程设计与应用角度,对问题的功能、性能等实现给出设计风格多样的,可满足不同设计优化目标的解决方案,在此基础上依据优化目标做出合理的选择。作为设计的重要组成部分测试,也要进行多样性解决方法的讨论。本书探讨了多样化测试与实现、快速测试与实现,及其选择方法。本书没有讨论硬件更深层次的约束,如可人工干预约束的布局布线 Logiclock 等工具,在挑战性能与资源极限的复杂应用情况下是要考虑的。

3.3 研究和应用设计工具

无论是资源问题还是优化问题,设计工具可以提供一些解决方法。要重视借助 EDA 工具的力量了解信息和对细节进行分析,考虑设计约束来更好的完成设计;借助第三方仿真工具 Modelsim 进行多种方法的仿真;借助 EDA 工具提供的在线工具(如测试工具 SignalTap Ⅱ 等)进行测试。

4. 本书的编写说明

本书编写的原则是提高信息量,简约规范,统一标准,紧凑编排,可读性强,方便学习与理解。本书部分采用了 FPGA 文档指南与规范说明,参见附录 A,在 6.4.2 小节做了举例说明。书中程序的注释角度以学习 VHDL 为主,一些注释是增量式的。

VHDL 格式自由,大小写不敏感。书中采用了缩进格式,统一将关键字、标准程序包提供的类型、常数大写字母开头和一般标识符小写。

书中仿真重点采用 Modelsim,版本是 Modelsim 6.5e(Windous XP)和 Modelsim 10,a(Window 7)。Modelsim 默认波形窗口的衬底为黑色,为了印刷的清晰,在工具栏中编辑参数,改为了白底黑字。在不以仿真说明为重点的段落里,有一些仿真结果来自 Quartus Ⅱ 8.0 平台自带的仿真器。设计结果大部分是在 Altera EP2C35F672C8 芯片上验证的。Quartus Ⅱ 主要采用版本是 Quartus Ⅱ 8.0, 10.0, 13.0。

VHDL 标准主要采用的是 1993 版,少量采用 2008 版,因为现在的 EDA 工具对 1993 版的支持普遍比较好。为了整个书籍的规范,一些在模板上修改而成的 1987 版测试程序等也归化到 1993 版或者 2008 版。了解标准的变化,从保留字上也可以反映出一些,参见附录 B;了解 1993 与 2008 版的特点参见附录 C。

本书主要是针对目前教学计划设立的 48 学时 VHDL 与数字系统设计课程和 8 学时通信系统原理课程实验的教学任务。

本书确切地说应该是基于 FPGA 的电子系统设计基础,FPGA 作为贯穿式教学的平台,在设计方面可以有许多延伸,书中作了些许尝试。

全书的编写了得到了北京工业大学教务处,实验学院以及信息工程系领导、同事的支持,更归功于中国高等教育学会对应用型本科教学“规划教材”的组织推动,以及西安交通大学出版社的承办。书中还参考了许多专家学者的著作和研究成果。在此向上面的组织与所有人表示衷心的感谢。

由于作者水平有限,加之书中的内容没有如期望的一样完全进入到教学实践中,书中可能有不妥之处,欢迎读者和同行批评指正。

目　录

1 概 述

本章的主要内容与方法：

(1)基于 FPGA 的电子系统设计的研究特点与方法。

(2)介绍借助 Quartus Ⅱ 模板资源学习。

(3)学习 Quartus Ⅱ 模板 VHDL 的案例。

(4)简单介绍 Quartus Ⅱ 中 1993 版和 2008 版的设置。

(5)VHDL 的基本结构。

(6)VHDL 的基本语句。

1.1 基于 FPGA 的设计特点与方法

现代工程是网络化的、系统化的、集成化的产品及相关过程的开发模式。这一模式使开发者从一开始就要考虑到质量、成本、开发时间及用户的需求、标准、环境问题等诸多方面因素，我们需要在顺序工程的基础上再进行并行工程的思考与实践。

FPGA 支持并行工程，采用自顶向下的协同式的设计，同时基于工程设计进行全过程、全要素、全盘考虑的需要，要开展 DFX(Design For…，可……设计)，其中 X 作为通配符可包括 A(Assembly，可装配)、R(Reliability，可靠性)、Y(Yield，可收益)、T(Test，可测试)以及可验证、可移植、可扩展、可维护、可配置、可复用、安全性、易用性等。设计不再仅满足可综合(Synthesis)可实现的范畴。比如，设计测试是设计描述的一部分，可靠性设计是设计描述的一部分，DFX 设计极大地丰富和立体化了设计描述。

自顶向下设计实现大型复杂系统的前提是丰富的资源保障，支撑是 EDA 工具，所以我们也要探讨引入资源和利用资源，探讨工具的运用。

设计多样性、设计理念与原则是相对应的。系统设计的实现简单的说要有系统规划，方案选择、模块划分，系统整合，测试与优化。因为基于 FPGA 的电子系统设计虽然可以有很多软件实现，但其本质实现是硬件的。在资源限制的情况下，要做工程各方面因素权衡，要从方案开始就对多样性极大关注，对设计载体、设计方法多样性广泛探究，对问题多种多样解决探讨实践，协调好不同的优化目标，一部分通过行程再造、一部分通过时空转换等方式直至问题的理想化或所谓优化解决。可以说没有比较就没有鉴别，没有基础就没有整合，没有多样性也就没有优化，设计是根据多样性针对应用需求给出的合适抉择。

如果仅仅是针对 FPGA 编写可综合的代码，在编程方面追求多样化会受到一些约束，而为了工程化的先虚拟再实际的理由，必定要用更丰富的语言来表现虚拟，即加入那些不可综合的语言，这使设计有了更大的空间，使设计进一步地降低成本和提高效率，使做 FPGA 设计更有趣味、更生态。更何况“虚拟”是设计不可或缺的重要组成部分。因而，在本书中提供的一些小程序，有些是为了说明可综合设计的多样性，有些是为了说明不可综合设计的多样性。

设计原则有速度与面积原则、系统原则、同步原则、硬件原则等。

速度与面积原则，要在满足速度要求的情况下尽可能节省面积解决；系统原则，要求不能仅仅从局部优化来考虑问题；同步原则，尽管当前讨论异步设计的声音又大了些，但是对FPGA设计而言，同步原则还是要重点关注的；硬件原则，代码的简约整齐只是设计需要兼顾的，硬件的优化合理实现才是评价的标准。

应探讨基于平台的设计，设计要跟随标准，要有好的设计规范和设计风格。

FPGA 的设计主要有两种方式，一是与工艺平台有关的设计，二是与工艺平台无关的设计。VHDL 可以容纳与沟通两种方式。

我们在设计中采用从个性化分析到标准化解决研究到回归个性化需求的方法。与平台无关的设计是标准化的解决研究，与平台有关的设计是回归个性化需求的一个方式。不管设计过程中采用何种风格，最终的设计都符合 FPGA 的原则，即正确的实例化底层单元模块，合理地使用固有的硬件结构，以达到最优化的设计效果。

从 VHDL 语言描述实体的行为方面，其结构体主要有以下三种描述方式：

①行为描述(behavioral)，用算法和行为描述设计。

②数据流描述(dataflow)，又称 RTL(Register Transfer Logic/Level)描述。

③结构化描述(structural)，采用层次化的方式描述设计。

三者的优缺点、描述特点和使用场合如表 1-1 所示。本书我们将给出案例来说明多种描述方式，同时根据算法、逻辑、不同的语法和优化目标等因素，从多种多样的设计中选定具体的实现。

表 1-1　三种描述方式的比较

描述方式	优缺点	描述特征	适用场合
结构化描述	连接网络关系清楚，电路模块划分清晰 电路不易理解、繁琐、复杂	结构化语句： Component Generic Map Port Map	电路层次化设计、顶层设计、模块连接
数据流描述 RTL 级描述	布尔函数定义明白 不易描述复杂电路 不易修改	面向可综合的设计	小门数的电路与模块设计
行为描述	电路特性清楚易读 综合效率相对较低	使用延时语句，包括惯性延时和传输延时 在多驱动处理时，采用决断函数 使用 Generic 语句对时序参数建模	大型复杂的电路设计系统仿真

书中用一个实体和多个结构体的程序来给出设计，通过 Congfiguration(配置)语句来指定起作用的结构体或者元件，借此来阐释设计比较的方法。

1.2　Quartus Ⅱ工程建立与语言模板(Template)

Quartus Ⅱ给出了一些设计语言的模板，如 VHDL，介绍了语言的结构，基本语法等，给出了一些典型的、常用的和基本的电路的描述，学习好这些案例，对了解 VHDL 的设计等会有帮助。

新建工程，进入 Quartus Ⅱ，File→New Project Wizard→填写工程的工作目录、工程名、顶层实体名→选择加入项目的文件，缺省→选择器件系列与具体型号→EDA 工具，缺省。

新建 VHDL 文件。File→New→Design File→VHDL File，进入了编辑窗口。输入 VHDL 文件，保存，编译(Compilation)，就可以得到一个最基本的工程了，工程名为 *.qpf，工程的设置文件为 *.qsf。

在编辑窗口，插入模板。Edit→Insert Template，方法如图 1-1 所示。

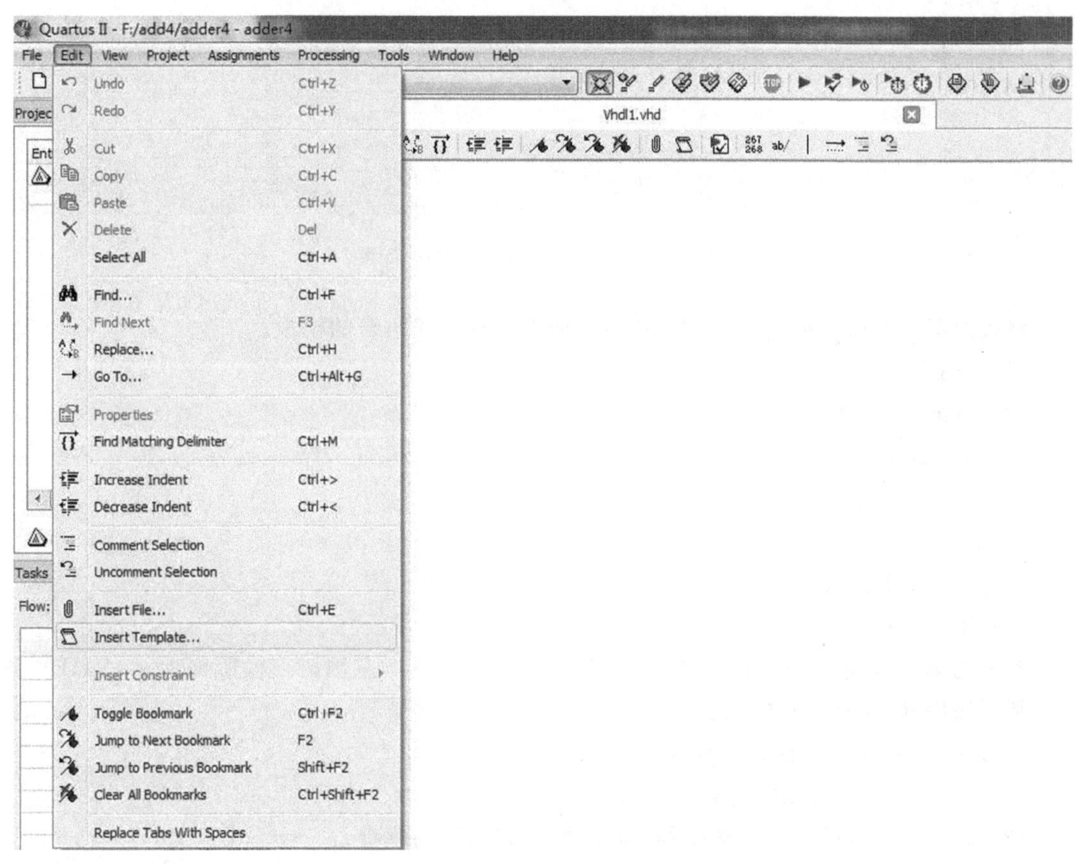

图 1-1　Edit 菜单中插入模板

选择 VHDL→Full Design→Arithmetic→Adders→Signed Adder，如图 1-2 右侧预览窗所示，可以看到一个 VHDL 有符号加法器的例程。

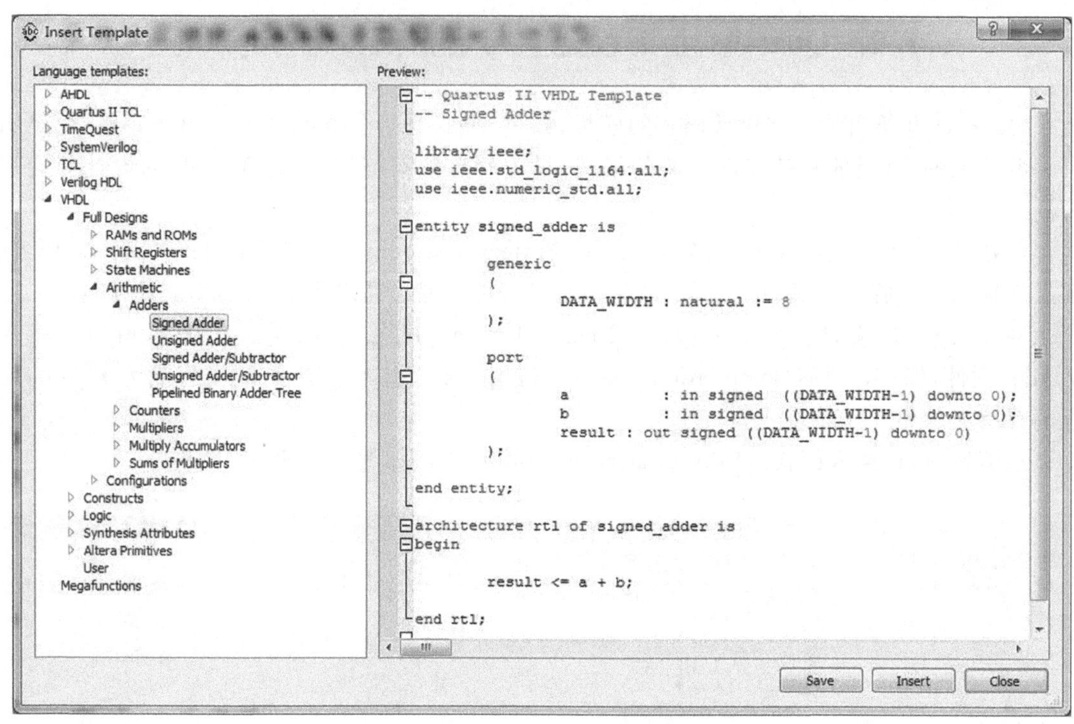

图 1-2　VHDL 模板设计案例选择

模板的第一层级，Quartus Ⅱ 10.0 中有如下 7 种描述语言的模板。

(1)AHDL。

(2)Quartus Ⅱ Tcl。

(3)TimeQuest。

(4)System Verilog。

(5)TCL。

(6)Verilog HDL。

(7)VHDL→。

模板的第二层级→VHDL，选择 VHDL 语言，如图 1-2 左侧语言模版窗所示，VHDL 语言模板层级展开，包括 6 种模板。

(1)Full Designs，VHDL 设计实例。

(2)Constructs，VHDL 的语言结构模板。

(3)Logic，逻辑单元，给出寄存器、锁存器、三态门等的模板。

(4)Synthesis Attributes，综合属性模板。

(5)Altera Primitives，Altera 的基本设计单元的模板，可调用缓冲器、寄存器、锁存器等。

(6)User 用户自定义模板。

1.2.1　查看模板给出的 VHDL 设计实例

模板的第三层级→Full Designs，VHDL 设计实例，层级展开，有 5 种常用设计单元模板。

(1)RAMs and ROMs，存储器。
(2)Shift Registers,移位寄存器。
(3)State Machines,状态机。
(4)Arithmetic,算术单元。
(5)Configuration,配置。
模板的第四层级→Arithmetic 算术单元,单元展开,有 5 种类别的运算器。
(1)Adders,加法器。
(2)Counters,计数器。
(3)Multipliers,乘法器。
(4)Multipliers and Accumulators,乘累加。
(5)Sum of Multipliers,乘加。
模板的第五层级→Adders 加法器,加法器展开,有 5 种常用的加法器。

(1)Signed Adder,有符号加法器,如图 1-2 所示,示例 1-1,文件命名 signed_adder. vhd。

(2)Unsigned Adder,无符号加法器。
(3)Signed Adder/Subtractor,有符号加减法器。
(4)Unsigned Adder/Subtractor,无符号加减法器。
(5)带流水线的二进制加法树。

在 VHDL 设计实例中,有许多案例,这里举一个模板给出的例子并加以注释,希望大家能够关注到这些资源。

示例 1-1 有符号 8 位加法器(8 Bit Signed adder)

```
--程序 1-1 signed_adder.vhd,文件名与实体名一致,后缀为 .vhd
--Quartus Ⅱ VHDL Template                    VHDL 模板,“--”代表 VHDL 的注释
--Signed Adder                                                  有符号加法器

Library Ieee;                                  --声明使用 Ieee 库
Use Ieee.Std_Logic_1164.All;                   --引用标准程序包 Std_Logic_1164
Use Ieee.Numeric_Std.All;                 --此程序包中含有符号、无符号及其相关函数

Entity Signed_adder Is                         --实体说明,实体名 signed_adder
Generic(Data_Width : Natural := 8);            --关键词 Generic 引导类属表
                                  --Natural(自然数,非负整数,Integers ≥ 0)
    Port(                                      --实体端口说明
    a: In Signed((Data_Width-1) Downto 0);     --2 进制 8 位被加数
    b: In Signed((Data_Width-1) Downto 0);     --2 进制 8 位加数
    result: Out Signed ((Data_Width-1) Downto 0) );
                                               --输出,和数
End Entity Signed_adder;                       --实体结束,1993 版 VHDL
```

```
Architecture rtl Of Signed_adder Is                    --名为 rtl 的结构体
Begin
    result <= a + b;                                   --有符号数相加
End Architecture rtl;                                  --结构体结束
```

程序说明：

(1) 因为进行有符号 Signed 与 Unsigned 无符号运算，需引用 Ieee. Numeric_Std 程序包，Ieee. Numeric_Std 内部将数据解释为 Std_Logic。与 Ieee. Numeric_Std 程序包相应的还有 Ieee. Numeric_Bit，其内部将数据解释为 Bit 类型。

有符号运算时，相应的数据类型是 Signed，综合时被解释为补码，最高位是符号位。无符号运算时，相应的数据类型是 Unsigned，综合时被解释为二进制数。

Ieee Std 1076.3 以及 Ieee 标准程序包提供了 Numeric_Std 程序包，相应的算术操作同时在该程序包中定义。算数操作符 abs，*，/，mod，rem，+，- 等在 Numeric_Std 进行了重载。

(2) 改变 Generic 类属中 Data_Width 的值，就改变了加法器的位数，如 8 改为 16，上述加法器就成为了 16 位加法器。

(3) Quartus Ⅱ VHDL 版本的设置。

有 3 种 VHDL 版本可选，1987，1993，2008 版。如设计采用 1993 版作为输入，设置如图 1-3所示，在 Quartus Ⅱ 主界面菜单栏选：Assignments→Settings→Analysis &Synthesis settings→VHDL Input。

从图 1-3 看到，Quartus Ⅱ 中还有许多参数可以设置。

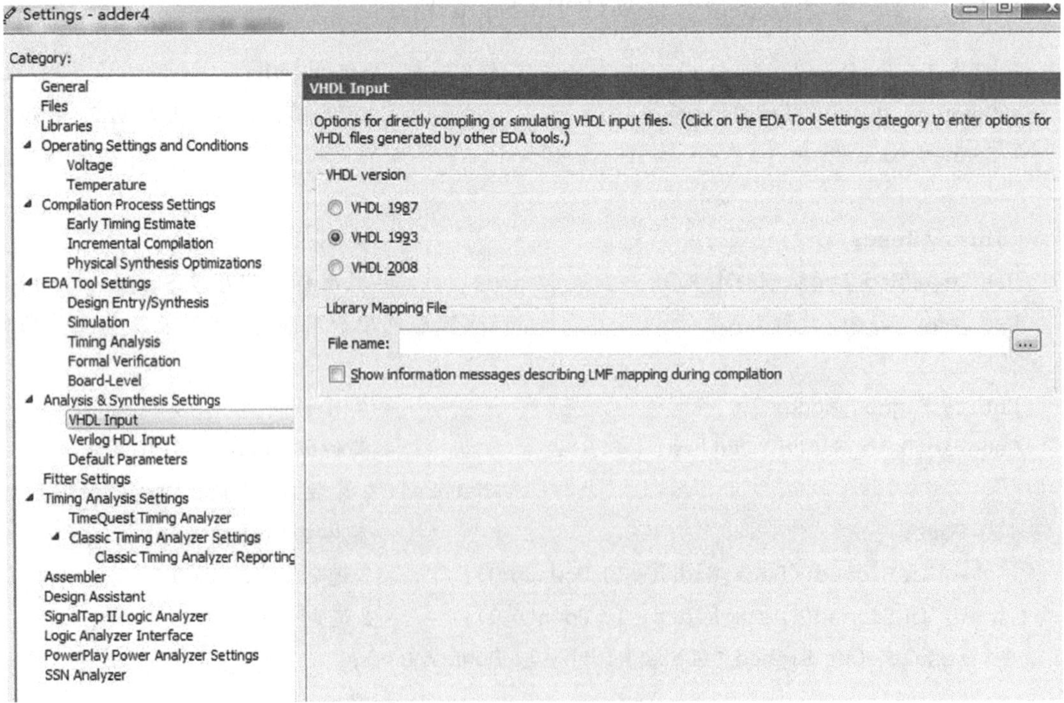

图 1-3 VHDL 输入版本的设置

比如要选择仿真工具,可以在 EDA Tool Settings 中选择 Simulation,可以进一步选择采用 Modelsim 仿真软件。

如果需在当前工程里加入其他文件夹里的文件,可以选择 File,浏览到需加入的文件,add 即可。

如果加入自己的库或一些特定的库,可以选择 Libraries。

(4)在插入模板右面的加法器模板中,点[−]→[+],加法器的层级收缩了,图 1-2 的效果变为了图 1-4 的效果。可以看到,一个最简单的 VHDL 可以由库、程序包调用、实体、结构体来构成。要想将设计展开,再点[+]→[−]。

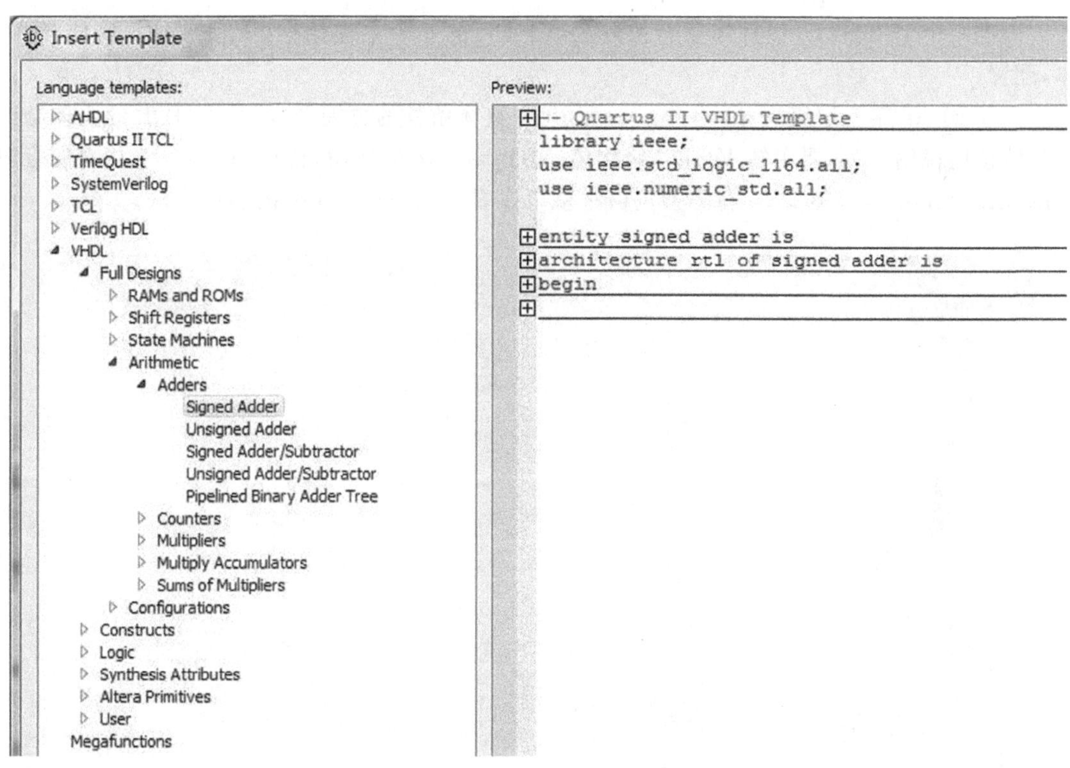

图 1-4 加法器的层级收缩

1.2.2 从 Quartus Ⅱ 模板了解的 VHDL 基本结构

Quartus Ⅱ VHDL 语言模板中有 VHDL 基本结构模板 Constructs。

VHDL 的 Constructs(基本结构)第一层级,包含 5 个条款,说明如下:

Design Unit 设计单元

Declaratins: 声明语句,使用前需要事先声明。

Concurrent Statemenst: 并行语句,直接在结构体中输写。

Sequential Statemens: 顺序语句,需要放在进程或子程序中。

Expressions: 表达式。

Design Unit 层级展开，即 VHDL 的 Constructs 第二层级，包含 9 个条款，说明如下：

Library clause：　库说明语句 Library。

Use clause：　程序包应用说明语句 Use。

Entity：　实体。

Architecture：　结构体。

Package：　程序包或称封装。

Package body ：　程序包体或称封装体。

Configuration_Declaraors：　配置声明。

Block Configuration：　块配置。

Component Configuration：　元件配置。

以上 9 个 Design Unit 具体条款实际是 VHDL 的 5 个主体结构，库，程序包，实体，结构体，配置。

Constructs 第三层级展开，即 Design Unit 中具体条款的再展开，就是 VHDL 的主体结构的具体使用说明。如，要了解 VHDL 结构体 Architecture 的使用，选择图 1－5 中左侧模板窗中的 Architecture，在图 1－5 右侧预览窗中就可以看到 Architecture 的使用方法。

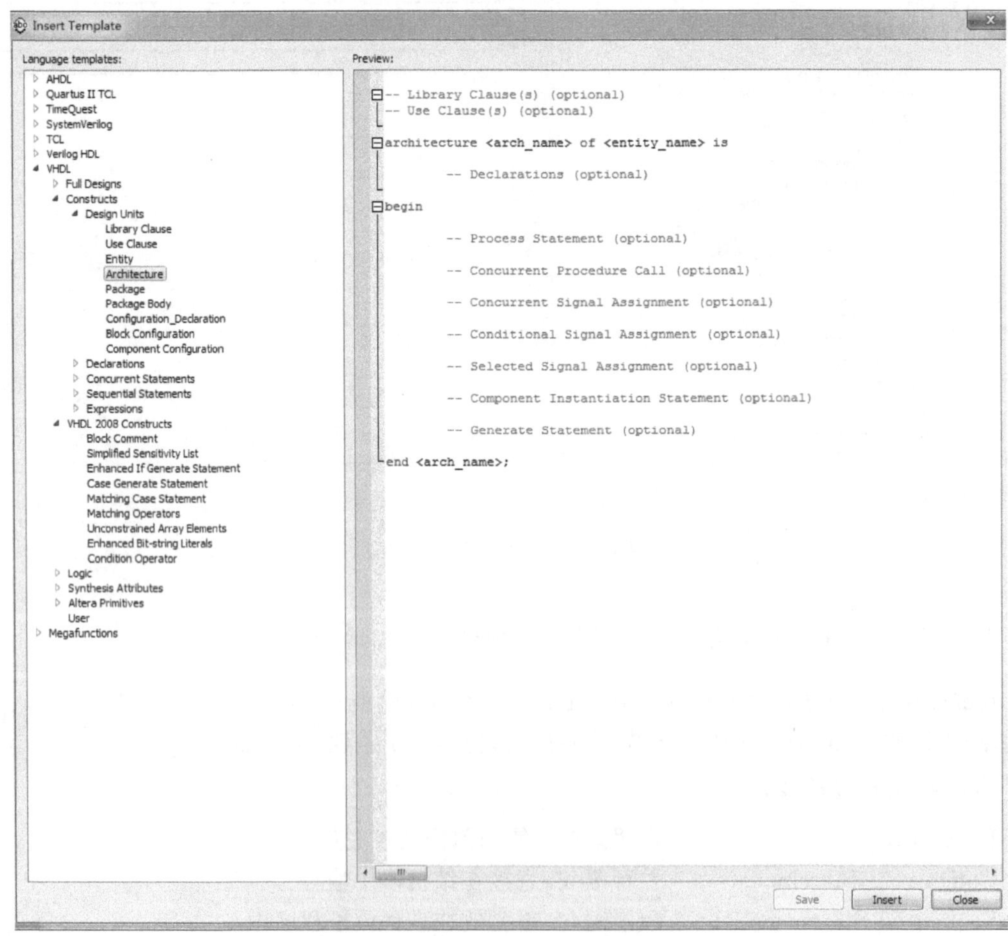

图 1－5　模板显示 VHDL 的基本结构

我们所看到的架构和语句可以直接点击图 1 - 5 显示窗口右下角 insert 将预览的内容插入到 VHDL 编辑窗口中，帮助我们编写程序。比如查看模板 Library clause，Use clause，Entity，Architecture条款，每一个条款都有英文解释，这里给出了上述 4 种条款汉语直译。

Library clause(库语句)说明

```
--A Library clause declares a name as a Library.
--It does not create the Library, it simply forward declares it
--库语句给一个库声明一个名字，它不是创建这个库，而是说明它
Library <Library_name>;
```

Use clause(程序包应用语句)说明

```
--Use clauses import declarations into the current scope
--If more than one use clause imports the same name into
--the same scope, none of the names are imported
--Use 语句给当前范围引入声明
--如果多于一个 Use 语句引入同样的名字在同一个范围，那些名字就没有引入

--Import all the declarations in a package          引用程序包中所有的声明
Use <Library_name>.<package_name>.All;
--Import a specific declaration from a package      引用程序包中特定的说明
Use <Library_name>.<package_name>.<object_name>;

--Import a specific Entity from a Library          从一个库里引用特定的实体
Use <Library_name>.<Entity_name>;

--Import from the work Library.  The Library is an alias for the Library
--containing the current design unit.
--引用工作库，这个库对于包含在当前设计单元的库是一个别名
Use work.<package_name>.All;

--Commonly imported packages:                               常被引用的程序包
--Std_Logic and Std_Logic_Vector types, and relevant functions
--程序包中包含标准逻辑、标准逻辑矢量类型及其相关函数
Use Ieee.Std_Logic_1164.All;

--Sigend and Unsigend types, and relevant functions
--程序包中包含有符号、无符号及其相关函数
Use Ieee.Numeric_Std.All;

--Basic sequential functions and concurrent procedures
```

```
--程序包中包含基本的顺序函数与并行过程
Use Ieee. VITAL_Primitives. All;

--Library of Parameterized Modules:
--customizable, device - independent logic functions
--程序包中包含用户可定义参数模块库LPM,其与器件逻辑功能无关
Use lpm. lpm_components. All;

--Altera Megafunctions                                  Altera宏函数说明
Use Altera_mf. Altera_mf_components. All;
```

Entity(实体)说明

```
Entity <Entity_name> Is                                         --实体模板
    Generic                                                     --类属说明
    (                                                           --类属参数表
    <name>: <type>   :=<default_value>;                         --可以多行说明
    …                                                           --VHDL 用分号结束语句
    );                                                          --最后一个分号在括号外
    Port(
        --Input ports                                           输入端口说明
        <name>: In  <type>;                                     --可以多个信号用多行说明
        <name>: In  <type> := <default_value>;                  --给输入信号缺省值
                                             --同类型信号可以用名字以逗号分隔在一行说明
        --Output ports                                          输出端口说明
        <name>: Out <type>;                                     --可以多个信号用多行说明
        <name>: Out <type> := <default_value>);                 --给输出信号缺省值
    End <Entity_name>;                                          --实体结束说明
```

Architecture(结构体)说明

```
 --Library Clause(s) (optional)                                 库语句(任选)
 --Use Clause(s) (optional)                                     包声明语句(任选)

Architecture <arch_name> Of <Entity_name> Is                    --结构体模板
--Declarations (optional)                                       声明语句(任选)
Begin                                                           --结构体开始
    --Process Statement (optional)                              进程语句(任选)
    --Concurrent Procedure Call (optional)                      并行过程调用(任选)
    --Concurrent Signal Assignment (optional)                   并行信号赋值语句(任选)
    --Conditional Signal Assignment (optional)                  条件信号赋值语句(任选)
    --Selected Signal Assignment (optional)                     选择信号赋值语句(任选)
```

```
    --Component Instantiation Statement (optional)        元件例化语句(任选)
    --Generate Statement (optional)                       生成语句(任选)
End <arch_name>;                                          --结构体结束说明
```

从上面说明可看出 VHDL 结构组成以及设计单元构成要素的使用方法、顺序等。

当我们需要了解更具体的内容，如 VHDL 如何声明数据对象、VHDL 有哪些并行语句和顺序语句以及他们的具体使用等也可以选中相应条目继续点击翻看。

查看结构说明中并行语句可以点击相应条款，如：

Conditional Signal Assignment 条件信号赋值语句；

Selected Signal Assignment 选择信号赋值语句。

Conditional Signal Assignment，条件信号赋值语句格式

```
<optional_label>:  <target> <=
    <value>  When  <condition>  Else
    <value>  When  <condition>  Else
    ...
    <value>;
```

注：<optional_label>，任选标号，所谓任选，就是可以指定也可以不指定，< >之中是需要我们指定的内容。这句话的含义是，当条件满足时，给目标赋值 1，否则赋值 1，以此类推。

Selected Signal Assignment，选择信号赋值语句格式

```
<optional_label>:  With  <expression>  Select
    <target>  <= <value>  When  <choices>
                 <value>  When  <choices>
    ...
    <value>  When Others;
```

注：这句话的含义是，根据表达式的不同选择，将对应的值赋值给目标。最后语句中有 Others 项，说明这个语句要求覆盖全部选项。

◇查看结构说明中顺序语句中的其中两条：

If Statement，If 语句格式

```
If <expression> Then                                      --expression:表达式
    --Sequential Statement(s)                             顺序语句
Elsif <expression> Then
    --Sequential Statement(s)                             顺序语句
Else
    --Sequential Statement(s)                             顺序语句
End If;
```

Case Statement，Case 语句格式

```
--All choice expressions in a VHDL Case statement must be constant
--and unique. Also, the case statement must be complete, or it must
--include an Others clause
--在 VHDL Case 语句中选择表达式必须是常数而且是唯一的，同时 Case 语句必须是完整的，或者必须包括 Others 语句
Case <expression> Is
    When <constant_expression> =>
        --Sequential Statement(s)                                顺序语句
    When <constant_expression> =>
        --Sequential Statement(s)                                顺序语句
    When Others =>
        --Sequential Statement(s)                                顺序语句
End Case;
```

可以查阅的条目还有很多，读者可以根据自己的需要查阅。

习　题

1－1　以 VHDL 有关内容为主，通览 Quartus Ⅱ设计模板的内容。

1－2　比较不同的硬件描述语言。

扩展学习与总结

1. 尝试对设计模板提供的 VHDL 例子进行仿真。

2. 小组分工查阅资料，对 FPGA 发展沿革，最新发展，FPGA 在不同领域的应用进行综述；对硬件描述语言的发展沿革，最新发展，以及应用进行综述；对于电子系统设计的发展沿革，最新发展、设计平台等进行综述，具体题目自拟。每人写出 4000 字综述报告，并开展小组交流与讨论。

2 多种形式的设计描述

本章的主要内容与方法：

(1)介绍课程的一些基本概念。

(2)顺序语句/并行语句的基本应用。

(3)并行语句 When Else,With Select,顺序语句 If,Case。

(4)VHDL 调用 Altera LPM 库中元件的方法。

(5)简单测试基准 Testbench 文件的编写。

(6)三种配置语句 Configuration 的应用。

(7)各种设计风格代码以及不同语句设计的综合结果比较。

(8)选择器的应用探讨。

2.1 本章主要概念

2.1.1 基于平台的设计方法

测试是设计的重要组成部分之一,我们不仅在电路设计中追求提高设计模块复用(Reuse)水平,还要开展基于平台的设计,即尽可能地将多种不同的设计来共享测试平台,提高设计质量与效率,降低开发生产与应用成本。

2.1.2 基于优化的 VHDL 设计

我们以一种基于优化的角度来说明 VHDL 设计。优化是面向不同目标的,比如从设计实现来讲占面积最小的,或者速度最快的,功耗最低的,可靠性高的等等;从设计时效来讲,用于快速原型设计的,用于 ASIC 验证的等等。

针对每一种优化目标,可能意味着有一种设计程序代码,配置(Configuration)的概念在设计面对多样化的选择和提高优化过程的效率来说是非常有用的。配置也帮助我们将多种不同的设计来共享测试平台 。

2.1.3 配置的概念与格式(Configuration)

配置(Configuration)相当于设计的部件/元件列表,它指定哪一个部件/元件适应设计的哪一部分。即 Configuration 语句的应用帮助我们在多方案、多风格设计中根据实际应用做出选择。

Configuration 主要有三种：

(1)块配置,其结果是实体与结构体组合。

(2)元件配置,其结果是实体与结构体与元件的组合。

(3)端口配置,可以对不同库里元件的性能做比较,也可以用于测试平台共享的一种说明方式。

本章 2.2 节对 4 选 1 电路给出多种结构体描述与注释，采用 Configuration 的默认形式指定了具体的实体与结构体对。在 Quartus Ⅱ模板的 VHDL 结构中有三条关于 Configuration 的条目。在模板的 Full Design 中，有关于配置的说明，我们把模板打开，用逐行注释的方式做了进一步说明。说明采用 VHDL 的注释符，即开头有“--”的行或文字表示注释。

```
Configuration declaration 配置声明：
  Configuration <configuration_name> Of <entity_name> Is
    For <architecture_name>
        --Use Clause (optional)
    --Block Configuration or Component Configuration (optional)
  End Configuration <configuration_name>;

  Configuration  <配置名>  Of  <实体名> Is                    --声明开始
  For  <结构体名>                                             --配置开始
    --Use 语句(可选项)
    --块配置或元件配置(可选项)
  End For;                                                    --配置结束
  End  Configuration <配置名>;                                --声明结束

--Note: A configuration declaration is used to configure one or more
  --instances of an entity. Quartus Ⅱ must be able to determine which
  --instance(s) to configure, or it will ignore the configuration-declaration.
  --Quartus Ⅱ is able to determine which instances to configure in
  --the following cases:
  --1. The configuration declaration pertains to the top-level entity.
  --2. The configuration declaration is named in a component configuration.
  --That is inside another, higher-level configuration-declaration.
  --3. The configuration declaration is named in a configuration specification.
  --注意:1 个配置声明是用于为实体配置一个或更多的例化
      --Quartus Ⅱ必须能够确定配置哪一个例化，否则它将忽略配置声明
      --Quartus Ⅱ在以下情况可以确定哪一例化做配置
      --1. 这个配置声明是有关顶层实体的
      --2. 这个配置以元件配置的名义声明，它包含在另一个更高级配置中
      --3. 这个配置以一个配置定义的名义声明
```

1. Block configurations 块配置

```
For <architecture_name, block_label, _or_ generate_label>          --块配置开始
--For <结构体名,块标号,_或_生成语句标号>
  --Use Clause (optional)
  --Use 语句(任选)
  --Block Configuration or Component Configuration (optional)
  --块配置与元件配置(任选)
End For;                                                            --块配置结束
```

2. Component configurations 元件/端口配置

```
--Component configurations go inside configuration declarations. See the
--full-design configuration templates for examples using component
--configurations.
--元件配置放在配置声明中
--参看模板 full-design 中 configuration 模板里应用元件配置的实例

For <instance_name>:<component_name>                                --配置开始
--For <例化名>:<元件名>
--Optionally specify either an entity or configuration (not both).
--任意指定实体或配置中的一个,不能都选
--Only use the semicolon if there is no port/generic binding to follow.
--如果没有端口和类属做如下绑定,则只能用分号
Use Entity <library_name>.<entity_name>(<optional_architecture_name>);
Use Configuration <library_name>.<configuration_name>;
--上面两句我们在应用时只能选 1 种方式应用
--Optionally specify port and generic bindings.
--任意指定端口/类属绑定
--Use these if the names of the ports/generics of the component don't match the
--names of the corresponding ports/generics of the entity being instantiated.
--如果元件端口/类属的名字与对应例化的实体相应的端口/类属不匹配,这样用:
    Generic Map(
    <instantiated_entity_generic_name> => <component_generic_name>,...)
    --(类属映射)是例化实体的类属名与元件类属名之间的映射
    Port Map(
    <instantiated_entity_input_name> => <component_input_name>,
    <instantiated_entity_output_name> => <component_output_name>,
    <instantiated_entity_inout_name> => <component_inout_name>,
          ...);
```

```
    --Block Configuration (optional)                     块配置语句任选
  End For;                                                 --配置结束
  --Port Map(端口映射)是例化实体的输入/输出名与元件输入/输出名之间的映射
```

2.1.4　类属说明(Generic)

Generic 类属,既可以说明常数,也可以说明物理量、时间量等,Generic Map 又称参数传递语句。

类属的应用可以使设计具有更多的灵活性。

类属可以将一个设计带入不同的参数,对结果快速进行比较。

设计复用(Reuse)是设计的一个重要方法。类属的应用可以使设计便于复用。

2.1.5　测试基准文件(Testbench)

测试基准文件 Testbench 将测试激励信号输入到待测试实例,通过第三方仿真软件 Modelsim 输出仿真数据或波形,其采用 VHDL 来编写,文件中有大量的只用于仿真不能综合的语句。测试模型如图 2-1 所示。目前 FPGA 的开发平台与仿真软件都支持不同语言的混合应用。

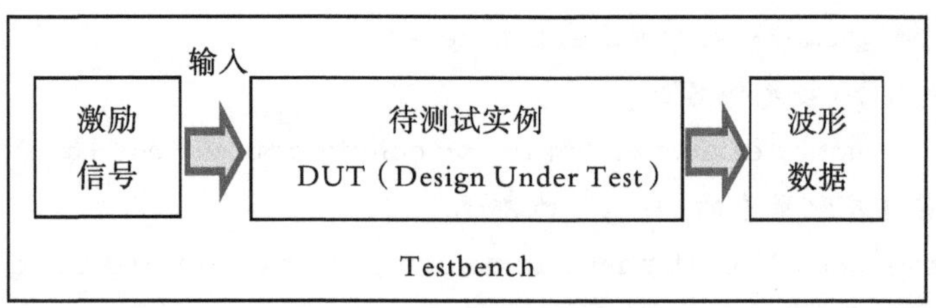

图 2-1　测试模型

2.1.6　设计验证的主要种类

Modelsim 仿真方式:

(1)只有输入激励驱动的测试:

· 输入波形方式产生激励驱动;

· 测试基准文件(Testbench)方式产生激励驱动;

· 特殊仿真,如 Modelsim 专用的用 Force 仿真命令等产生激励驱动。

(2)完全测试平台,应用 STD 库中的 Textio 程序包,输入激励和结果从文件中读取,仿真输出结果以及比较。

(3)快速测试平台,如 Testbench 用矩阵表给出输入激励和结果,仿真输出结果以及比较。

(4)混合测试,组合几种测试技术。

这一章我们只涉及到第一种测试平台,在第 3、4 章中还有进一步的讨论。

2.2 多种形式的多路选择器

设计目的与要求：

(1)初步了解 VHDL 顺序与并行的特点。通过一个实体多个结构体的方式，采用 2 种并行选择语句 When Else 与 With Select 语句，2 种顺序选择语句 If,Case 语句，给出 VHDL 多种方式描述的 4 选 1 电路的程序代码。

(2)多样设计与高效的优化选择。编写配置(Configuration)语句，指定当前具体应用的结构体，对各种方式的综合实现进行比较。

(3)提高设计的灵活性。参数化设计，尝试类属(Generic)的应用。

(4)编写 VHDL 测试激励文件 Testbench 采用第三方软件 Modelsim 进行仿真比较。

(5)探讨基于 FPGA 结构的选择器设计优化，如用香农定理，将 6 位选择器逻辑化为多数 FPGA 需要的 4 输入端模式，实现面积优化。

(6)探讨多路选择器的设计应用。

(7)讨论参数化的选择器的设计仿真与实现。

2.2.1 多种四选一电路的 VHDL 设计

示例 2-1 四选一电路实体设计

```
--程序 2-1 mux41a.vhd
Library Ieee;                             --声明使用 Ieee 库
Use Ieee.Std_Logic_1164.All;              --声明引入 Std_Logic_1164 包中的所有项

Entity  mux41a  Is                        --声明实体名 mux41a,此实体对应 7 个结构体
  Port(a,b: In Std_Logic;                 --标准逻辑类型,需要 Std_Logic_1164 程序包
       c,d:In Std_Logic;                  --a,b,c,d 为 4 个被选择信号
       s:In Std_Logic_Vector(1 Downto 0); --s 为 2 位标准逻辑输入矢量选择信号
       y:Out Std_Logic);                  --1 位选择器输出
End Entity mux41a;                        --结束实体端口的描述,VHDL93 版标准
--End mux41a;                             --VHDL 87 版标准
```

/＊4 选 1 VHDL 顶层程序文件要以实体名命名，以 .vhd 为后缀，文件名为 mux41a.vhd 实体名注意要符合 VHDL 的标识符规范，如实体名是字母开头的数字与字母组合。

Std_Logic 标准逻辑类型含九值：'1'，'0'，'H'，'L'，'X'，'Z'，'U'，'W'，'-'，含义分别为强'1'、'0'，弱'1'、'0'，不定态，高阻，未知、弱未知，无关。Std_Logic，Std_Logic_Vector 类型便于仿真，但是 VHDL 标准程序包不支持这个类型，Std_Logic_1164 程序包支持，所以需要说明使用 Std_Logic_1164 程序包。＊/

示例 2-1a When Else 条件信号赋值语句(并行语句)

```
Architecture  muxone  Of  mux41a  Is        --声明第一个结构体名为 muxone
Begin
    y <= a  When  s = "00"  Else            --当 s = "00"条件成立,a 传输到 y
         b  When  s = "01"  Else            --否则当 s = "01"条件成立,b 传输到 y
         c  When  s = "10"  Else
         d  When  s = "11"  Else
         --Null ;                            --空操作,慎用,会引入锁存器
         --'Z' ;                             --Std_logic 类型的高阻
         --'X' ;                             --Std_logic 类型的不确定态
         Unaffected;                         --VHDL 保留字
End  Architecture muxone;                    --结束对第一个结构体的描述
```

示例 2-1b With Select 选择信号赋值语句(并行语句)

```
Architecture  muxtwo  Of  mux41a  Is        --声明第二个结构体名为 muxtwo
Begin                                        --结构体描述开始
    With  s  Select                          --并行语句,选择信号赋值语句
      y <= a When  "00",                     --注意','分隔,而不是';'
           b When  "01",
           c When  "10",
           d When  "11",                     --With Select 要覆盖条件
           Unaffected When Others;           --Std_Logic 有 9 值所以要说明 Others
--Unaffected 保留字,表示不执行任何操作,也可以用'-'
End  Architecture muxtwo ;                   --结束对第二个结构体的描述
```

示例 2-1c If 条件语句(顺序语句)

```
Architecture  muxthree  Of  mux41a  Is      --声明第三个结构体名为 muxthree
Begin
    Process (a,b,c,d,s) Is                   --顺序语句放在进程中
    Begin
        If s = "00"  Then  y <= a;      --如果满足条件 s = "00",执行跟着的语句
        ElsIf s = "01"  Then  y <= b;
                                       --否则如果满足条件 s = "01",执行跟着的语句
        ElsIf s = "10"  Then  y <= c;        --多位数据用双引号括起
        ElsIf s = "11"  Then  y <= d;        --ElsIf 不用写 End If
        --Else y <='X' ;                     --1 位数据用单引号括起
        Else y <='-' ;                       --Std_logic 类型的"-"无关项
        End  If;                             --结束多条件判断语句
```

```
    End  Process;                             --进程结束语句
End  Architecture muxthree;                   --结束对第三个结构体的描述
```

示例 2-1d　Case 条件语句(顺序语句信号赋值法)

```
Architecture  muxfour  Of  mux41a  Is         --声明第四个结构体名为 muxfour
Begin                                         --结构体开始

    Process(a,b,c,d,s)Is                      --括号中是敏感表,设计输入
    Begin                                     --进程开始
        Case s Is                             --Case 引导顺序语句
            When "00"  =>y <= a;              --s 的值为“00”,将 a 传输到 y
            When "01"  =>y <= b;
            When "10"  =>y <= c;              --每个值不能重复
            When "11"  =>y <= d;              --要覆盖所有值
            When Others=> Null;               --Std_Logic 有 9 值,要对其他值处理
        End  Case;                            --Case 语句结束
    End Process;
End Architecture muxfour;                     --结束对第四个结构体的描述
```

示例 2-1e　Case 条件语句(顺序语句变量计算法)

```
Architecture muxfive Of mux41a Is             --声明第五个结构体名为 muxfive
Begin
    Process(a,b,c,d,s(0),s(1)) Is             --进程声明中有敏感表
                  --信号 s: Std_Logic_Vector(1 Downto 0),可以表示为 s(0),s(1)
    Variable muxval : Integer Range 3 Downto 0;
                                              --变量 muxval 在进程中起作用
                                              --声明整数时要给出范围
    Begin
        muxval := 0;                          --变量赋初值
        If(s(0) = '1') Then                   --s(0)是 s 数组的最低位信号
            muxval := muxval + 1;             --变量 muxval 计算赋值立即起作用
        End If;
        If(s(1) = '1') Then
            muxval := muxval + 2; --这种计算选择值的方式在输入选择更多时方便
        End If;
        Case muxval Is
            When 0 => y <= a;                 --相当于 s0=0,s1=0
            When 1 => y <= b;                 --相当于 s0=1,s1=0
            When 2 => y <= c;                 --相当于 s0=0,s1=1
```

```
            When 3 => y <= d;                    --相当于 s0=1,s1=1
            When Others =>Null;                  --Case要全面覆盖,每个值必须是唯一的
        End Case;
    End Process;
End Architecture muxfive;                        --结束对第五个结构体的描述
```

示例 2-1f　Case 条件语句(顺序语句信号计算法)

```
Architecture muxsix Of mux41a Is                 --声明第六结构体名为 muxsix
--此结构体与 2-1e 比较,用于学习,可仿真,不能综合,不推荐使用
Signal muxval:Integer Range 3 Downto 0;          --信号说明,整数类型范围 3 到 0
Begin
    Process Is                                   --进程声明中无敏感表,要配合 Wait 语句
    Begin
        muxval <= 0;                             --说明成 Signal,要加等待时间
        Wait For 0 ns;                           --综合器往往会忽略掉这个延时
                                 --Quartus Ⅱ综合器不支持这个语句,可 Modelsim 仿真
        If(s(0) = '1') Then
            muxval<= muxval + 1;                 --Signal 赋值不能立即起作用
            Wait For 0 ns;                       --加等待时间
        End If;
        If(s(1) = '1') Then
            muxval <= muxval + 2;                --信号的运算与赋值
            Wait For 0 ns;                       --加等待时间
        End If;
        Case muxval Is
            When 0 => y <= a;                    --s0=0,s1=0
            When 1 => y <= b;                    --s0=1,s1=0
            When 2 => y <= c;                    --s0=0,s1=1
            When 3 => y <= d;                    --s0=1,s1=1
            When Others => Null;                 --必须说明表达式没有用到的值
        End Case;
        Wait On a,b,c,d,s(0),s(1);               --执行敏感表
    End Process;
End Architecture muxsix;
--以并行为主要特征的语言,又必须有顺序的处理,所以系统本身设有延时
```

示例 2-1g　利用 LPM 参数模块生成的元件

/＊在原理图窗口,双击,元件插入或在 Tools 菜单点击 Magaward Plag - in Manger→Gates→LPM_MUX,输入自定义文件名称,lpmux41.vhd。VHDL 后缀 .Vhd,Verilog 后缀 .v。

打开 lpmux41.vhd 可以看到系统生成的元件的端口信息，按照这个信息写 Component 声明。以下结构体即利用 LPM 参数模块生成的元件 lpmux41 写成的 4 选 1。*/

```
Architecture muxseven Of mux41a Is                  --声明第七个结构体名为 muxseven
    Component lpmux41 Is                            --声明 LPM_MUX 元件 4 选 1 名为 lpmux41
        Port (data0,data1,data2,data3: In Std_Logic ;       --4 输入信号
              sel: In Std_Logic_Vector (1 Downto 0);        --2 选择信号
              result: Out Std_Logic );                      --1 输出信号
    End Component lpmux41;
    Begin
    u1: lpmux41 Port Map(a,b,c,d,s,y);              --LPM_MUX 元件 lpmux41 例化
                               --位置对应法，顺序要与 Component 中 Port 信号说明一致
End Architecture muxseven;
```

2.2.2　配置应用(块配置)

```
Configuration mux41a_cfg Of mux41a Is               --配置语句说明，可附在主体程序之后
    For muxtwo                                      --指定实体 mux41a 对应结构体 muxtwo
    End For;
End Configuration mux41a_cfg;                       --配置语句说明结束
```

2.3　多种四选一电路设计的语法讨论

2.3.1　块配置讨论

块配置是配置语句的默认形式。采用配置的原因是一个实体可以有多个结构体，而在综合时实体与结构体是 1 对 1 的。

Configuration 语句段落可放在程序代码中、测试 Testbench 文件中或单独存放。仿真时以测试 Testbench 文件中的说明起作用，换言之，最后的说明可以覆盖前面的说明起作用。

Configuration 语句段落也可以缺省，缺省时为指定最后一个结构体。

2.3.2　有优先级的选择

(1)If 语句与 When Else 相对应。If 为顺序语句，When Else 为并行语句，两者都有优先级，而且可以隐含条件。从综合结果来看，这两个语句综合结果是一致的。

(2)慎用 Null。If 与 When Else 语句的条件 Others 值为 Null 时，综合结果出现了一个 Latch，这并不是我们预期的，所以 If 与 When Else 语句用 Null 时需要注意或者慎用。

(3)注意‘Z’的应用条件。If 与 When Else 语句的条件 Others 值为‘Z’时，输出会夹带 1 个三态门，在设计中用的多时往往会适配不到芯片中，因为‘Z’通常只能出现在芯片的 IO 端口的单元中。

(4)注意 If 语句中的‘X’。纯粹的 4 选 1 的 Others 值是当值为不确定‘X’或者无关态‘-’时出现的。要注意 If 语句中的‘X’。

①If(sel='1')　Then
　y <= '0';
 	
Else
　y <= '1';
End If;
	②If(sel ='0')　Then
	　y <= '1';
	
	Else
	　y <= '0';
	End If;
		If(sel='1') Then
		　y <='0';
		Elsif(sel='0') Then
		　y <='1';
		Else　y <='X';
		End If;

当 sel ='X'时， y <='1';	当 sel ='X'时， y <='0';	修改后

①与②描述的电路应该一样，在后仿真时没问题，因为 sel 一般不会出现‘X’。但是前仿真时，当 sel = ‘X’时，y 会得出不同结果，违反了电路原理，因而对于前仿真而言需要更严谨的描述。因为后仿真时，仿真软件往往具有缺省的初始值，前仿真则往往不具有缺省的初始值。

2.3.3　无优先级的选择

Case 与 With Select 相对应，没有优先级。Case 为顺序语句，With Select 为并行语句。示例因信号采用 9 值标准逻辑类型，两个语句都需要 Others 语句，即要求值覆盖条件，同时条件对应的值是唯一的。从综合结果来看，这两个语句综合结果是一致的。

2.3.4　信号 Signal 与变量 Variable 的讨论

信号 Signal、变量 Variable、常数 Constant、文件 File 是 VHDL 的数据对象。信号 Signal 是内部连线，在结构体中全局有效，赋值在进程结束时变化，信号赋值 y<=a；变量 Variable 在进程内部暂存并作用，赋值立即变化，变量赋值 y：=a；信号 Signal、变量 Variable、常数 Constant 在结构体或进程声明时置初值都采用“：=”。

变量 Variable 原本只用于子程序如函数、过程或进程，在 93 版以后，因为有了共享变量，Shared Variable，所以可以和信号 Signal 一起说明了。

2.4　选择器的设计优化

2.4.1　If 和 Case 语句的速度与面积平衡

Case，With Select 语句与 If，When Else 语句相比，前者综合的结果更简洁易验证，如表 2-1所示，所以如果没有要求优先级的实现，应该尽可能选择用 Case，With Select 语句描述方式。

表 2-1 If 和 Case 语句的比较

If 语句特点	Case 语句特点
If 语句是逐项条件顺序比较的过程 If-Then-Else 结构,综合出来的电路是串行的,增加了时延,路径长 可以实现单元复用,面积可以更小 If 语句综合电路有优先级	Case 语句的条件性是独立的、排它的 Case 语句可读性较好,易于维护,有一定的并行性,速度快
cmp cmp cmp min	cmp cmp cmp min

2.4.2 If 语句与设计的局部调整优化

If 语句优化前后对比如表 2-2 所示。

表 2-2 If 语句优化前后对比

优化前	优化后

◇条件 3 为关键路径
◇条件 1,2,3 不同时成立
If 条件 1 Then
do instruction1;
Elsif 条件 2 Then
do instruction2;
Elsif 条件 3 Then
do instruction3;
End If;
	◇调整顺序,逆序
	
	If 条件 3 Then
	do instruction3;
	Elsif 条件 2 Then
	do instruction2;
	Elsif 条件 1 Then
	do instruction1;
	End If;

◇如果条件 2-4 没有优先级
If 条件 1 Then 信号置 1;
Elsif 条件 2 Then 信号置 0;
Elsif 条件 3 Then 信号置 1;
Elsif 条件 4 Then 信号置 0;
End If;
	◇合并 IF 语句
	If 条件 1 Or 条件 3 Then
	信号置 1;
	Elsif 条件 2 Or 条件 4 Then
	信号置 0;
	End If;

2.4.3 用更少输入端选择器实现面积优化

在实际 FPGA 设计优化时，我们要关注 FPGA 的结构，多数 FPGA 的基本单元为 4 位输入的查找表 LUT，采用香农分解可以将 6 输入端的 4 选 1 转换为 4 输入单位的单元来实现，可以获得面积优化。

假设输入为 a，b，c，d，选择端为 s0，s1，输出为 y，则 4 选 1 电路的输出函数为

$$y = s1's0'a + s1's0b + s1s0'c + s1s0d \quad (2-1)$$

实际设计时考虑将如式(2-1)所示的输出函数分解为 3 个函数，如式(2-2)，(2-3)，(2-4)所示，即 4 选 1 电路改为 3 个 2 选 1 电路 MUX 来实现。

$$y1 = s0'a + s0b \quad (2-2)$$

$$y2 = s0'c + s0d \quad (2-3)$$

$$y = s1'y1 + s1y2 \quad (2-4)$$

6 选 1 的香农变换如图 2-2 所示。

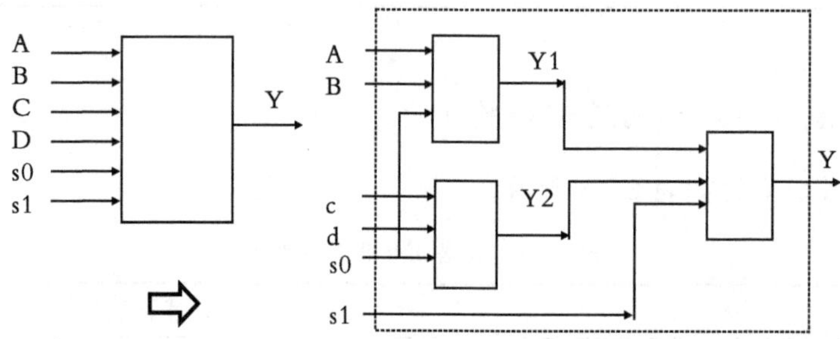

图 2-2　6 选 1 的香农变换

对于任意 6 个输入端函数，假如输入为 a，b，c，d，e，f，输出为 y。

$$y(a,b,c,d,e,f)$$

$$= a'y(0,b,c,d,e,f) + ay(1,b,c,d,e,f)$$

$$= a'b'(0,0,c,d,e,f) + ab'y(1,0,c,d,e,f) + a'by(0,1,c,d,e,f) + aby(1,1,c,d,e,f) \quad (2-5)$$

$$= a'b'y1 + a'by2 + ab'y3 + aby4 \quad (2-6)$$

式(2-6)与(2-1)实际上是等同的。

一般 FPGA 的基本单元都既有 LUT 也有 MUX，如果不进行香农变换，任意 6 个输入函数采用 LUT 实现，要采用 7 个 LUT；进行香农变换，仅采用 4 个 LUT。

2.4.4 VHDL 语句与优化

FPGA 顺序语句与并行语句、FPGA 速度与面积优化问题等就像一对孪生兄弟不可分离，FPGA 设计要不断处理好顺序与并行，速度与面积等的问题，这也是 FPGA 设计的语言与其他语言不同的地方。

VHDL 语句在 4 选 1 的不同实现中可以看到，有顺序语句 If，Case。其中，顺序中的顺序用 If，顺序中的并行用 Case。有并行语句 When Else，With Select，其中并行中的顺序用 When

Else,并行中的并行用 With Select,甚至一些保留字的使用也分顺序和并行,如 Unaffected,Null 含义等同,前者用于并行语句,后者用于顺序语句。

VHDL 的这种特点无处不在。评价 VHDL 学习是否入门了,在于当面对所有的设计都能将顺序语句、结构与并行语句、结构自如恰当地针对不同优化目标应用,进一步提高则要能够实现描述风格等的自如变换。

优化是针对结构与应用的,有时是反逻辑化简的,如 2.4.3 小节香农变换,有时还要逻辑复制,提高驱动。

2.5 四选一各种描述综合结果比较

四选一电路的各种描述综合结果如表 2-3 所示。

表 2-3 示例 2-1 四选一电路的各种描述综合结果一览

语句	RTL Viewer 结果
If When Else 其他条件 为 Null 时	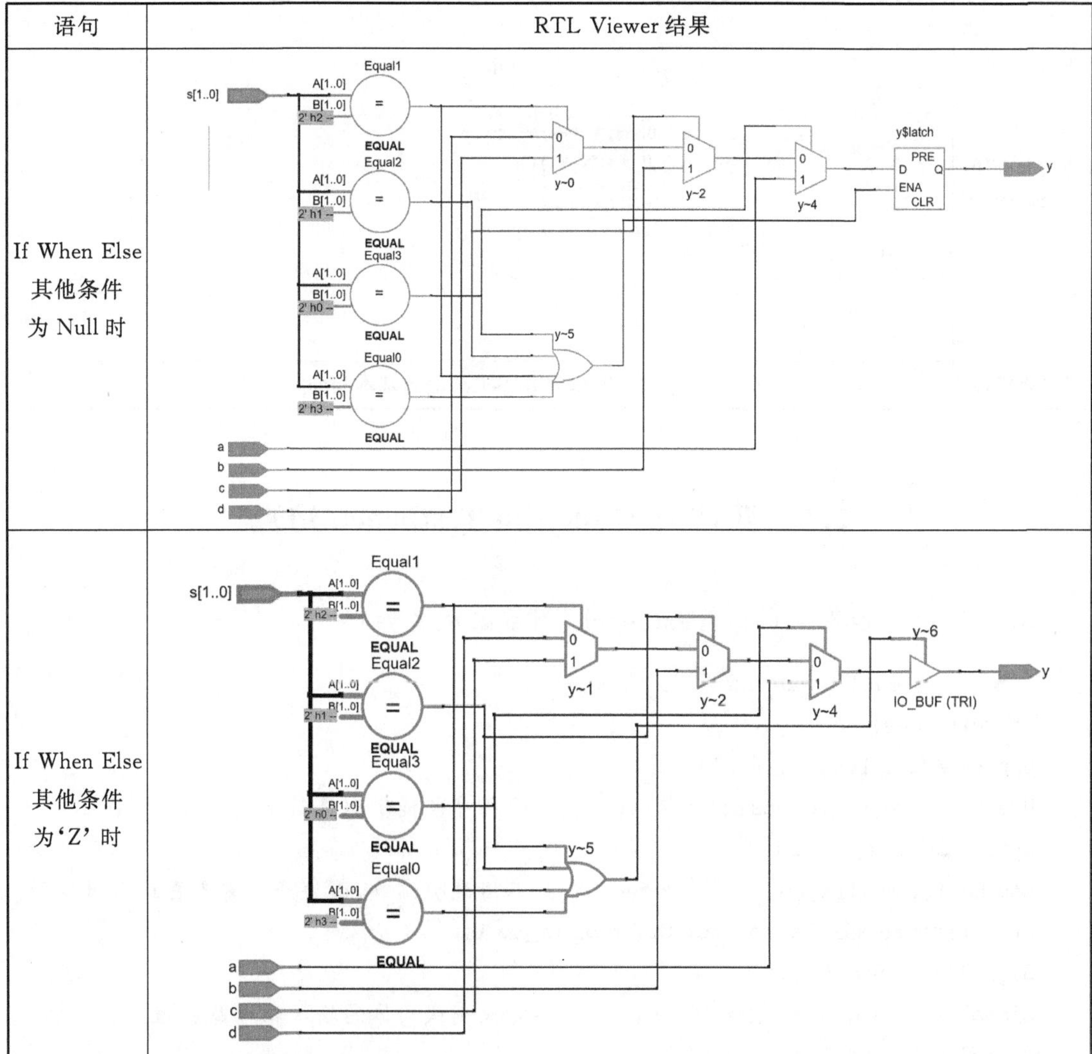
If When Else 其他条件 为‘Z’时	

续表

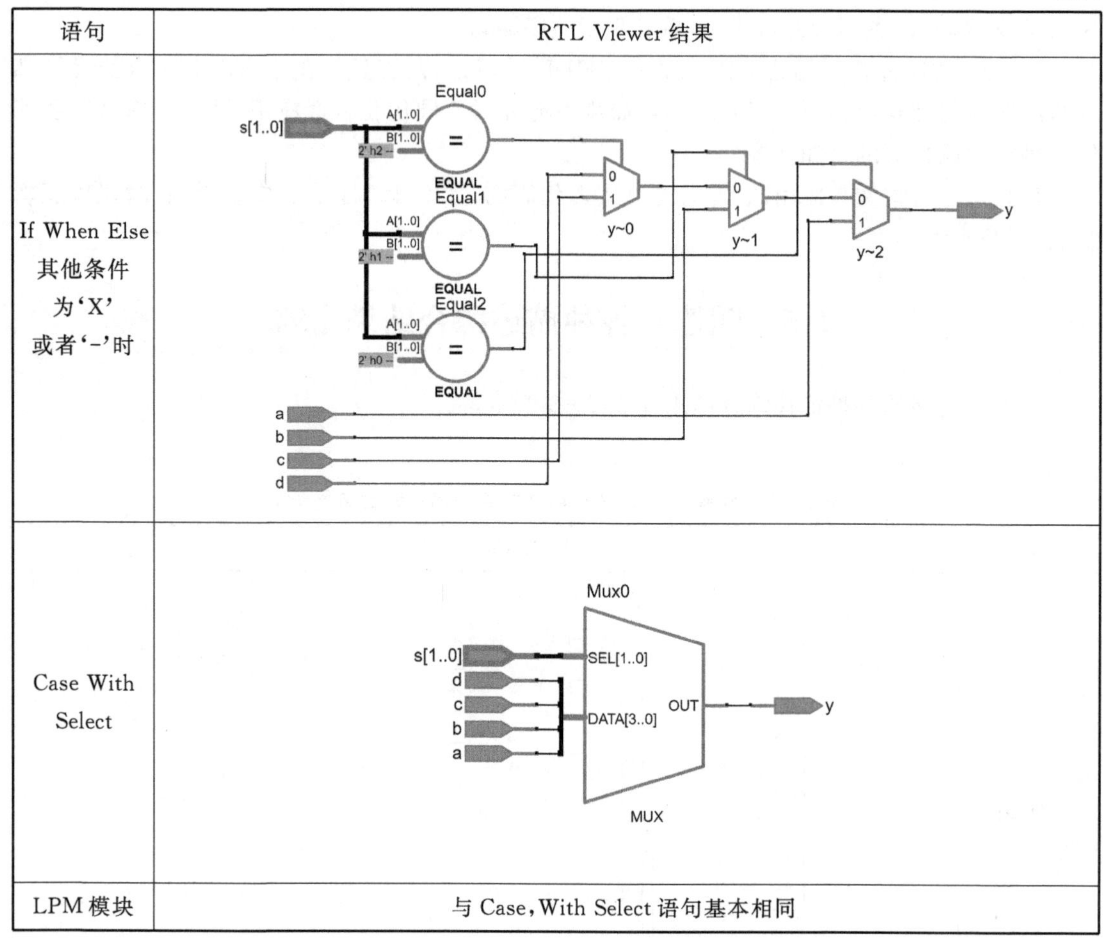

语句	RTL Viewer 结果
If When Else 其他条件为‘X’或者‘-’时	
Case With Select	
LPM 模块	与 Case，With Select 语句基本相同

2.6　四选一 Modelsim Testbench 仿真

示例 2-2　四选一电路 Testbench（测试基准文件）

```
--程序文件 mux41a_tb. vhd
Library Ieee;
Use Ieee. Std_Logic_1164. All;
Use Ieee. Std_Logic_Unsigned. All;       --此程序包提供运算符重载函数
Entity mux41a_tb  Is
End Entity mux41a_tb;                    --不用说明端口，因为并不需要真正引出端子
Architecture mux41a_tb_arch Of  mux41a_tb Is
Signal y :  Std_Logic  ;
Signal a :  Std_Logic := '0';            --对被测试实体的输入信号赋初值
Signal b :  Std_Logic := '0';
```

```
Signal c :  Std_Logic :='0';
Signal d :  Std_Logic :='0';
Signal s :  Std_Logic_Vector(1 Downto 0):="00" ;   --选择运算初值
    Component mux41a Is                            --元件说明
        Port (y: Out Std_Logic ;                   --端口与 mux41a 实体端口一致
              a, b, c, d: In Std_Logic ;           --4 个输入
              s: In Std Logic Vector(1 Downto 0) ); --2 个选择端
    End Component mux41a;
    --For All:mux41a Use Entity Work.mux41a(muxone);
                                           --采用 mux41a 中的 muxone 结构体
Begin
    DUT : mux41a              --元件例化,(DUT,Design Under Test)被测单元
    Port Map (y => y, a => a, b => b, c => c, d => d, s => s );

    Process(a,b,c,d,s) Is
    Begin
        a <=Not a After 50 ns;             --给出被测试实体输入端口信号的变化
        b <=Not b After 100 ns;                   --在 100 ns 后,b 取非
        c <=Not c After 200 ns;
        d <=Not d After 400 ns;
        If s<= 3 Then
            s<= s+1  After 800 ns;         --此语句含不同类型运算,需要用重载
        Else s<="00";
        End If;
    End Process;
End Architecture mux41a_tb_arch;

Configuration mux41a_cfg Of mux41a Is              --配置说明语句
    For muxone                              --采用 mux41a 实体的结构体 muxone
    End For;
End Configuration mux41a_cfg;
--这段配置说明与结构体说明后注释的 For All 语句的指定含义相同
```

2.7 仿真注意事项与仿真程序说明

2.7.1 仿真文件的编写注意事项

(1)尽可能完备测试,追求错误覆盖率,仿真易于观察、验证。

(2)尽可能给周期信号或有代表性的人工码等,通信、测量等测试要求高的电路还要给出

随机码或者伪随机码。

(3)对于复杂电路要做功能(Functional)、时序(Timing)甚至静态时序仿真。

(4)在输入激励文件编写配置 Configuration 语句,加快优化选择以及提高测试代码复用。

(5)注意 Generic 语句的使用,加快优化选择以及提高测试代码复用。

2.7.2 Modelsim 使用说明

(1)Modelsim 可以根据被测模块的 VHDL 描述自动生成 Testbench 的总体结构框架。对于这个四选一测试程序,自动生成 Testbench 的总体结构框架中,原实体的端口转变成 Testbench 中的信号,我们要对输入信号设初值,然后给出输入信号的变化。

(2)执行 Modelsim 仿真 Simulate 命令,测试程序将调用测试单元给出输出仿真结果。如图 2-3 所示,为四选一程序 Modelsim 功能仿真结果,即在 Modelsim Wave 窗口看到的结果。具体操作见第 4 章说明。

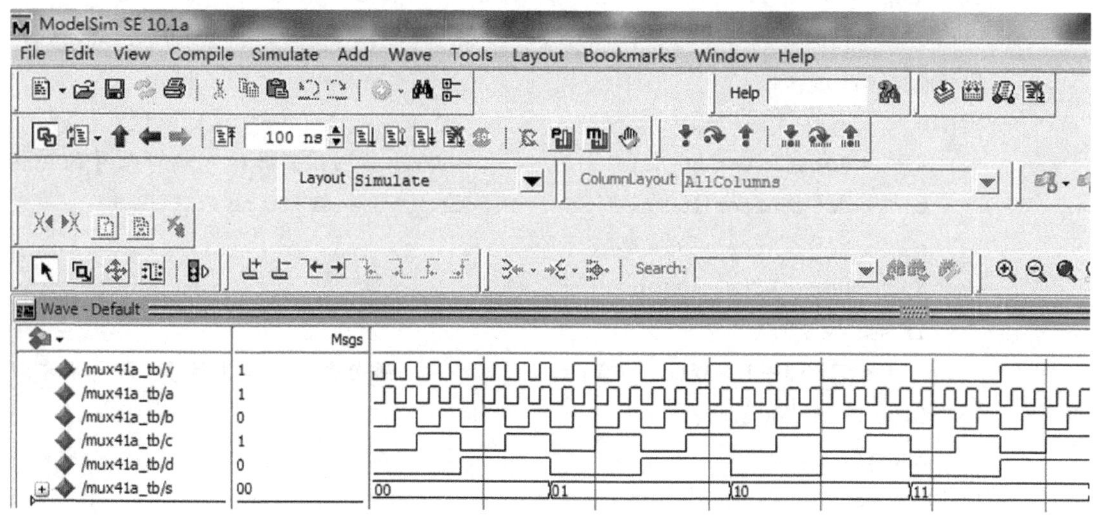

图 2-3 四选一 Modelsim 功能仿真结果

2.7.3 Testbench 编写说明

(1)Testbench 中信号的延迟语句,如 After 50 ns 是不可综合的语句,仅用于仿真。程序中的 50 ns 作为基本的延时单位时间,在通常的设计中一般会采用 Generic 语句来指定为一个 Delay 来表示。

如 1:设计实体语句说明中说明 Generic(Delay:Delay_length:=50 ns);

这句话也可以写做 Generic(Delay:Time:=50 ns);

后面 After 50 ns 可以写为 After Delay;

After 100 ns 可以写为 After 2 * Delay;

如 2:在元件例化前可以重申,但时延为 100 ns;

Generic(Delay:Delay_length:=100 ns);

后面 Generic 说明的参数可以覆盖前面的说明。

(2)如果没有 If 语句,也可以不采用 Process,因为结构体也具有类似进程在输入激励变化时循环的特点。

(3)仿真具有多结构体的程序时,Configuration 语句放在测试基准文件中比较方便,只需在 Testbench 中,改换 Configuration 语句指定的结构体,就可以选择出最适合性能要求的程序段。如果源代码与 Testbench 都说明了 Configuration,仿真时 Testbench 说明的起作用。

(4)可以不用 Configuration 语句,而是在元件说明处或是 Testbench 被测元件说明之后采用 For 短语引导的例化说明来绑定实体与结构体对。如:

```
For All:mux41a Use Entity Work. mux41a(muxone);
```

For 语句虽然也可以指定多个结构体中具体采用的结构体,但是开发和设计比较复杂的电路时,还是采用 Configuration 比较好。

(5)与电路追求 RTL 级描述不同,Testbench 的编写更倾向于行为级描述。前者可以获得更精微的电路调控能力等,后者则更有利于提高测试效率。

2.8　选择器的设计应用

2.8.1　选择器作为 ROM 应用

选择信号看做地址,被选的输入信号看做 ROM 的内容,给出不同的地址,就可以选出不同的内容。

2.8.2　选择器的总线应用

1. 端口的三态设计

早期总线往往采用 74LS373 芯片,此芯片中有三态门,而在 FPGA 只有在端口的单元有三态门,内部的电路需要用带使能的选择结构取代三态门结构。

示例 2-3　八位四选一三态电路设计

```
--If 语句应用中常见的错误举例与修改
Library Ieee;
Use Ieee. Std_Logic_1164. All;
Entity tristate Is
    Port(in3,in2,in1,in0 : In Std_Logic_Vector(7 Downto 0);
         en : In Std_Logic_Vector(1 Downto 0);
         dout : Out  Std_Logic_Vector(7 Downto 0));
End Entity tristate ;
Architecture wrong_tri Of tristate Is
Begin
    Process(en,in3,in2,in1,in0) Is                          --错误进程
    Begin
        If en = "00" Then dout <= in0 ;
```

```
        Else dout <=(Others=> 'Z'); End If ;
        If en = "01" Then dout <= in1 ;
        Else dout <=(Others=> 'Z'); End If ;
        If en = "10" Then dout <= in2 ;
        Else dout <=(Others=> 'Z'); End If ;
        If en = "11" Then dout <= in3 ;
        Else dout <=(Others=> 'Z'); End If ;
    End Process;
End Architecture wrong_tri;
--由于激励信号的同时性,对于同一输出信号,只有最靠近进程结束的语句起作用
--同一进程最好只放一个 If 语句(嵌套的除外)

Architeture right_tri Of tristate Is
Begin
    Process(en,in3,in2,in1,in0)  Is                        --正确进程
    Begin
        If  en = "00" Then dout <= in0 ;                   --嵌套 If 语句
        Elsif  en = "01" Then dout <= in1 ;
        Elsif  en = "10" Then dout <= in2 ;
        Elsif  en = "11" Then dout <= in3 ;
        Else dout <=(Others=> 'Z');
        End If ;
    End Process;
End Architecture right_tri;
```

2. 将 1~16 位的计算值送到具有三态门控制的 8 位总线

系统常常会将计算数据送到总线上去,总线为 8 位,数据的位数却是变动的。当计算的数字是 16 位时,要将它分为高 8 位和低 8 位,如果计算的数字是 12 位,要先在最高位前面补零,然后再分高 8 位和低 8 位,如果希望数据能通过三态缓冲器控制输出,就需要设计带三态输出缓冲器的参数化数据选择器。

示例 2-4　带三态输出控制的参数化选择器

```
--程序文件 muxparam.vhd
Library Ieee;
Use Ieee.Std_Logic_1164.All;
Entity muxparam Is
    Generic( N:Positive := 12);                       --参数指定,Positive 正整数
    Port(din: In Std_Logic_Vector( N-1 Downto 0);     --12 位数据输入
         sel:In Std_Logic;                            --高 8 位与低 8 位选择
         oe_bar:In Std_Logic;                         --低电平使能
```

```
        dout:Out Std_Logic_Vector( 7 Downto 0));      --数据位自动对齐的输出
End Entity muxparam;
Architecture muxparam_arch Of muxparam Is
Signal din_s: Std_Logic_Vector( 15 Downto 0) :=(Others=>'0');
Begin
    g0: For i In 0 To 15 Generate                         --For Generate 循环语句
        g1: If i<= N-1 Generate                           --并行语句 If Generate 语句
            din_s(i) <= din(i);                           --循环变量 i 不用声明
        End Generate g1;
    End Generate g0 ;
    dout <= din_s(15 Downto 8) When oe_bar='0' And sel ='0' Else
            din_s( 7 Downto 0) When oe_bar='0' And sel ='1' Else
            (Others=>'Z');                                --When Else 嵌套语句
End Architecture muxparam_arch;
```

程序语句说明：

(1)不要见到 If 和 Case 就以为是顺序语句，If Generate，Case Generate 是并行语句。

(2)数据位自动对齐的功能非常有用，比如：Std_Logic 与 Std_Logic_Vector 是不能直接运算的，可是有时需要这样算，通常在计算时也不能保证不同输入的数位数一样。我们要能够自动对齐，如果系统不能提供程序包支持，为了方便使用，要将上面的程序变成函数，通过调用函数实现对齐。幸运的是，Std_Logic_Unsigned 程序包可以针对不同位数的输入进行运算。

(3)如图 2-4 参数化选择器仿真结果，设计输入是设参数为 12 的 12 位数字，前面需要补 0，具体的 2 进制数值用 16 进制表示，是输入从 A51 开始不断递增的 12 位数据，当 sel 为‘0’，输出 8 高位 0A，其中 0 位是自动补的。当 sel 为‘1’时，输出 8 低位 51 等。

(4)如图 2-5 参数化选择器仿真结果，设计输入是设参数为 16 位的 16 数字，前面不需要补 0，设计输入用 16 进制数，具体数值是输入从 1200h 开始不断递增的数据，当 sel 为‘0’，输出高位 A5，当 sel 为‘1’时，输出低位 01 等。

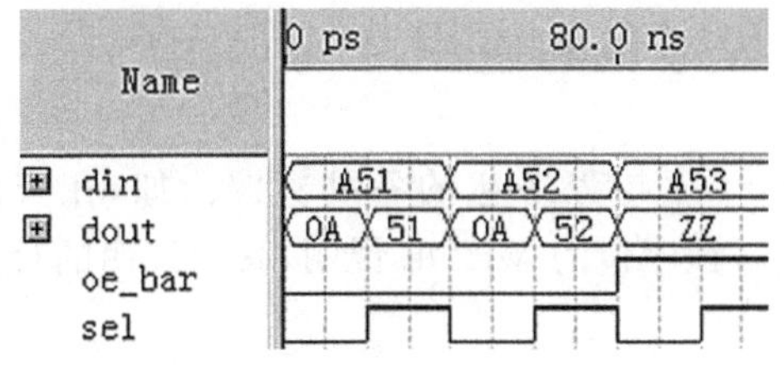

图 2-4 参数化选择器 N=12 的功能仿真结果

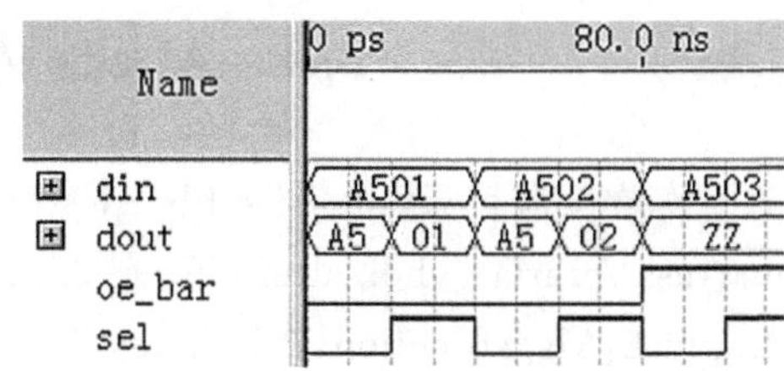

图 2-5 参数化选择器 N=16 的仿真结果

示例 2-4 在分别带入参数 12 和 16 时的综合结果如图 2-6、2-7 所示。两种结果内部元件资源占用是一致的，而参数为 12 时，占用布线资源较少。从此例可以看出设计中如果信号能够置初值，有利于综合后资源的减少，而且有时是必不可少的。

三态门通常只在 FPGA 芯片端口才有，不能在内部出现。有的综合工具可以将三态门从内部挤到端口。

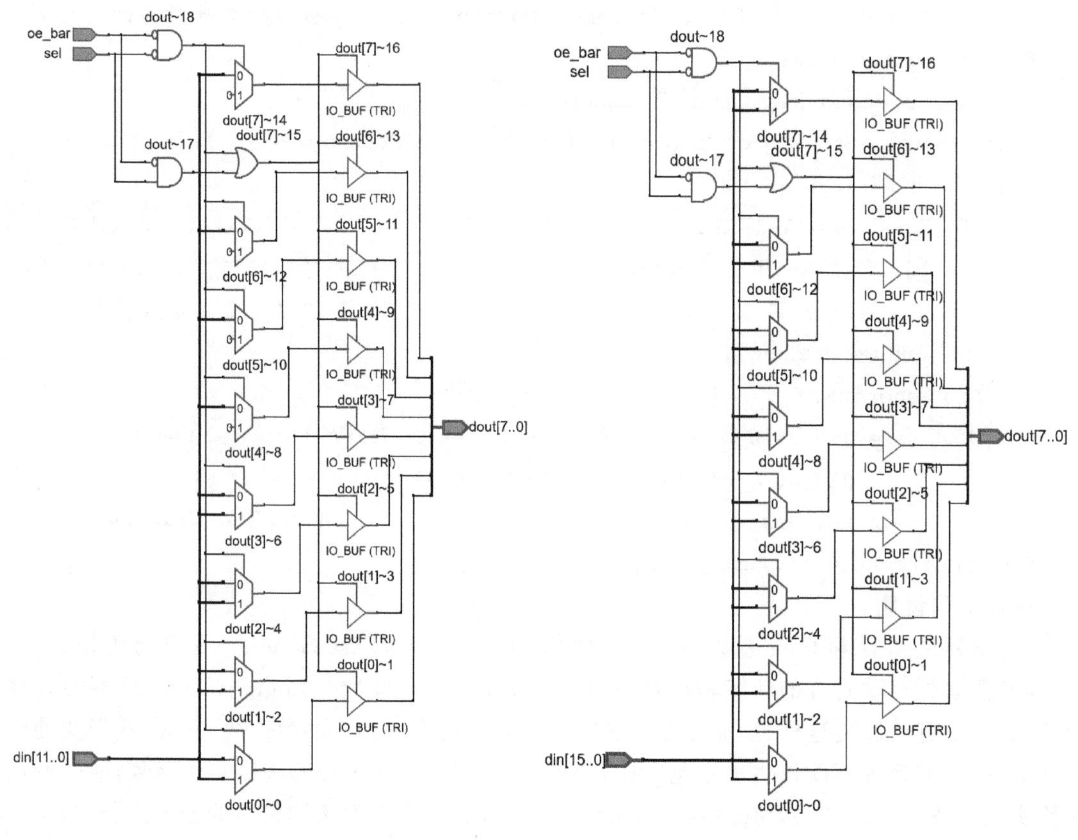

图 2-6　N=12 的综合结果　　　　图 2-7　N=16 的综合结果

2.8.3　选择器的测试应用

在四选一 testbench 中，我们通过给 4 输入端计数以及选择端计数，带入四选一元件后，得到了步进的方波，这种步进的方波常常会在测试应用中出现，参见图 2-3。

2.9　Altera 库与程序包的应用

Altera 在安装目录/quartus/libraries 有 Altera 库与程序包的各种资源。如，在安装目录/quartus/libraries/vhdl/altera 下，有 Altera 库与程序包的 VHDL 说明，其中说明的程序包有 Maxplus2，Magafunction，Primitives 等。

(1)Maxplus2 主要是时序电路模块和运算单元模块；

(2)Primitives 主要是基本功能单元；

(3)Magafunction 主要是宏功能模块。

/quartus/libraries/primitives，/maxplus2 等目录下，还有库中各种元件图可以用。

二选一元件 a_21mux 的说明在/quartus/libraries/vhdl/altera/maxplus2. vhd 文件中，原来 maxplus2 中的一些 74 系列电路的说明也在其中。在这里分别给出利用 Altera 程序包对元件直接调用以及进行结构体声明元件后调用的例子。

示例 2－5 利用程序包对元件说明

```
--程序文件 amux21.vhd
Library Ieee;
Use Ieee.Std_Logic_1164.All;
Library Altera;                                  --打开 Altera 库
Use Altera.Maxplus2.All;                         --使用 Altera.maxplus2 程序包

Entity amux21 Is
    Port(s,a,b:In Std_Logic;
         y:Out Std_Logic);
End Entity amux21;

Architecture amux21_arch Of amux21 Is
Begin
    u1: a_21mux Port Map(s,a,b,y);               --调用 Maxplus2 程序包中的 21mux 器件
End Architecture amux21_arch;
```

Altera.Maxplus2 程序包中含有元件二选一的说明，其名称为 a_21mux，打开程序包后，程序就可以直接对元件进行调用。

在这个程序里元件例化采用了位置对应的方式，位置对应的方式要求实参与元件说明的形参位置完全对应。

示例 2－6 利用结构体中元件说明

```
--程序文件 t7400.vhd
Library Ieee;
Use Ieee.Std_Logic_1164.All;
--Library Altera;                                  --关闭 Altera 库
--Use Altera.Maxplus2.All;                         --关闭 Altera.Maxplus2 程序包

Entity t7400 Is
    Port(a,b :In Std_Logic;
         c :Out Std_Logic);
End Entity t7400;

Architecture t7400_arch Of t7400 Is
    Component a_7400 Is                            --元件声明
        Port (a_2: In Std_Logic;
              a_3: In Std_Logic;
              a_1: Out Std_Logic);
```

```
    End Component a_7400;                          --元件声明结束

Begin
    u1: a_7400 Port Map(a,b,c);                    --元件例化
                                                   --元件说明按位置对应
End Architecture t7400_arch;
```

由于没有打开程序包，因此引用元件需在结构体的说明部分对元件进行说明。这种使用方式仅限于在 Altera 公司提供的 EDA 软件环境中进行。

习　题

2-1　写出示例 2-4 参数化数据选择器和三态输出缓冲器的 Test bench，用 Modelsim 仿真。

2-2　给出示例 2-4 另外一种结构体描述方式，采用类属定义的方式，指定 N 为 13 或 10，并在 Modelsim 程序中用 Configuration 语句进行指定并仿真。

扩展学习与总结

1. VHDL 语言结构组成部分及其作用是什么？
2. 程序包由哪几个部分组成？其功能和书写格式是什么？怎么使用？
3. 如何定义和使用配置？
4. 实体的类属参数与端口说明有何作用？如何使用？
5. 什么是元件例化？如何定义元件和调用元件？如何映射参数和引脚？
6. 了解 VHDL 的文字规则。

3 多种运算单元设计

本章的主要内容与方法：

(1)结合运算单元不同风格的设计深入学习 VHDL 语言。

· 学习如何自定义库与程序包；

· 学习重要的子电路方法、元件例化、函数、过程等；

· 深入学习编程语句，Generic，Block，Generate，Configuration 等。

(2)加减运算的进一步讨论。

· 采用 LPM 模块的设计；

· BCD 加法器的设计；

· 超前进/借位加减法的设计；

· 带寄存器以及流水线的设计；

· ALU 设计与数据通道。

(3)简要介绍设计原语。

3.1 主要概念说明

1. 结构化的设计(Hierarchical Coding)

结构化的设计是 VHDL 结构体描述的风格之一。

(1)结构化设计的目的：

①模块化，降低复杂性，使问题更加清晰，易于为设计者理解。

②提高系统的可综合性。

③形成知识产权 IP(Intellectual Property)的方法之一。

(2)结构化设计的方法：采用元件定义和元件例化来实现层次化，类似于电路的网络表，将各元件通过语言描述进行连接，元件隐藏了详细内容，用黑盒子(Black box)的方式，使在某一级只有元件的输入输出是可见的。

(3)结构化编码的注意事项：

①设计层次不宜太深，3～5 层即可。

②顶层模块最好仅仅包括所有模块的组织和调用。

③所有的 I/O，如输入、输出、双向端口的描述全局时钟、全局复位/置位、三态缓冲在顶层。

2. 类属 Generic 语句和生成 Generate 语句

采用类属 Generic 进行设计是在系统设计中所推荐的，Generic 可以从实体外向实体或者元件传递参数。采用 Generic 的设计方法可以方便地改变电路的结构与规模，不仅可以指定常数，也可以指定仿真时序参数以及物理参数等。

借助 Generic 带入时序参数的仿真方法是 Generic 非常重要的应用。

并行语句生成语句是实现单元复用的重要语句。本章的案例中加法器的位数由类属 Generic 决定，通过 If Generate，Generate 生成语句进行了元件的例化。

3. Block 块结构

Block（块结构）格式与实体格式极为类似，只是将相应的 Entity 改换为 Block。类似格式的关键字还有 Component。Block 放在结构体语句中，每一个 Block 相当于一个子电路原理图。

Block 在 VHDL 中具有的一种划分机制，允许设计者合理地将一个模块分为数个区域，在每个块都能对其局部信号、数据类型和常量加以描述和定义。任何能在结构体的说明部分进行说明的对象都能在 BLock 说明部分中进行说明。

Block 将结构体中的并行语句进行组合，可改善并行语句及其结构的可读性，还可以利用 Block 的保护表达式关闭某些信号。Block 不产生低层次的元件模块结构，并发执行。VHDL 综合器将略去所有的 Block 块语句说明。

采用 Block 语句这种平铺结构，可以加快仿真与综合。

3.2 基本加/减运算单元设计

基本运算是必须给予关注的，因为电子系统设计需要有高效灵活的运算支持，我们通过基本加/减运算单元设计来学习本章的内容与方法。

设计要求：

(1)基本运算部分：

①采用元件例化、函数、过程、块、包、库等多种方法实现 8 位基本加/减运算单元设计，其中库包括 LPM 库以及自定义库，函数包括自定义重载函数。

②采用一个实体多个结构体的方式给出多种方法的加/减 VHDL 程序代码。

③学习指定当前具体应用的结构体以及底层的结构体，探讨不同的实现形式。采用默认配置(Configuration)，元件配置语句。采用 For 说明语句。

④对各种设计进行仿真与实现比较。

(2)运算的拓展部分：

①流水线实现 32 位加法器，流水线与非流水线比较。

②先行进位加/减法器设计，以及先行进位与行波链加法器比较。

③BCD 加法器的设计。

④探讨采用 LPM(可定义参数宏模块)与原语实现设计。

⑤ALU 设计及 ALU 数据通道的设计。

(3)仿真与综合：

①应用类属(Generic)进行参数化的任意位加法器的设计与仿真。

②对各种方式的实现进行综合，采用第三方软件 Modelsim 编写 VHDL 测试激励文件 Testbench 进行仿真比较。

3.2.1 多种加法器介绍与设计

任意指定位加法器可以采用 Generic 类属语句通过参数指定位数实现，类属语句 Generic

给出常数，如常数为 8，可构成 8 位加法器，常数为 16，可构成 16 位加法器。

为了表达设计的多样性，顶层设计采用 1 个实体和 8 种结构体来描述，设计最终结构体的运用可由配置 Configuration 或 For 语句来说明。

示例 3-1:8 位加法与参数化加法器程序采用了元件配置，它不仅给实体指派顶层具体的结构体，同时指派结构体调用元件所采用的结构体。

层次化加法器或称结构化加法器，采用行波链加法器(Ripple Carry Adder)，连接如图 3-1 所示。每一位加法对应 1 位加法器，整个加法器是将前级(低位)的进位输出连到后级(高位)的进位输入。

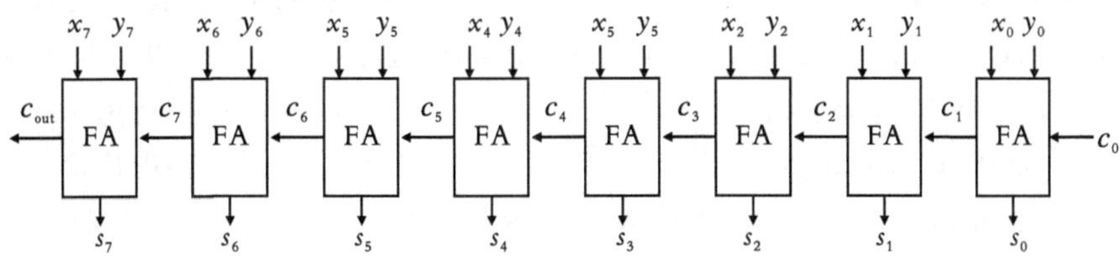

图 3-1　8 位行波链加法器

我们先对 9 个结构体示例 3-1a～3-1i 做一个简要的介绍。

(1)程序 3-1a，第 1 个结构体描述的是 8 位结构化加法器。

在结构体中对 1 位加法器元件进行 8 次直接例化调用。

(2)程序 3-1b，8 位结构化加法器。

与上一个结构体的区别是引用了 Work 库，比较了循环，其中采用的 8 个全加器中具体是哪一种全加器(fadder)由结构体中的 For 例化语句决定。

(3) 程序 3-1c，8 位结构化加法器。

这个电路是对程序 3-1a 的修改，通过 Generate 语句的运用循环对元件 1 位加法器 fulladder 进行例化。

(3)程序 3-1d，函数定义法。

函数可以在程序包里定义，也可以在结构体中定义，这里我们给出了函数在结构体中定义的实例，所以加法实现只需 1 个文件即可。

(4)程序 3-1e，过程定义法，采用类属指定位数的方式。

过程可以在程序包里定义，也可以在结构体中定义，我们给出了过程在结构体中定义的实例，加法实现只需 1 个文件即可。

(6) 程序 3-1f，过程定义法，应用 VHDL 的数组范围属性自动判断位宽的方式。

应用 VHDL 的数组范围属性自动判断位宽，并采用断言 Assert 语句判断如果加数与被加数的位宽不等，引导 Report 语句提示和 Severity 语句给出错误级别。加法实现只需 1 个文件即可。

(7) 程序 3-1g，重载运算函数定义法。

引用 Std_Logic_Unsigned 程序包中的重载运算函数方式实现行为级加法运算。加法实现只需 1 个文件即可。

(8) 程序 3-1h，3-1i，自定义程序包与重载运算函数法。

采用自定义程序包与重载运算函数，不用 Std_Logic_Unsigned 程序包。

示例 3-1 参数化/结构化加法器，实体 addern

```
Library Ieee;                                    --Ieee 库声明
Use Ieee.Std_Logic_1164.All;                     --引用程序包 Std_Logic_1164
Use Ieee.Std_Logic_Unsigned.All;                 --引用程序包 Std_Logic_Unsigned
Use Work.All;                                    --引用当前工程 Work 库中的元件

Entity addern Is                                 --声明加法器实体 addern
    Generic(N: Positive:=8);                     --类属 Generic 定义，N 为正整数 8
    Port(a,b: In Std_Logic_Vector( N-1 Downto 0);        --N 为被加数，加数
         ci: In Std_Logic;                               --进位输入
         co: Out Std_Logic;                              --进位输出
         s: Out Std_Logic_Vector(N-1 Downto 0));         --N 为和数
End Entity addern;
```

示例 3-1a 元件例化法，addern 结构体 top1

```
Architecture top1 Of addern Is                   --结构化非参数化的 8 位加法

    Component fulladder Is                       --1 位加法器元件声明
        Port(x,y,cin: In Std_Logic;              --元件端口说明
             cout,sum: Out Std_Logic);
    End Component fulladder;

Signal c: Std_Logic_Vector(7 Downto 1);          --内部连线，进位级联信号
Begin
    fa0: fulladder Port Map(x=>a(0),y=>b(0),cin=>ci,cout=>c(1),sum=>s(0));
                                                 --此元件例化是非参数化的
    fa1: fulladder Port Map(x=>a(1),y=>b(1),cin=>c(1),cout=>c(2),sum=>s(1));
    fa2: fulladder Port Map(x=>a(2),y=>b(2),cin=>c(2),cout=>c(3),sum=>s(2));
    fa3: fulladder Port Map(x=>a(3),y=>b(3),cin=>c(3),cout=>c(4),sum=>s(3));
    fa4: fulladder Port Map(x=>a(4),y=>b(4),cin=>c(4),cout=>c(5),sum=>s(4));
    fa5: fulladder Port Map(x=>a(5),y=>b(5),cin=>c(5),cout=>c(6),sum=>s(5));
    fa6: fulladder Port Map(x=>a(6),y=>b(6),cin=>c(6),cout=>c(7),sum=>s(6));
    fa7: fulladder Port Map(x=>a(7),y=>b(7),cin=>c(7),cout=>co,sum=>s(7));
    --这里采用名称映射法，此法实际信号可与元件端口说明中的信号说明顺序不同
End Architecture top1 ;
```

示例 3－1b 元件例化法，addern 结构体 top2

```
Architecture top2 Of addern Is                           --结构化非参数化的 8 位加法

Signal c：Std_Logic_Vector(8 Downto 0)；

--绑定元件例化给实体/结构体对可以采用下面的方法
--在这个例子中，fa0，fa1，fa2 绑定在同一个结构体 fulladder 不同子结构体上
--以下 3 句例化语句也可以在 Configuration 中采用元件配置的方式来指定
For fa0 ：fulladder  Use Entity Work.fulladder(faone)；
For fa1 ：fulladder  Use Entity Work.fulladder(fatwo)；
For fa2 ：fulladder  Use Entity Work.fulladder(fathree)；
--如果没有下句综合结果 fa3－fa7 与 fa2 同，即沿用上面最后一句的说明
For Others：fulladder  Use Entity Work.fulladder(faone)；
--这句说明 fa3－fa7 采用 faone 结构
--For All：fulladder  Use Entity Work.fulladder(faone)；
--如用"For All："则所有结构体采用 faone 结构
--对较简单的设计采用这种方法，对于较复杂的设计采用 Configuration 比较好

Begin
    c(0)<=cin；
    fa0：fulladder Port Map(a(0)，b(0)，c(0)，c(1)，s(0))；
    fa1：fulladder Port Map(a(1)，b(1)，c(1)，c(2)，s(1))；
    fa2：fulladder Port Map(a(2)，b(2)，c(2)，c(3)，s(2))；
    fa3：fulladder Port Map(a(3)，b(3)，c(3)，c(4)，s(3))；
    fa4：fulladder Port Map(a(4)，b(4)，c(4)，c(5)，s(4))；
    fa5：fulladder Port Map(a(5)，b(5)，c(5)，c(6)，s(5))；
    fa6：fulladder Port Map(a(6)，b(6)，c(6)，c(7)，s(6))；
    fa7：fulladder Port Map(a(7)，b(7)，c(7)，c(8)，s(7))；
    --For i In N－1 Downto 0 Ccncratc
    --u1：fulladder                                    --fulladder 例化，例化语句
    --Port Map(a(i)，b(i)，cin=>c(i)，cout=>c(i+1)， s(i))；
    --End Generate；                                   --这 4 句可替代 fa0 到 fa7 语句
    c(0)<= c(8)；
    --元件例化采用位置对应法，这种方法带入信号顺序必须与元件说明顺序完全对应
    --注释的并行循环生成语句虽可替代 fa0 到 fa7 语句却难以做元件配置了
    --fulladder.vhd 在 Work 库中，实体前有"Use Work.All；"则可不说明元件直接引用
End Architecture top2 ；
```

示例 3－1c　循环元件例化法，addern 结构体 top3

```
Architecture top3  Of addern Is                          --结构化参数化的 N 位加法

    Component fulladder Is                               ---一位加法器元件说明
        Port(x,y,cin: In Std_Logic;
             cout,sum: Out Std_Logic);
    End Component fulladder;

Signal c: Std_Logic_Vector(N Downto 0);                  --N+1 位
Begin
    Gen: For i In N-1 Downto 0 Generate        --生成语句(并行),无需声明循环变量 i
        t0: If i = 0 Generate            --条件生成语句,例化第一级, 例化语句须标志
            u1: fulladder                                --fulladder 例化,例化语句
            Port Map(x=>a(i),y=>b(i),cin=>ci,cout=>c(i+1),sum=>s(i));
        End Generate t0;                                 --第一级单元例化结束
        t1: If i > 0 And  i<N-1 Generate                 --生成语句须标志
            u1 : fulladder                               --例化中间级单元
            Port Map(x=>a(i),y=>b(i),cin=>c(i),cout=>c(i+1),sum=>s(i));
        End Generate t1;                                 --中间级单元例化结束
        t2: If i =N-1  Generate                          --并行语句,可以有 Else
            u1 : fulladder                               --例化最后级单元
            Port Map(x=>a(i),y=>b(i),cin=>c(i),cout=>co,sum=>s(i));
        End Generate t2;                                 --最后级单元例化结束
    End Generate Gen;
End Architecture top3;
```

示例 3－1d　函数法，addern 结构体 addf

```
Architecture addf Of addern Is                           --参数化的 N 位函数加法

Signal s1: Std_Logic_Vector(N Downto 0);
Function addn(a,b: Std_Logic_Vector( N-1 Downto 0);carry: Std_Logic)
    Return Std_Logic_Vector Is                           --函数说明
    Variable cout: Std_Logic;                            --进位输出
    Variable cin: Std_Logic:= carry;                     --进位输入
    Variable sum: Std_Logic_Vector(N Downto 0):=(Others=>'0');       --和数
    Begin                                                --函数中可以是顺序语句
        loop1: For i In 0 To N-1 Loop                    --0 到 N-1 的循环 loop1
            cout:=(a(i) And b(i)) Or (a(i) And cin) Or (b(i) And cin);
```

```
          sum(i):=a(i) Xor b(i) Xor cin;
        End Loop loop1;                    --循环 loop1 描述结束
        sum(N):= cout;
        Return sum;                        --这种 Return 返回专用于函数,必须返回值
    End Function addn;                     --函数说明结束
    Begin                                  --结构体描述开始
    s1 <= addn(a,b,ci);                    --函数调用带回一个值,实参为输入类型
    s <= s1(N-1 Downto 0);                 --N 位加法和数
    co <= s1(N);                           --N 位加法进位输出
End  Architecture addf;
```

示例 3-1e　过程法 1,addern 结构体 addp1

```
Architecture addp1 Of addern Is                       --参数化的 N 位过程实现的加法
    Procedure addvec                                  --声明过程与参数
    ( a: In Std_Logic_Vector(N-1 Downto 0);           --输入参量,被加数
      b: In Std_Logic_Vector(N-1 Downto 0);           --输入参量,加数
      cin: In Std_Logic;                              --输入参量,进位输入
      Signal sum: Out Std_Logic_Vector;               --输出参量,和数
      Signal cout: Out Std_Logic;                     --输出参量,进位输出
      N: In Positive) Is                              --Positive 为正整数
    Variable c: Std_Logic;
    Begin                                             --开始过程体中的执行语句
        c:=cin;
        loop1: For i In 0 To N-1 Loop                 --0 到 N-1 的循环
            sum(i)<=a(i) Xor b(i) Xor c;
            c:=(a(i) And b(i)) Or (a(i) And c) Or (b(i) And c);
        End Loop loop1;
        cout<=c;
    End Procedure addvec;                             --过程说明结束

Begin                                                 --结构体描述开始
    addvec(a,b,ci,s,co,N);                            --过程调用,填入输入和输出实参
End Architecture addp1;                               --结构体描述结束
```

示例 3-1f　过程法 2,addern 结构体 addp2

```
--用数组属性自动判断位数的加法过程
--当加数与被加数不等长会通过断言报告,错误级别为出错,断言语句不可综合
Architecture addp2 Of addern Is                          --应用属性的过程加法
    Procedure addvec2(add1,add2: Std_Logic_Vector;       --端口类型缺省指输入
```

```
                    cin: In Std_Logic;
                    Signal sum: Out Std_Logic_Vector;
                    Signal cout: Out Std_Logic) Is
    Variable c: Std_Logic:=cin;
    Alias n1: Std_Logic_Vector(add1'Length-1 Downto 0) Is add1;          --别名
    Alias n2: Std_Logic_Vector(add2'Length-1 Downto 0) Is add2;
    Alias s:  Std_Logic_Vector(add2'Length-1 Downto 0) Is sum;
    Begin
        Assert ((n1'Length=n2'Length) And (n1'Length=s'Length))
                                                      --断言语句
        Report "vector Lengths must be equal!"        --报告信息
        Severity Error;                               --错误级别 Error
        loop1: For i In n1'Reverse_Range Loop
                                                 --反排序范围值从低位 LSB 开始循环
            sum(i)<=n1(i) Xor n2(i) Xor c;
            c:=(n1(i) And n2(i)) Or (n1(i) And c) Or (n2(i) And c);
        End Loop loop1;
        cout<=c;                                      --进位输出
    End Procedure addvec2;                            --过程说明结束
Begin
    addvec2(a,b,ci,s,co);                             --过程调用
End Architecture addp2;
```

示例 3-1g　重载函数法,addern 结构体 oadd

```
Architecture oadd Of addern Is                        --参数化的重载函数加法
Signal s1:Std_Logic_Vector(N Downto 0);
Begin
    s1<='0'&a+b+ci;          --& 为并置符,这里表示在数值 a 前拼接'0',扩展出进位
                                  --引用了 Ieee 库 Std_Logic_Unsigned 程序包
    s<=s1(N-1 Downto 0);
    co<=s1(N);                    --拆位,最高位为进位输出,其余为和数
End Architecture oadd;
```

示例 3-1h　自定义重载函数法,addern

重载函数的应用通常要求说明 Ieee 库 Std_Logic_Unsigned 程序包,但对于自行设计的如本例要将命名为 Std_Logic_overload. vhd 的文件先编译。Std_Logic_overload 包中说明了重载函数。运行前要在实体用 Work 库自定义的程序包来替代 Std_Logic_Unsigned 程序包。

```
--自定义重载函数加法
Architecture adderfo_arch Of addern Is
```

```
Signal s1,q1:Std_Logic_Vector(N Downto 0);
Signal carry:Std_Logic_Vector(N-1 Downto 0);
Signal q2:Std_Logic;
Constant zeros: Std_Logic_Vector(N-2 Downto 0):=(Others=>'0');

    Function "+" (Add1, Add2: Std_Logic_Vector)
        Return Std_Logic_Vector Is
        --返回两个 Std_Logic_Vector 类型相加的运算结果
        Variable sum:  Std_Logic_Vector(N-1 Downto 0);
        Variable sum1: Std_Logic_Vector(N Downto 0);
        Variable c: Std_Logic := '0';                    --Carry In
        Alias n1: Std_Logic_Vector(N-1 Downto 0) Is Add1;
        Alias n2: Std_Logic_Vector(N-1 Downto 0) Is Add2;
        Begin
        For i In 0 To N-1 Loop
            sum(i) := n1(i) Xor n2(i) Xor c;
            c := (n1(i) And n2(i)) Or (n1(i) And c) Or (n2(i) And c);
        End Loop;
        sum1:=c&sum ;
        Return (sum1);
    End Function  "+";
Begin

    carry<=zeros&ci;                        --前面加 0 对齐数据的位数
    s1 <=a+ b;                              --自定义重载加法
    q1 <= s1(N-1 Downto 0)+carry;           --被加数、加数、进位输入相加
    q2<= q1(N) Xor s1(N);
    co<= q2;
    s<=q1(N-1 Downto 0);
End Architecture adderfo_arch;
```

示例 3-1i　自定义程序包和自定义重载函数法

```
Library Ieee;
Use Ieee.Std_Logic_1164.All;

Package Std_Logic_overload Is                       --程序包头说明
    Function "+" (add1, add2: Std_Logic_Vector)     --包头的声明公共可见
    Return Std_Logic_Vector;                        --返回值类型 Std_Logic_Vector
End Package Std_Logic_overload;                     --程序包头说明结束
```

```
                          --这个函数返回两个 Std_Logic_Vector 类型相加的运算结果
Package Body Std_Logic_overload Is                      --程序包体说明
    Function "+" (add1, add2: Std_Logic_Vector)         --函数说明
    Return Std_Logic_Vector Is                       --函数返回类型 Std_Logic_Vector
    Variable sum: Std_Logic_Vector(add1'Length-1 Downto 0);
    Variable c: Std_Logic := '0';
                                                        --包体中的声明为内部不可见
    Alias n1: Std_Logic_Vector(add1'Length-1 Downto 0) Is add1;          --别名
    Alias n2: Std_Logic_Vector(add2'Length-1 Downto 0) Is add2;
    Begin
        For i In sum'Reverse_Range Loop                 --用信号范围属性获得循环值
            sum(i) := n1(i) Xor n2(i) Xor c;            --逻辑"异或"运算
            c := (n1(i) And n2(i)) Or (n1(i) And c) Or (n2(i) And c);
        End Loop;                                       --逻辑"与""或"运算
        Return (sum);                                   --函数返回和数
    End Function "+";                                   --函数说明结束
End Package Body Std_Logic_overload;                    --包体说明结束

/* 实体说明与其他结构体相同,但需要调用自定义程序包,
Use Work.std_logic_overload.All;为了体现自定义重载,关闭程序包
--Ieee.Std_Logic_Unsigned */
Architecture oadds Of addern Is                         --参数化的重载函数加法
Signal s1,q1:Std_Logic_Vector(N Downto 0);
Signal carry:Std_Logic_Vector(N-1 Downto 0);
Signal q2:Std_Logic;
Constant zeros: Std_Logic_Vector(N-2 Downto 0):=(Others=>'0');
Begin
    carry<=zeros&ci;                                    --前面加 0 对齐数据的位数
    s1 <=a+ b;                                          --自定义重载加法
    q1 <= s1(N-1 Downto 0)+carry;                       --被加数、加数、进位输入相加
    q2<= q1(N) Xor s1(N);
    co<= q2;
    s<=q1(N-1 Downto 0);
End Architecture oadds;
```

(1)运行例 3-1h addern 的 oadds 这个结构体的注意与准备。

要先在实体程序包说明的位置加上自定义的程序包 std_logic_overload,即在实体前加上"Use Work. std_logic_overload. All;"语句。关闭(注释上)Std_Logic_Unsigned 程序包说明。

示例 3－1i 自定义的重载函数包 std_logic_overload 没有考虑不同数位数对齐的问题，因而运算进位位要用前面加 0 对齐的处理。函数、过程与进程一样，其中采用顺序语句描述。

程序包如果仅仅是定义数据类型数据对象等内容，只需包首即可，而要包含子程序则一定要有包体。

(2)可以不说明 Work 库。Work 库与 Std 库在 VHDL 中是缺省的。要注意，Work 中的文件在 Quartus Ⅱ主菜单栏点击 Assignments→Settings→Files 中应该有，当然通常在工程中正确编译的程序会自动添加在其中的。有时文件结构乱了，可以整理一下 Files。

(3)add1 ' Length 中“' ”是属性，' Length 可以得到数组的长度。

sum ' Reverse_Range，' Reverse_Range 属性的应用得到的是反排序标识数组 sum 范围的数据。

(4)程序中的语法，Assert 断言语句。

```
    Assert Boolean - expression                      --断言
      Report string - expression                     --报告
    [Severity severity - level;]                     --错误级别
```

Severity level 有 4 种可能的值：

Note：注意。在仿真时传递信息。

Warning：警告。此时仿真过程可继续，但结果是不可预知的。

Erorr：错误。仿真过程不能继续执行。

Failure：致命错误。发生了致命错误，仿真过程必须停止。

注：Assert 断言语句是不可综合的。

以上是示例 3－1 实体 addern 的 9 个结构体，以下程序是示例 3－1 程序中调用的元件 1 位加法器 fulladder 的 VHDL 描述。

示例 3－1j　全位加器 fulladder 实体(addern 的元件)

```
--全位加器程序文件 fulladder.vhd
Library Ieee;
Use Ieee.Std_Logic_1164.All;
Use Ieee.Std_Logic_Unsigned.All;

Entity fulladder Is                          --声明 1 位加法器的实体 fulladder
    Port(x,y,cin: In Std_Logic;              --被加数、加数、进位输入
         cout,sum: Out Std_Logic);           --进位输出，和数
End Entity fulladder;                        --1 位加法器实体端口描述结束
```

示例 3－1k　全位加器 fulladder 结构体 RTL 级描述

```
--全位加器结构体,RTL 级描述
Architecture faone Of fulladder Is           --1 位加法器的第 1 个结构体 faone
Begin
```

```
    sum<= x Xor y Xor cin;
    cout<=(x And y) Or (x And cin) Or (y And cin);            --逻辑符号有优先级
End Architecture faone;
```

示例 3－1l　全位加器 fulladder 结构体 Block RTL 级描述

```
--全位加器结构体,应用 Block RTL 级描述
Architecture fatwo Of fulladder Is                 --1 位加法器的第 2 个结构体 fatwo
Signal so1,co1,co2 : Std_Logic;                    --中间连线
Begin
    h_adder1 : Block                               --第 1 个 Block 说明
    Begin
        so1<= x Xor y;                             --半加器
        co1<= x And y;
    End Block h_adder1;                            --Block 说明结束
    h_adder2: Block                                --第 2 个 Block 说明
    Signal so2 : Std_Logic;
    Begin
        so2 <= so1 Xor cin;
        co2 <= so1 And cin;
        sum <= so2;                                --和数
    End Block h_adder2;
    or2 : Block                                    --第 3 个 Block 说明
    Begin
        cout <= co2 Or co1;                        --进位输出
    End Block or2;
End Architecture fatwo;
```

示例 3－1m　fulladder 结构体行为级描述

```
--全位加器结构体,行为描述
Architecture fathree Of fulladder Is               --1 位加法器的第 3 个结构体 fathree
Signal so1 : Std_Logic_Vector(1 Downto 0);
Begin
    so1<='0'&x+y+cin;          --不同位数的加法,需要 Std_Logic_Unsigned 包
    sum<=so1(0);                                   --1 位加法器的和数
    cout<=so1(1);                                  --1 位加法器进位输出
End Architecture fathree;
```

示例 3－1n　全位加器 fulladder 结构体结构性描述

```
--全位加器结构体,结构性描述
```

```
Architecture fafour Of fulladder Is              --1 位加法器的第 4 个结构体 fafour
    Component halfadder Is                       --半加器元件说明
        Port(x,y: In Std_Logic;
             cout, sum: Out Std_Logic);
    End Component halfadder;

Signal d,e,f: Std_Logic;
Begin
    u1: halfadder Port Map(x=>x,y=>y,cout=>d,sum=>e);
                                                 --元件半加器的例化
    u2: halfadder Port Map(x=>e,y=>cin,cout=>f,sum=>sum);
    cout<=d or f;                                --逻辑"或"运算
End Architecture fafour;
```

示例 3－1o　半加器 halfadder RTL 级描述

```
--半加器程序文件 halfadder.vhd
Library Ieee;
Use Ieee.Std_Logic_1164.All;

Entity halfadder Is                              --声明半加器实体 halfadder
    Port (x,y: In Std_Logic;                     --被加数、加数
          cout, sum: Out Std_Logic);             --进位输出、和数
End Entity halfadder;                            --半加器实体端口描述结束

Architecturehalfadder_arch Of halfadder Is       --半加器结构体
Begin
    sum <= x Xor y;                              --逻辑"异或"运算
    cout <= x And y;                             --逻辑"与"运算
End Architecture halfadder_arch;
```

3.2.2　配置实现与配置方法

1. 默认配置法,fulladder 配置

```
Configuration addern_cfg1 Of addern Is           --配置 addern 的说明
    For top1                                     --指定 addern 的结构体为 top1
    End For;
End Configuration addern_cfg1;  --可改变结构体名,对比 addern 不同的结构体

Configuration fulladder_cfg Of fulladder Is      --配置实体 fulladder 的说明
    For fathree                                  --指定 fulladder 的结构体为 faone
```

```
    End For;
End Configuration fulladder_cfg;                    --配置说明结束
```

2. 元件配置法 1,addern 实体结构对应法

```
Configuration addern_cfg2 Of addern Is              --配置 addern 的说明
    For top1                                        --指定 addern 的结构体为 top1
        For fa0 :fulladder Use Entity Work.fulladder(faone);
        End For;                   --采用 top1 中的 fulladder 3 个结构体中的 faone
        For fa1 :fulladder Use Entity Work.fulladder(fatwo);
        End For;                   --采用 top1 中的 fulladder 3 个结构体中的 fatwo
        For fa2:fulladder Use Entity Work.fulladder(fathree);
        End For;
        For fa3 :fulladder Use Entity Work.fulladder(faone);
        End For;
        For fa4 :fulladder Use Entity Work.fulladder(fatwo);
        End For;
        For fa5 :fulladder Use Entity Work.fulladder(fathree);
        End For;
        For fa6 :fulladder Use Entity Work.fulladder(faone);
        End For;
        For fa7 :fulladder Use Entity Work.fulladder(fatwo);
        End For;
    End For;
End Configuration addern_cfg2;
```

3. 元件配置法 2,addern 调低层配置文件法

```
Configuration addern_cfg3 Of addern Is              --配置 addern 的说明
    For top3                                        --指定 addern 的结构体为 top3
        For Others:fulladder Use Configuration Work.fulladder_cfg; --调低层配置
        End For;  --top3 中的全加器采用 Work 库中配置文件 fulladder_cfg 中的说明
    End For;
End Configuration addern_cfg3;
```

注:

(1)根据 Configuration 语句的应用说明,addern 的 Configuration 只能使用实体结构对应法或者调用低层配置文件法之一。

(2)For Others 中用 Others 替代了具体的例化元件的标号。

3.2.3 加法器程序说明

(1)示例 3-1c 程序采用了 If Generate 语句,2008 版的支持这个语句有 Else,同时还有 Case Generate 语句。

(2)程序 3－1g,即加法 addern 的 第 7 个结构体采用 Std_Logic_1164 和 Std_Logic_Unsigned 程序包,其运算缺省的是无符号运算(Unsigned 8-bit Adder)。与程序 1－1 有符号加法器运算比较,这个运算含有进位,进位是通过在运算前增加了'0'连接实现的,'0'&a 表示在 8 位数组的最高位前加'0',成为了 9 位数组,这里 & 代表拼接之意。

增加 Std_Logic_Unsigned 程序包说明。VHDL 作为强类型语言,通常要求运算的数据不仅要输入/输出类型一致,而且要求位宽一致。Std_Logic_Unsigned 含有运算符重载函数,使得标准类型、标准矢量类型可以与整数等一起运算,运算可以位宽不同。对于这个程序的意义就是运算数据的实际位宽不同。

(3)程序中 Generic 的运用是值得推荐的,它可以实现实体间的参数传递,参数的重置,其参数化的特点使得程序的通用性、灵活性等更好,其不仅能如本例中设置位宽,将其设为时间参数,还可以仿真不同的时序带给设计的不同。比如,了解和比较用不同器件库的设计差异。

(4)示例 3－1 中被调用的一位加法器说明:

①一位加法器的结构体 1(faone)、2(fatwo)、4(fafour)采用的是 RTL 描述风格。

②结构体 2(fatwo)、4(fafour)两种方式可以完全对应,一个是展开的用 Block 来划分描述,一个是采用元件例化进行调用的。

③一位加法器的结构体 2(fatwo)采用了 Block 格式描述,提了高加法器的可读性,事实上,采用 Block 的方式仿真与综合的速度也更快。

④一位加法器的结构体 3(fathree)采用的是行为级的结构体描述风格。要注意的是,这里涉及到了不同类型和不同位宽的加法运算,所以应用了 Std_Logic_Unsigned 程序包的重载函数。

(5)一个程序是否优化不在于它的代码看起来是否简洁,一个看似不简洁的程序也许是一个简洁优化的电路,最终的判断要关注综合后逻辑资源报告、RTL Viewer 视图以及时序分析结果直至硬件下载测试应用。目前设计到的例子都比较短小,因而基本上仅从 RTL Viewer 视图结果来看就可以说明了。如表 3－1 所示综合结果。

(6)示例 3－1 中采用了元件配置和默认的块配置来决定实体与结构体对。块配置直接指定了实体与结构体对,如 addern 的配置 addern_cfg1;元件配置在此基础上可以进一步指定结构体中元件所调用的单元,即更低层级的实体与结构体对,如 addern 的配置 addern_cfg2。

在示例中元件配置给出来 3 种方法,第一种是在结构体说明去用 For 短语引导,见示例 3－1b;第二种是在 Configuration 语句中,用 Use 引导直接指定实体与结构体对,如 addern_cfg2;第三种是 Use 引导指定底层 Configuration 设定,如 addern_cfg3。

元件配置还常常用于端口配置。如果例化的元件与配置中元件的端口名不一致,可以做端口映射,具体方法是在指定实体与结构体对语句后,写端口映射语句 Port Map,同样 Port Map 语句也是可以前置类属参数映射 Generic Map。

端口配置的使用可以进一步提高设计模块和测试模块的复用能力。

表 3-1　addern 各层级描述综合结果一览

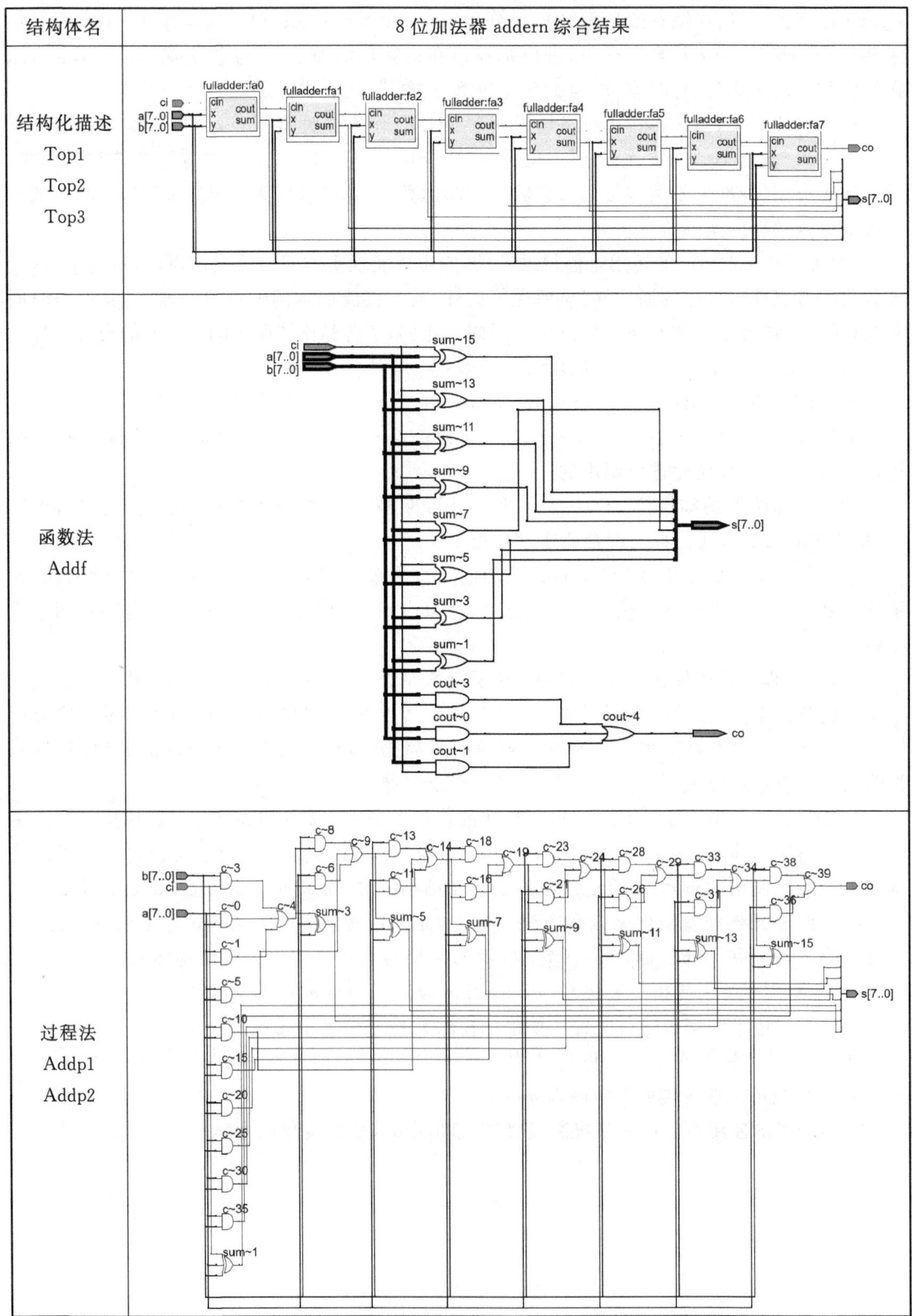

结构体名	8 位加法器 addern 综合结果
结构化描述 Top1 Top2 Top3	
函数法 Addf	
过程法 Addp1 Addp2	

续表

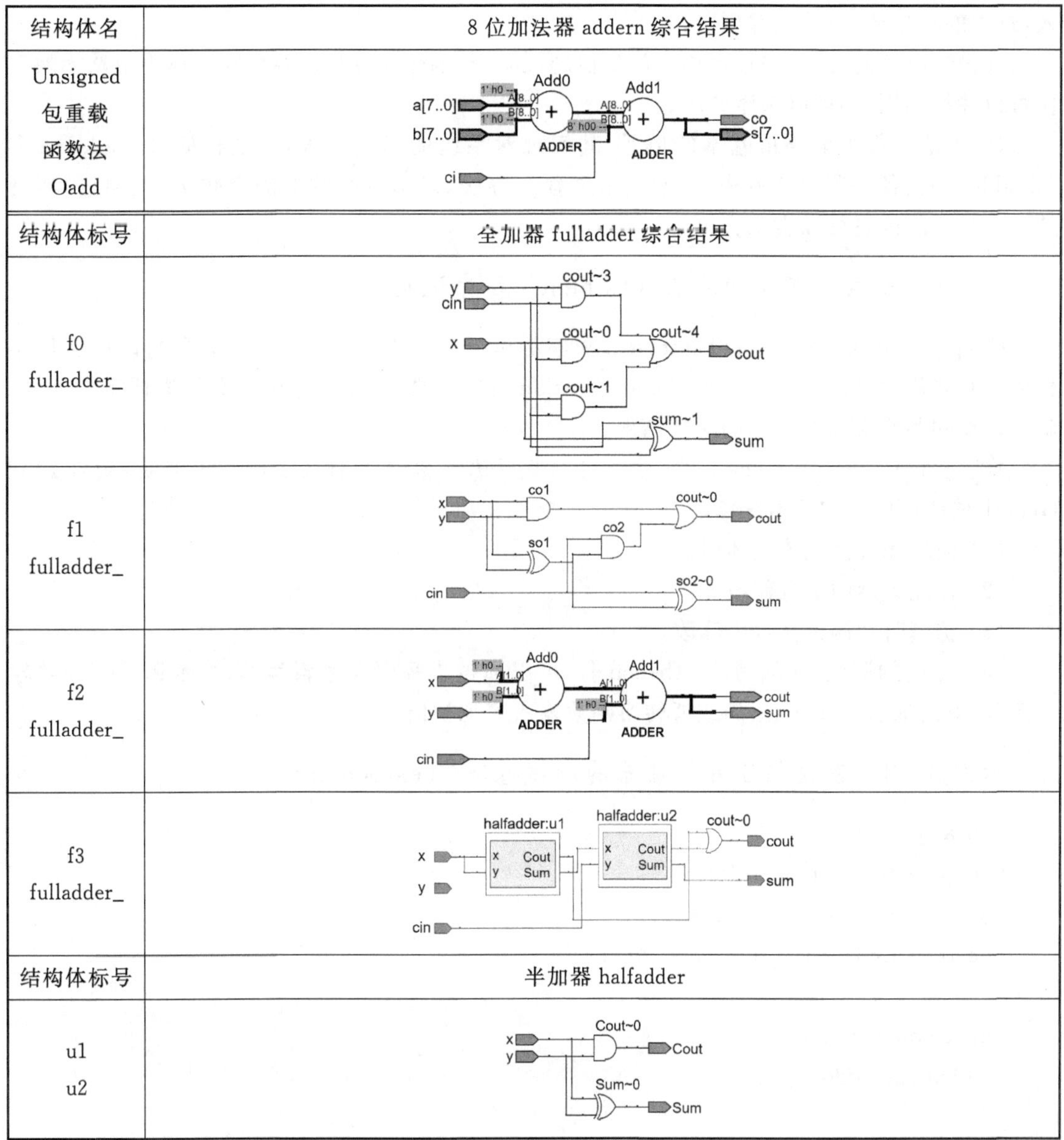

结构体名	8 位加法器 addern 综合结果
Unsigned 包重载 函数法 Oadd	
结构体标号	全加器 fulladder 综合结果
f0 fulladder_	
f1 fulladder_	
f2 fulladder_	
f3 fulladder_	
结构体标号	半加器 halfadder
u1 u2	

(7)加法器行为描述的 RTL Viewer 结果是加法器符号单元及其级联。

加法器结构化的描述 RTL Viewer 结果是加法单元模块及其级联,选中其中一个模块点击,可以看到底层的 RTL 描述。底层加法器 RTL 描述的 RTL Viewer 结果是门电路组成的。

直接调用 8 次一位加法器实现的电路比用循环调用的电路运行速度略快,所以在程序中有时更简洁、更直观(硬件层面)的语法性能会更好。

综合器可以进行速度、面积、所见即所得的优化等,具体采用的优化可以在 Quartus Ⅱ中根据设计需要进行设置,然后编译程序,再用 RTL Viewer 工具查看优化后的综合结果。当然,优化过程中,最先也是最基础的考虑是设计本身,如原理、算法、逻辑、设计结构,代码风格,其次是约束,最后才考虑工具。

(8)函数实现的加法比过程实现的加法占用资源少,这也很容易理解,函数的参数只有输入,过程的参数输入/输出都可以。

在设计应用时,函数与过程往往放到包的说明中,函数与过程的好处在于作为电路模块不仅可以并行调用,还可以顺序调用。

(9)加法器作为电路最基本的单元,其实现对系统的影响很大,很值得研究。addern 和 fulladder 的其他一些描述方法,比如应用进程的方法,采用专用的加法级联缓冲(原语)实现的。有兴趣的读者可以尝试。

3.2.4 参数化任意加法器 Testbench 程序说明

Congfiguration 语句放在 Testbench 程序测试平台中非常方便,对于不同的设计实体结构体对只要修改 Testbench 中的 Congfiguration 反复仿真,无需对原设计文件进行重新编译,就可以获知各个实体结构体的时序结果。

推荐在设计中采用 Generic 类属语句。类属语句默认实体 Generic 的声明,要注意到 Generic的值可以在如下情况修改:

(1) 使用元件例化该实体时。

(2) 在配置声明部分修改。

(3) 仿真时在测试文件中修改。

为了全面的测试,在编写 Testbench 时,可以通过进程以及逻辑运算、算术运算等来给输入序列,有时通过 Others 语句或者转换函数输入序列初值。

示例 3-2 参数化任意加法器的测试基准(Testbench)

```
Library Ieee;
Use Ieee.Std_Logic_Arith.All;
Use Ieee.Std_Logic_1164.All;
Use Ieee.Std_Logic_Unsigned.All;

Entity addern_tb Is
End Entity addern_tb;

Architecture addern_tb_arch Of addern_tb Is
Signal a:Std_Logic_Vector(N-1 Downto 0):=Conv_Std_Logic_Vector(0,N);
                                            --a 的初值是“0000……0”,N 个‘0’
Signal b:Std_Logic_Vector(N-1 Downto 0):=Conv_Std_Logic_Vector(1,N);
                                            --b 的初值是“0000……1”
Signal ci:Std_Logic :='0';                  --进位输入初值为‘0’
Signal co:Std_Logic;
Signal s:Std_Logic_Vector(N-1 Downto 0);

    Component addern Is                     --N 位加法器元件说明
```

```
        Generic (N:Positive:= 8);            --Positive,大于 0 的正整数,8 位加法器
        Port(a,b:In Std_Logic_Vector(N-1 Downto 0);
             ci:In Std_Logic;
             co:Out Std_Logic;
             s:Out Std_Logic_Vector(N-1 Downto 0));
    End Component addern;

Begin
    DUT:addern Generic Map(N=>8)--Generic Map 重新指定位数,如指定 8 位
    Port Map(a=>a,b=>b,ci=>ci,co =>co,s=>s);

    Process(a,b,ci) Is                    --利用进程、逻辑运算与算术运算给出输入序列
    Begin
        ci <=Not ci After 100 ns,'0' After 200 ns ;
        a<=a+1 After 50 ns;                       --采用了操作符重载
        b<=b+1 After 50 ns;                       --After 仅用于仿真,不可综合
    End Process;                          --进程可以无限循环,所以输入激励可连续不断
End Architecture addern_tb_arch;
```

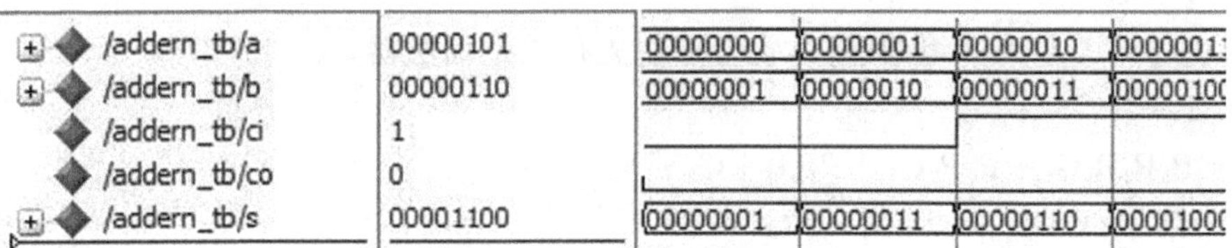

图 3-2　8 位加法器 Modelsim 仿真结果

程序说明：

(1)Generic Map(N => 8) 是类属表的按名称映射，或称参数传递语句，它可在使用元件例化实体时，修改原实体如 addern 中说明的 Generic 数据。第一个 N 是 addern 指定的形参，第二个 N 是 addern_tb 指定的实参。加法器的位数可以在仿真 addern_tb 指定。

(2)加法器采用按位置映射的方法，如果做 4 位加法仿真，上述语句也可以写作 Generic Map(4)。

(3)在初始化值的时候，如果各位都是'0'，可以由(Others=>'0')来指定。

(4)采用转换函数将整数转换为指定位宽的 Std_Logic_Vector 类型的值时很方便。如，Conv_Std_Logic_Vector(1,N)是采用转换函数，将整数 1 转换为 Std_Logic_Vector 类型的数"00000001"。

3.3　先行进位加法器

加法器布尔表达式：

$$S_0 = A_0 \oplus B_0 \oplus C_0 \tag{3-1}$$

$$C_1 = (A_0 \oplus B_0)C_0 + A_0B_0 \tag{3-2}$$

$$S_1 = A_1 \oplus B_1 \oplus C_1 \tag{3-3}$$

$$C_2 = (A_1 \oplus B_1)C_1 + A_1B_1 \tag{3-4}$$

$$S_2 = A_2 \oplus B_2 \oplus C_2 \tag{3-5}$$

$$C_3 = (A_2 \oplus B_2)C_2 + A_2B_2 \tag{3-6}$$

$$S_3 = A_3 \oplus B_3 \oplus C_3 \tag{3-7}$$

$$C_4 = (A_3 \oplus B_3)C_3 + A_3B_3 \tag{3-8}$$

先行进位加法器定义两个函数：

产生函数：$G_i = A_iB_i$ (3-9)

传递函数：$P_i = A_i \oplus B_i$ (3-10)

将(3-9)(3-10)代入上面(3-1)～(3-8)式

$$C_1 = P_0C_0 + G_0 \tag{3-11}$$

$$C_2 = P_1P_0C_0 + P_1G_0 + G_1 = P_1C_1 + G_1 \tag{3-12}$$

$$C_3 = P_2P_1P_0C_0 + P_2P_1G_0 + P_2G_1 + G_2 = P_2C_2 + G_2 \tag{3-13}$$

$$C_4 = P_3P_2P_1P_0C_0 + P_3P_2P_1G_0 + P_3P_2G_1 + P_3G_2 + G_3 = P_3C_3 + G_3 \tag{3-14}$$

……

$$C_n = P_{n-1}C_{n-1} + G_{n-1} \tag{3-15}$$

$$S_n = P_n \oplus C_n \tag{3-16}$$

如果进行以 4 位为一组的级联，设计成组先行进位加法器，设

$$P_G = P_3P_2P_1P_0 \tag{3-17}$$

$$G_G = P_3P_2P_1G_0 + P_3P_2G_1 + P_3G_2 + G_3 \tag{3-18}$$

其中：S_n 为第 n 位和数；C_n 为第 n 位进位；

A_n 为第 n 位被加数；B_n 为第 n 位加数；

G_G 为 4 位先行进位单元的产生函数；

P_G 为 4 位先行进位单元的传递函数。

推导 C_4，C_8，C_{12}，C_{16} 得到：

$$C_4 = P_{G0}C_0 + G_{G0} \tag{3-19}$$

$$C_8 = P_{G1}P_{G0}C_0 + P_1G_{G0} + G_{G1} = P_{G1}C_1 + G_{G1} \tag{3-20}$$

$$C_{12} = P_{G2}P_{G1}P_{G0}C_0 + P_{G2}P_{G1}P_{G0} + P_2G_1 = P_{G2}C_2 + G_{G2} \tag{3-21}$$

$$C_{16} = P_{G3}P_{G2}P_{G1}P_{G0}C_0 + P_{G3}P_{G2}P_{G1}G_{G0} + P_{G3}P_{G2}G_{G1} + P_{G3}G_{G2} + G_{G3} = P_{G3}C_3 + G_{G3} \tag{3-22}$$

公式(3-11)(3-12)(3-13)(3-14)与公式(3-19)(3-20)(3-21)(3-22)形式相同，如果把前者的进位逻辑称为组内进位逻辑，后者的进位逻辑称为组间进位逻辑，可以看到，组内进位逻辑与组间进位逻辑是相同的。

4 位先行进位器的 4 级连接如图 3-3 所示，下半部分组合电路所示的是进位逻辑，上半部则是 4 个带有 P 和 G 引出端的 4 个全加器。

16 位双重先行进位加法器的 4 级连接如图 3-4 所示，这种连接的加法器称为成组先行进位加法器(Block Carry Look Ahead，BCLA)。16 位 4 级先行进位也可以由 4 片 ALU 元件

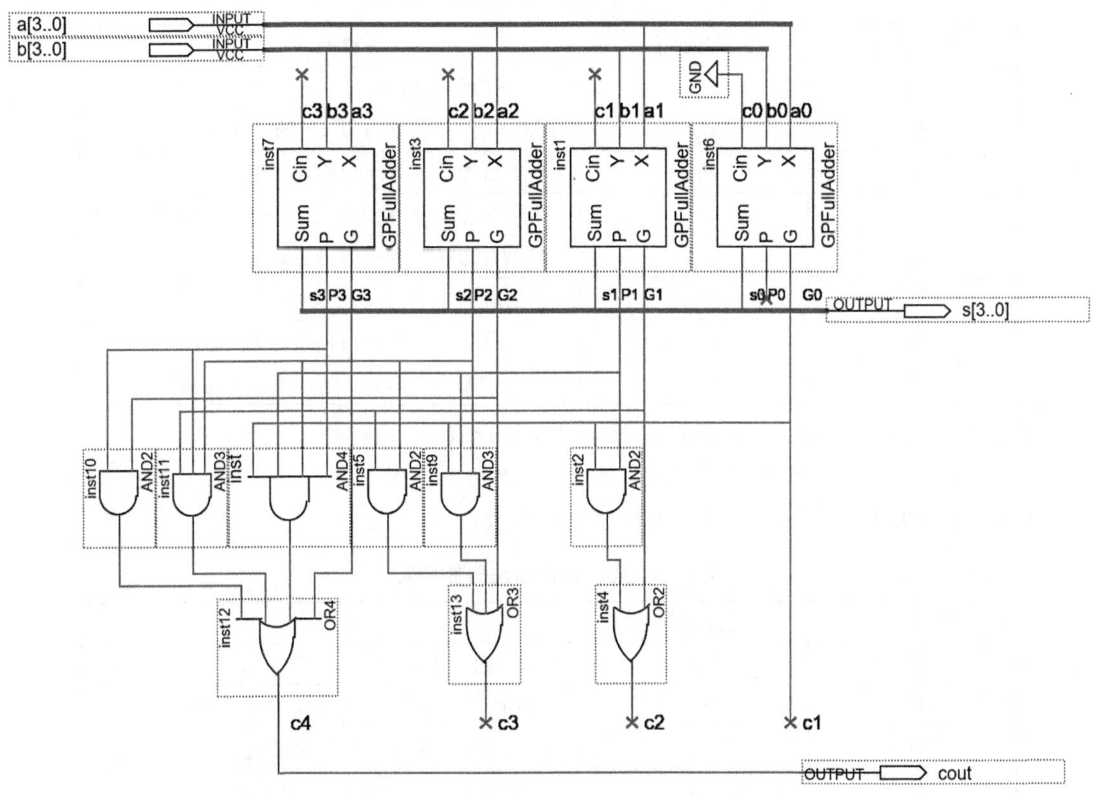

图 3 - 3　4 位先行进位加法器的 4 级电路连接

74181 与 4 片进位逻辑元件 74182 级联得到。

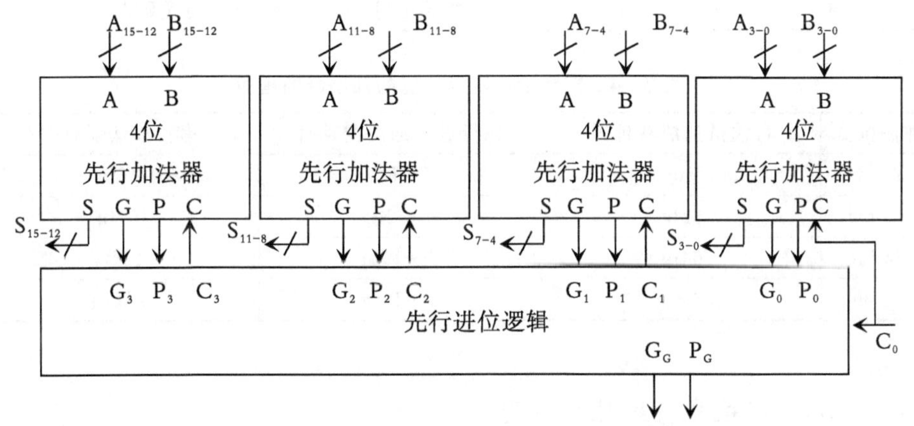

图 3 - 4　16 位先行进位加法器的 4 级连接

涟波进位加法器延时(如表 3 - 2 所示)与先行进位加法器延时(如表 3 - 3 所示),不同加法器的延时比较如表 3 - 4 所示。

表 3-2　涟波进位加法器信号延迟

输出信号	所需时间	说明
S_0	$2t_{pd}$	2 个 Xor 门传递延迟
C_1	$3t_{pd}$	1 个 Xor 门、一个 And 及 Or 的传递延迟
S_1	$4t_{pd}$	C1 产生后加一个 Xor 门传递延迟
C_2	$5t_{pd}$	C1 产生后加一个 And 门及 Or 的传递延迟
S_2	$6t_{pd}$	C2 产生后加一个 Xor 门传递延迟
C_3	$7t_{pd}$	C2 产生后加一个 And 门及 Or 的传递延迟
S_3	$8t_{pd}$	C3 产生后加一个 Xor 门传递延迟
C_4	$9t_{pd}$	C3 产生后加一个 And 门及 Or 的传递延迟

对于 N 位元加法器，C_n的传递延迟：$2n+1, n=1\sim N$

S_n的传递延迟：$(n+1)\times 2, =0\sim(N-1)$

显然，这种加法器随着位元的增加延迟会非常大。

表 3-3　先行进位加法器信号延迟

输出信号	所需时间	说明
S_0	$2t_{pd}$	2 个 Xor 门传递延迟
C_1	$3t_{pd}$	1 个 Xor 门、一个 And 及 OR 的传递延迟
S_1	$4t_{pd}$	C1 产生后加一个 Xor 门传递延迟
C_2	$3t_{pd}$	一个 Xor 门、And 门及 Or 的传递延迟
S_2	$4t_{pd}$	C2 产生后加一个 Xor 门传递延迟
C_3	$3t_{pd}$	一个 Xor 门、And 门及 Or 的传递延迟
S_3	$4t_{pd}$	C3 产生后加一个 Xor 门传递延迟
C_4	$3t_{pd}$	一个 Xor 门、And 门及 Or 的传递延迟

表 3-4　实际设计不同位宽的加法器的延时

加法位宽	行波链加法器延时	先行进位加法器延时	串行加法器延时
4 bit	8tg	5～6tg	16tg
16 bit	32tg	7～8tg	64tg
32 bit	64tg	9～10tg	128tg
64 bit	128tg	9～10tg	256tg

示例 3-3a　四位先行进位加法器

```
--4 位先行进位加法器的顶层文件 cla4.vhd
Use Work. All
Entity cla4 Is
    Port(a,b: In Bit_Vector( 3 Downto 0);
        ci: In Bit;
```

```
        s: Out Bit_Vector( 3 Downto 0);
        co,pg,gg: Out Bit);
End Entity cla4;

Architecture cla4_arch Of cla4 Is
Signal g,p: Bit_Vector(3 Downto 0);
Signal c: Bit_Vector(3 Downto 1);
Begin
    carrylogic: clalogic Port Map(g,p,ci,c,co,pg,gg);
 --先行进位逻辑元件例化
    FA0: gpfulladder Port Map(a(0),b(0),ci,g(0),p(0),s(0));
 --1 位加法元件例化
    FA1: gpfulladder Port Map(a(1),b(1),c(1),g(1),p(1),s(1));
    FA2: gpfulladder Port Map(a(2),b(2),c(2),g(2),p(2),s(2));
    FA3: gpfulladder Port Map(a(3),b(3),c(3),g(3),p(3),s(3));
End Architecture cla4_arch;

Entity gpfulladder Is
--1 位加法模块
    Port(x,y,cin: In Bit;
 --输入信号,Bit类型包括'0','1'
        g,p,sum: Out Bit);
 --进位引出端,求和引出端和输出端
End Entity gpfulladder;

Architecture gpfulladder_arch Of gpfulladder Is
Signal p_int: Bit;
Begin
    g <= x And y;
 --x 与 y 逻辑运算产生函数
    p <= p_int;
    p_int <= x Xor y;
 --x 异或 y 传递函数
    sum <= p_int Xor cin;
End Architecture gpfulladder_arch;

Entity clalogic Is
--先行进位逻辑模块
    Port(g,p: In Bit_Vector(3 Downto 0);
        ci: In Bit;
        c: Out Bit_Vector(3 Downto 1);
        co,pg,gg: Out Bit);
End Entity clalogic;

Architecture clalogic_arch Of clalogic Is
```

```
Signal gg_int,pg_int: Bit;
Begin
    c(1)<=g(0) Or (p(0) And ci);
    c(2)<=g(1) Or (p(1) And g(0)) Or (p(1) And p(0) And ci);
    c(3)<=g(2) Or (p(2) And g(1)) Or (p(2) And p(1) And g(0)) Or
            (p(2) And p(1) And p(0) And ci);
    pg_int<=p(3) And p(2) And p(1) And p(0);
    gg_int<=g(3) Or (p(3) And g(2)) Or (p(3) And p(2) And g(1)) Or
             (p(3) And p(2) And p(1) And g(0));
    co <=gg_int Or (pg_int And ci);
    pg<=pg_int;
    gg<=gg_int;
End  Architecture clalogic_arch;
```

程序说明：

(1)库声明与程序包引用。Bit等类型以及常用的逻辑运算等含在Std库里，VHDL中对Std库和Work库都是可以缺省说明的，因而这个程序可以不声明库。

经常可以看到一些本不需要引用的程序包在程序中引用了，如果在这个程序中引用了Ieee并引入Std_Logic_1164程序包，也没有关系，因为编译器或综合工具通常会自动地过滤掉一些没有用到的东西。

(2)占用资源与性能。先行进位加法通过定义传递函数和产生函数，将传统行波链加法中串行级联的进位逻辑拆出来，进行并行逻辑组合，串行变为并行，速度加快了，但也会有一些面积的代价。

示例3-3b　四位先行进位加法器测试基准文件

```
Entity cla4_tb  Is                                        --不需要说明端口
End Entity cla4_tb;

Architecture cla4_tb_arch Of cla4_tb Is
Signal ci:Bit:='0';                                       --进位输入初值为'0'
Signal a:Bit_Vector(3 Downto 0) :="0000";                 --被加数
Signal b:Bit_Vector(3 Downto 0) :="0000";                 --加数
Signal gg:Bit;
Signal pg:Bit;
Signal co:Bit;                                            --进位输出
Signal s:Bit_Vector(3 Downto 0);                          --和数

    Component cla4 Is                                     --先行进位加法器元件
        Port(ci:In Bit;
             a,b:In Bit_Vector(3 Downto 0);
```

```
            gg,pg,co:Out Bit;
            s:Out Bit_Vector(3 Downto 0) );
    End Component cla4;
Begin                                                --先行进位加法器元件例化
    DUT:cla4  Port Map( ci => ci,a => a, b => b,
                gg => gg,pg => pg,co => co, s => s);
    ci<='1'After 50 ns;                              --延时 50 ns 后,ci 赋值 为'1'
    a<= "0000" After 50 ns,"0001" After 100 ns,"0010" After 150 ns;
    b<= "0001" After 50 ns,"0010" After 100 ns,"0100" After 150 ns;
End Architecture cla4_tb_arch;
```

示例 3-4 任意指定位超前进位加减法器

加法器设计采用超前进位的方法(公式 3-15,3-16),减法运算通过加补码的方式由加法器实现。补码就是取反加 1 的运算。

```
Library Ieee;
Use Ieee.Std_Logic_1164.All;

Entity preaddsub Is                                  --任意指定位超前进位加减法器
    Generic(N:Positive:= 4 );                        --指定为四位加法器
    Port(opa,opb: In Std_Logic_Vector(N-1 Downto 0);
        c_i:In Std_Logic;                            --进位输入
        addsub_i: In Std_Logic;                      --加减法选择
        c_o: Out Std_Logic;                          --进位输出
        result: Out Std_Logic_Vector(N-1 Downto 0)); --和数
End Entity preaddsub;

Architecture preaddsub_arch Of preaddsub Is
Begin

    Process(opa,opb,c_i,addsub_i) Is
    Variable s_c: Std_Logic_Vector(N-1 Downto 0);
    Variable p,g: Std_Logic_Vector(N Downto 1);
    Variable c: Std_Logic_Vector(N Downto 0);
    Begin
        If addsub_i='1'Then                          --加法
            p:=opa Xor opb;                          --加法传递函数
            g:=opa And opb;                          --加法产生函数
            c(0):=c_i;
            For i In 1 To N Loop
```

```
            c(i):=g(i) Or (p(i) And c(i-1));        --加法进位运算
        End Loop;
        c_o<=c(N);
        s_c:=p Xor c(N-1 Downto 0);
    Else                                            --减法
        p:=opa Xor (Not opb);                       --减法传递函数
        g:=opa And (Not opb);                       --减法产生函数
        c(0):=Not c_i;
        For i In 1 To N Loop
            c(i):=g(i) Or (p(i) And c(i-1));        --减法借位运算
        End Loop;
        c_o<=Not c(N);
        s_c:=p Xor c(N-1 Downto 0);
    End If;
    result<=s_c;
  End Process;
End Architecture preaddsub_arch;
```

二进制加减法分如下 8 种情况仿真,仿真结果如图 3-5 所示。

加法无进位输入,无进位输出,1001+0101;

加法有进位输入,无进位输出,1001+0101+0001;

加法无进位输入,有进位输出,1001+1001;

加法有进位输入,有进位输出,1001+1001+0001;

减法无借位输入,无借位输出;1001-0101;

减法有借位输入,无借位输出;1001-0101-0001;

减法无借位输入,有借位输出;0101-1001;

减法有借位输入,有借位输出;0101-1001-0001;

观察图 3-5,可以验算减法的补码运算,即一个减数可以对其取反码再加 1 变成补码,补码与原来的被减数相加即可得到的被减数减去减数的结果。

如:16 进制运算 5-8=0101-1000=0101+0111+1=1101,通过 c_i 为 1 实现加 1。

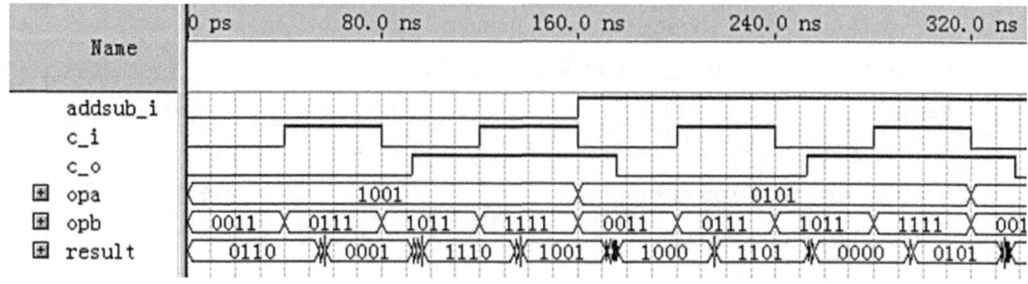

图 3-5　加减法仿真

3.4　BCD 加法器

BCD 码(Binary Coded Decimal),即二—十进制编码。四位二进制码元可以有 16 种组合 0～9,A～F。BCD 码只用四位二进制码的 10 种组合表示0～9,后 6 种组合不用。其原理如图 3-6 所示。

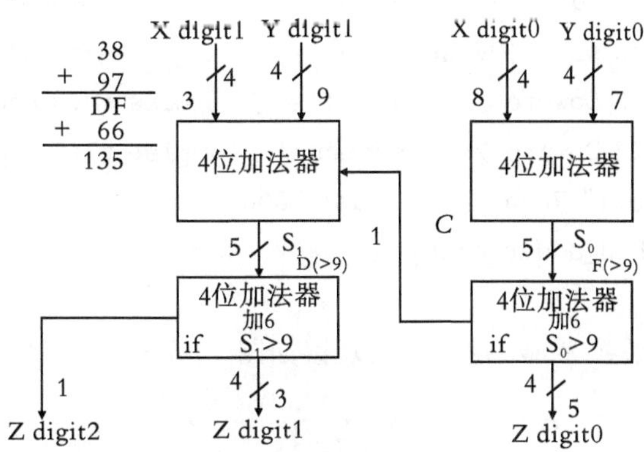

图 3-6　BCD 加法原理

这种编码在数据大于 9 时就需要加 6 进位。多位 BCD 码加法采用的是行波链的连接方式,即前级加法的进位输出接到后级加法的进位输入。

示例 3-5 BCD 加法程序采用了别名 Alias 语句,使设计达到了化归的效果,有利于复用。

示例 3-5　BCD 加法器

```
--这个例子是 3 位 BCD 加法器,采用行为描述
Library Ieee;
Use Ieee.Std_Logic_1164.All;
Use Ieee.Std_Logic_Unsigned.All;                              --运算符重载程序包

Entity bcd_adder Is
    Port(x,y: In Std_Logic_Vector(7 Downto 0);                --被加数、加数
         z: Out Std_Logic_Vector(11 Downto 0));               --BCD 输出
End Entity bcd_adder;

Architecture bcd_adder_arch Of bcd_adder Is
Alias xdig1: Std_Logic_Vector(3 Downto 0) Is x(7 Downto 4);   --别名
Alias xdig0: Std_Logic_Vector(3 Downto 0) Is x(3 Downto 0);
Alias ydig1: Std_Logic_Vector(3 Downto 0) Is y(7 Downto 4);
Alias ydig0: Std_Logic_Vector(3 Downto 0) Is y(3 Downto 0);
Alias zdig2: Std_Logic_Vector(3 Downto 0) Is z(11 Downto 8);
```

```
Alias zdig1: Std_Logic_Vector(3 Downto 0) Is z(7 Downto 4);
Alias zdig0: Std_Logic_Vector(3 Downto 0) Is z(3 Downto 0);
Signal s0, s1: Std_Logic_Vector(4 Downto 0);
Signal c: Std_Logic;
Begin
    s0 <= '0'& xdig0 + ydig0;                       --扩展出进位位求和
    s1 <= '0'& xdig1 + ydig1 + c;                   --运算符重载,位宽不同的加法
    c <= '1' When s0 > 9 Else '0';                  --十六进制与十进制转换
    zdig0 <= s0(3 Downto 0) + 6 When s0 > 9 Else s0(3 Downto 0);
    zdig1 <= s1(3 Downto 0) + 6 When s1 > 9 Else s1(3 Downto 0);
    zdig2 <= "0001" When s1 > 9 Else "0000";
End Architecture bcd_adder_arch;
```

BCD adder 仿真结果在 Wave 窗口中的观察如图 3-7 所示。

	Msgs		
/bcd_adder_tb/x	80	20	80
/bcd_adder_tb/y	88	80	88
/bcd_adder_tb/z	168	100	168

图 3-7　BCD 加法器 Modelsim 仿真结果

程序说明:

(1)BCD 码的输入输出要用十六进制观察,如果显示不是十六进制数据,需要修改显示数制为十六进制,方法是选中数据信号名按右键,Radix→Hexadecimal。

(2)别名 Alias 的主要用途是对已有的对象定义一个替换名,Alias 语句可以在 Architecture、Entity、Process、Subprogram、Package 说明部分、Package Body 语句中使用。它常为数组或者指令的每个字段提供命名的机制,并且可由别名直接去引用这个字段。

3.5　流水线与非流水线加法器

流水线(Pipeline)结构:典型的同步设计,用串行的方法提高性能。

实现方法是把大的布尔逻辑拆分成 N 个小的布尔逻辑,其间插入触发器,输入与输出皆通过触发器。极端的流水线使系统像是寄存器链,组合逻辑淹没在触发器的建立时间和保持时间中,带来了方便的时序预测以及毛刺的消失。需要说明的是,流水线提高性能的作用主要体现在系统级应用。系统级应用时,这里的小的布尔逻辑可能成为比较大的逻辑甚至模块,这时可能更要关注 Pipeline 结构模块间的配平问题。

示例 3-6 流水线与非流水线加法器

```
Library Ieee;
Use Ieee.Std_Logic_1164.All;
Use Ieee.Std_Logic_Unsigned.All;

Entity addercn Is                                  --加法器实体部分,对应两个结构体
    Generic(N:Positive:=4);      --类属 Generic 定义常数 N 为正整数 4,即 4 位加法
    Port (cin,clr_n: In Std_Logic;                 --进位输入、清零信号
          clk: In Std_Logic;                       --增加时钟输入端
          a,b : In Std_Logic_Vector (N-1 Downto 0);  --被加数、加数
          s: Out Std_Logic_Vector (N-1 Downto 0);    --和数
          cout: Out Std_Logic);                    --进位输出
End Entity addercn;
```

示例 3-6a 端口带寄存器的可清零的加法器

```
--实体 addercn 的结构体 one
Architecture one Of addercn Is
Signal sint,aa,bb,cc: Std_Logic_Vector (N Downto 0);
Begin

    Process(clr_n,clk) Is
    Begin
        If clr_n='0' Then                                    --异步清零
            aa<=(Others=>'0');bb<=(Others=>'0');       --aa,bb 初值"0000"
        Elsif clk'Event And clk='1' Then                     --时钟上升沿
            aa<='0'&a;                 --在 If 不完整结构中,信号每次传输产生触发器
            bb<='0'&b;                      --扩展出进位位
            cc<=cin;                        --同步原则需要端口都加触发器
            sint<=aa+bb+cc;                 --加数、被加数、进位位之和
            s<= sint(N-1 Downto 0);         --因为数据需要满足建立时间和保持时间
            cout<= sint(N);                 --所以触发器有去毛刺的作用
        End If;
    End Process;
End Architecture one;
```

示例 3-6 结构体 one 与结构体 two 的时序仿真如图 3-8 和图 3-9 所示,结果都无明显的毛刺,也没有夹杂错误的数字。结构体 two 加了流水线结果略好于不加流水线,这说明综合器对连加的处理能力也是很好的。

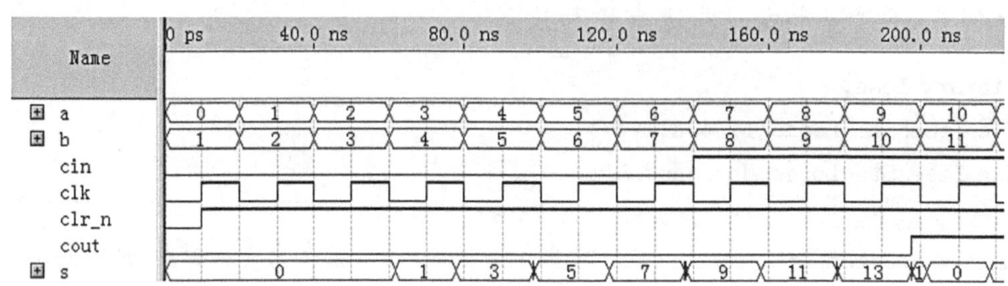

图 3 - 8　示例 3 - 6a 结构体 one 仿真

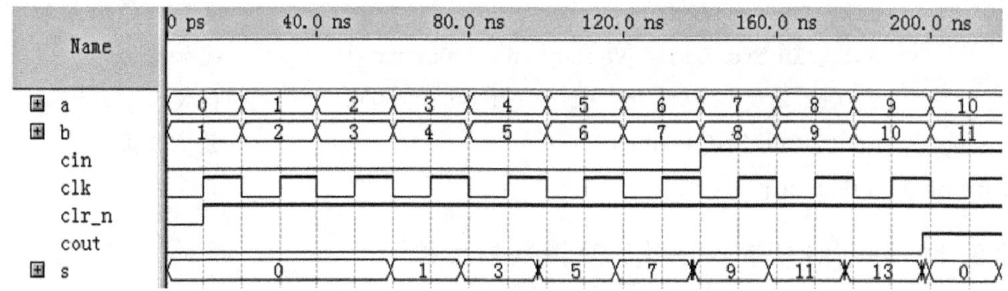

图 3 - 9　示例 3 - 6a 结构体 two 仿真

程序 3 - 6 结构体 one 综合后 RTL Viewer 的结果(图 3 - 10)看起来不是特别理想，通常没有必要在输出连接 2 个寄存器，连加中间是组合电路，在芯片速度很快时也可能有毛刺带来的风险。

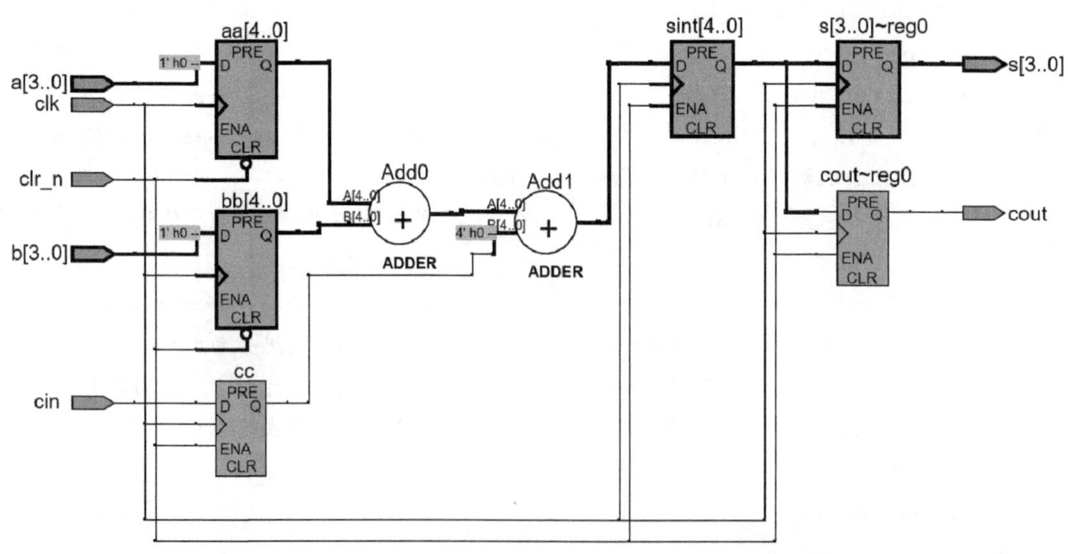

图 3 - 10　示例 3 - 6a 结构体 one RTL Viewer 的结果

示例 3-6b 带流水线的可清零的加法器

```
Architecture two Of addercn Is                          --实体 addercn 的结构体 two
Signal sint1,sint: Std_Logic_Vector (N Downto 0);
Signal aa,bb: Std_Logic_Vector (N Downto 0);
Signal cc1,cc2: Std_Logic;
Begin

    Process(clr_n,clk) Is
    Begin
        If clr_n='0'Then                                --异步清零
            aa<=(Others=>'0');                          --aa<="0000";
            bb<=(Others=>'0');
        Elsif clk'Event And clk='1'Then                 --时钟上升沿
            aa<='0'&a;
            bb<='0'&b;
            cc<=cin;
            sint1<=aa+bb;
            cc2<=cc1;
                                        --流水线要将大的布尔运算拆分成小的布尔运算
            sint<= sint1+cc2;           --不完整 If 语句中信号赋值将产生 D 触发器
        End If;
        s<=sint(N-1 Downto 0);                          --放在 If 语句之外,不产生寄存器
        cout<= sint(N);
    End Process;
End Architecture two;

Configuration adder4_cfg Of addercn Is                  --配置语句说明
    For two                                             --指定实体 addercn 的结构体为 two
    End For;
End Configuration adder4_cfg;
```

从 RTL Viewer 看到(图 3-11),程序 3-6 结构体 two 优于结构体 one。结构体 two 占用资源更少,有更少的组合电路竞争冒险的风险,结构体 two 直觉上也简约、畅达、标准。

结构体 two 的加法部分即典型的流水线划分结构的设计方法,这种方法将大的布尔逻辑连“加”化为小的布尔逻辑分步“加”,利用触发器来分割,利用触发器的建立时间和保持时间来完成布尔逻辑,使得系统以时钟的节拍来工作,既有利于时序预测,减少毛刺,也有利于系统级性能的提高。

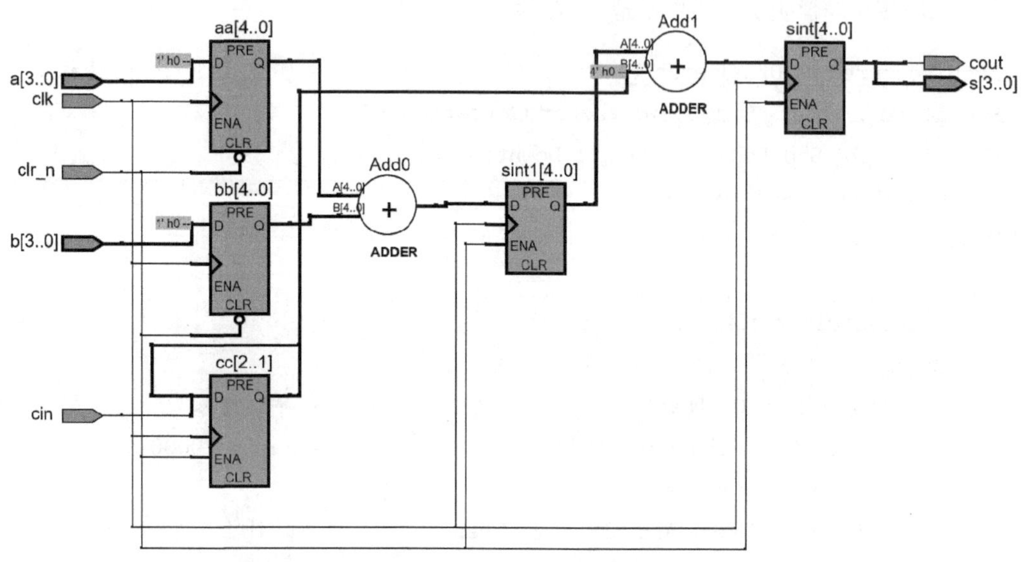

图 3-11　示例 3-6b 结构体 two RTL Viewer 的结果

示例 3-7　32 位流水线加法器设计

```
--32 位加法器程序文件 add32.vhd
Library Ieee;
Use Ieee.Std_Logic_1164.All;
Use Ieee.Std_Logic_Unsigned.All;

Entity add32 Is                                           --实体声明,对应两个结构体
    Port(clk,cin : In Std_Logic;                          --时钟与进位位
         x,y: In Std_Logic_Vector(31 Downto 0);           --加数与被加数
         q: Out Std_Logic_Vector(31 Downto 0));           --和数
End Entity add32;
```

示例 3-7a　32 位流水线加法器结构体 1

```
Architecture add32_arch1  Of add32 Is                     --add32 结构体 1
Signal tmp1,tmp2,tmp3,tmp4,tmp5:Std_Logic_Vector(8 Downto 0);
Signal tmp6,tmp7,tmp8: Std_Logic_Vector(8 Downto 0);
Signal tmp9,tmp10,tmp11,tmp12: Std_Logic_Vector(8 Downto 0);
Signal ci1,ci2:Std_Logic;
Begin

    Process(clk) Is
    Begin
```

```
            If clk'Event And clk='1'Then                           --时钟上升沿触发
                tmp1<='0'&(x(7 Downto 0));  tmp2<='0'&(y(7 Downto 0));
                tmp3<='0'&(x(15 Downto 8)); tmp4<='0'&(y(15 Downto 8));
                tmp5<='0'&(x(23 Downto 16));tmp6<='0'&(y(23 Downto 16));
                tmp7<='0'&(x(31 Downto 24));tmp8<='0'&(y(31 Downto 24));
                ci1<=cin;ci2<=ci1;                                  --输入加寄存器
                tmp9<=tmp1(8 Downto 0)+tmp2(8 Downto 0)+ci2;
                                                                    --运算加寄存器,扩展进位
                tmp10<=tmp3(8 Downto 0)+tmp4(8 Downto 0)+tmp9(8);
                tmp11<=tmp5(8 Downto 0)+tmp6(8 Downto 0)+tmp10(8);
                tmp12<=tmp7(8 Downto 0)+tmp8(8 Downto 0)+tmp11(8);
                q(7 Downto 0)<=tmp9(7 Downto 0);                    --输出加寄存器
                q(15 Downto 8)<=tmp10(7 Downto 0);
                q(23 Downto 16)<=tmp11(7 Downto 0);
                q(31 Downto 24)<=tmp12(7 Downto 0);
            End If;
        End Process;
End Architecture add32_arch1;
```

示例 3-7b　32 位流水线加法器结构体 2

```
Architecture add32_arch2 Of add32 Is                                --add32 结构体 2
    Component CARRY Is                                              --专用进位缓冲器
        Port(a_in : In Std_logic;
             a_out : Out Std_logic);
    End Component CARRY;
    Alias d130: Std_Logic_Vector(7 Downto 0) Is x(31 Downto 24);
    Alias d120: Std_Logic_Vector(7 Downto 0) Is x(23 Downto 16);
    Alias d110: Std_Logic_Vector(7 Downto 0) Is x(15 Downto 8);
    Alias d100: Std_Logic_Vector(7 Downto 0) Is x(7 Downto 0);
    Alias d230: Std_Logic_Vector(7 Downto 0) Is y(31 Downto 24);
    Alias d220: Std_Logic_Vector(7 Downto 0) Is y(23 Downto 16);
    Alias d210: Std_Logic_Vector(7 Downto 0) Is y(15 Downto 8);
    Alias d200: Std_Logic_Vector(7 Downto 0) Is y(7 Downto 0);
    Alias out3: Std_Logic_Vector(7 Downto 0) Is q(31 Downto 24);
    Alias out2: Std_Logic_Vector(7 Downto 0) Is q(23 Downto 16);
    Alias out1: Std_Logic_Vector(7 Downto 0) Is q(15 Downto 8);
    Alias out0: Std_Logic_Vector(7 Downto 0) Is q(7 Downto 0);
Signal ci,ci1,ci2: Std_Logic;
Signal tmp0,tmp1,tmp2,tmp3: Std_Logic_Vector(8 Downto 0);               --暂存中间结果
```

```
Signal d13,d12,d11,d10,d23,d22,d21,d20: Std_Logic_Vector(7 Downto 0);
Begin
                                         --Instantiating carry,例化缓冲器 carry
    ca : carry
    Port Map (a_in => ci, a_out =>ci1);
    Process(clk) Is
    Begin
        If clk'Event And clk='1' Then
            d13 <= d130; d12 <= d120; d11 <= d110; d10 <= d100;
            d20 <= d200; d21 <= d210; d22 <= d220; d23 <= d230;
            ci <=cin;  ci2 <= ci1;                              --输入加寄存器
            tmp0 <='0'&d10+d20;  tmp1 <='0'&d11+d21;  --运算加寄存器
            tmp2 <='0'&d12+d22;  tmp3 <='0'&d13+d23;
            out0 <=tmp0(7 Downto 0)+ci2;
            out1 <=tmp1(7 Downto 0)+tmp1(8);
            out2 <=tmp2(7 Downto 0)+tmp2(8);                  --输出加寄存器
            out3 <=tmp3(7 Downto 0)+tmp3(8);--cout<=tmp3(8);
        End If;
    End Process;
End Architecture add32_arch2;

--运用 Configuration 对 2 个结构体进行比较分析
Configuration add32_cfg Of add32 Is
    For add32_arch2
    End For;
End Configuration add32_cfg;
```

在运算类设计中，进位是一个很重要也是很容易出错的问题，此流水线的设计也是这样。进位在哪一级介入？进位介入以后如何保证流水线结构综合结果的严整？这个示例做了这样的探讨，结构体 2 为此还尝试应用了系统提供的进位缓存器。

32 位流水线加法器 RTL Viewer 观察结果如图 3 - 12 所示。

设计实现说明：

(1)采用流水线设计毛刺问题基本不存在。

(2)结构体 add_arch1 的 RTL Viewer 结果划分与对称性不够好，中间级性能不够好。

(3)结构体 add_arch2，进一步运用流水线，结果的对称性改善，速度与性能有改善，除此之外，add_arch2 程序中采用了别名语句 Alias，程序的规整性有提高。

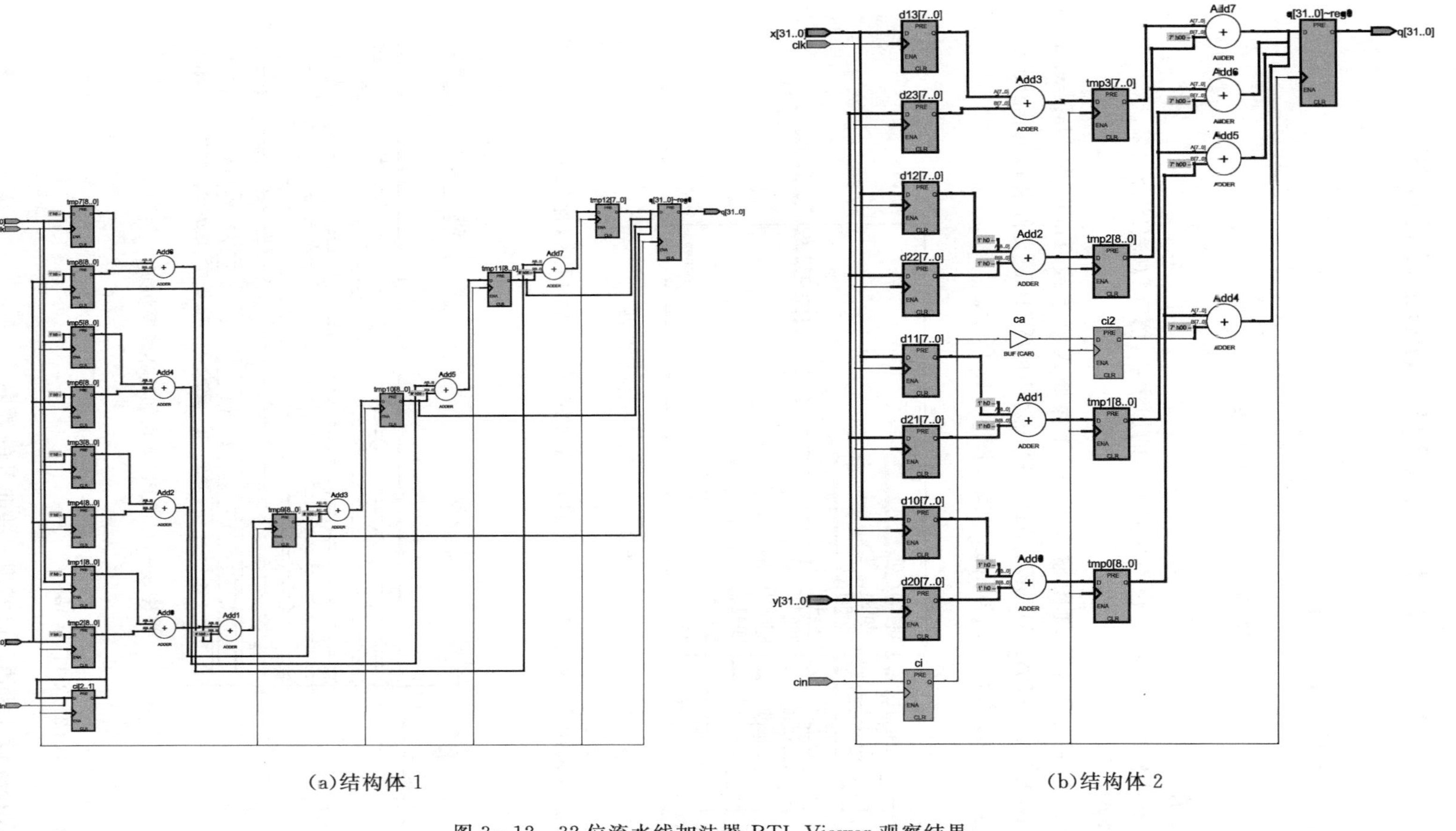

(a)结构体 1 (b)结构体 2

图 3-12 32 位流水线加法器 RTL Viewer 观察结果

3.6　LPM 模块应用与 Altera 设计原语

3.6.1　LPM 模块应用

示例 3-8　采用 LPM 模块设计加法器

设计要求：

采用 LPM 模块设计 8×8 位不加流水线的加法器，仿真，观察综合结果。

设计提示：

调用 LPM 模块可在原理图输入窗口，元件插入菜单或 Tools 中 Magaward Plag - in Manger→Arithmetic→PARALLEL_ADD，输入对应的 VHDL 文件名称 lpmadd. vhd，选择不加流水线，生成的 8×8 位加法器的原理图如图 3-13 所示。

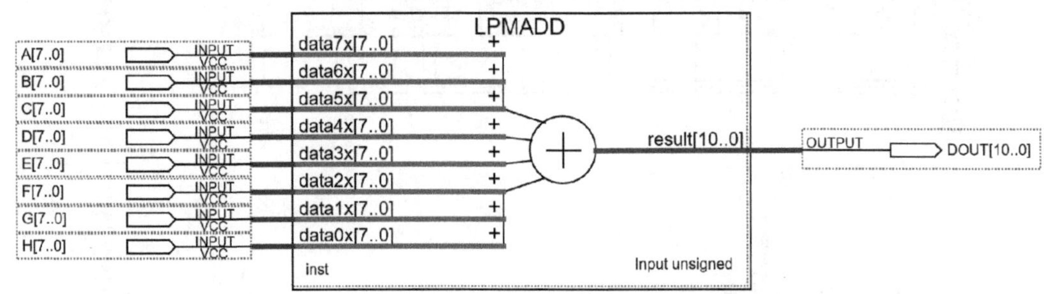

图 3-13　LPM 8 位加法器

8 位加法器采用 LPM 模块设计仿真的结果如图 3-14 所示，比较程序 3-2 仿真图的准确性要好许多。

Name	0 ps		20.0 ns		40.0 ns		60.0 ns		80.0
a	0	1	2	3	4	5	6	7	
b	1	2	3	4	5	6	7	8	
c	2	3	4	5	6	7	8	9	
d	3	4	5	6	7	8	9	10	
e	4	5	6	7	8	9	10	11	
f	5	6	7	8	9	10	11	12	
g	6	7	8	9	10	11	12	13	
h	7	8	9	10	11	12	13	14	
dout	28		36		44	52	60	68	

图 3-14　LPM 8 位加法器时序仿真

3.6.2　Altera 设计原语

调用 Altera 设计原语可以实现一些非凡的效果。

利用 RTL Viewer 看一下示例 3-8 综合后的电路结果，如图 3-15 所示。可以看到，

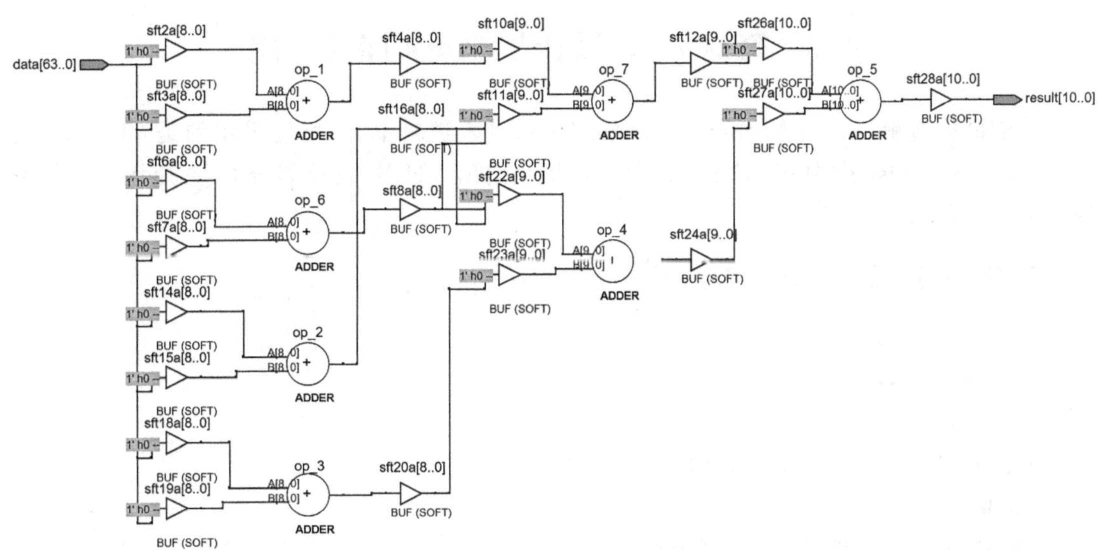

图 3 - 15　LPM 8 位加法器 RTL Viewer

LPM 模块实现的加法并不是通过触发器减少毛刺，而是通过 Altera 库中的一种称为 soft 的缓冲器实现的。这些缓冲器不是寻常的描述语言可以描述出来的，这些缓冲器要采用所谓“原语”来表述。

原语与具体的设计平台有关。厂家提供的库里有些单元就是所谓“原语”单元。

仅就缓冲器 BUF 而言，Altera 提供了许多种，有用于时钟的、进位的、级联的、普通 IO 的、LVDS 差分高速输入/输出的等等。

比如设计需要一个有些延时的 BUF，如果在设计中自己用两级反相器搭建会被综合器认为无效元件综合掉，而插入 Altera 提供的一个 Lcell 作为缓冲器是可以不被综合掉的。这种“原语”单元也可以在原理图中从 Altera 库中调出来，然后在 VHDL 程序中被引用。

在 Altera 的模板中给出了 VHDL 引用 Lcell 的方法：

```
--Add the library and use clauses before the design unit declaration
--在设计单元声明前增加如下 Library 和 Use 语句：
Library Altera_mf;
Use Altera_mf.Altera_mf_components.All;
--Instantiating LCELL  例化缓冲器 LCELL
<instance_name> : LCELL
Port Map (a_in => <data_in>, a_out => <data_out>);
```

通常我们不推荐期望调用固定的可延时单元在系统中来实现延时功能，因为这种单元与工艺、生产批次、应用环境等许多因素有关，其量化的指标不是很稳定、可靠。实际延时功能通常都是在寄存器应用的基础上实现的。

Lcell 的内部电路实现就是一个传输门，可以在系统保持时间不够时加延时。

使用 Lcell 还要注意将 Quartus Ⅱ工具的“Ignore Lcell Buffers”选项关闭。

3.7 简易 ALU 设计与 Work 库的应用

我们最初接触 ALU(Arithmetic Logic Unit,算术逻辑单元)可能采用的是 74LS181 元件,Altera maxplus2 库中有 74LS181 元件,这里我们用 VHDL 设计如表 3-5 所示的功能更简单的 ALU。

示例 3-9 简易 ALU 参考程序

```
--简易 ALU 程序文件 alu.vhd
Library Ieee;
Use Ieee.Std_Logic_1164.All;
Use Ieee.Std_Logic_Unsigned.All;
Use Work.alu_lib.All;
--Work 可以不声明,用 Work 的包需要说明

Entity alu Is
    Port( a, b: In Std_Logic_Vector(15 Downto 0);          --16 位操作数
          sel : In t_alu;                                   --4 位功能选择
          c : Out Std_Logic_Vector(15 Downto 0));           --16 位运算结果

End Entity alu;
Architecture alu_arch Of alu Is

Begin

    aluproc: Process(a,b,sel) Is
    Begin
        Case sel Is
            When alupass => c <= a;
            When andop => c <= a And b;
            When orop => c <= a Or b;
            When xorop => c <= a Xor b;
            When notop => c <= Not a;
            When plus => c <= a + b;
            When alusub => c <= a - b;
            When inc => c <= a + 1;
            When dec => c <= a - 1;
            When zero => c <= (Others=> '0');
            When Others => c <= (Others=> '0');
```

表 3-5 简易 ALU 功能表

sel 输入	操作
0000	C=A
0001	C=A And B
0010	C=A Or B
0011	C= Not A
0100	C= A Xor B
0101	C=A+B
0110	C=A−B
0111	C=A+1
1000	C=A−1
1001	C=0

```
        End Case;
    End Process aluproc;
End Architecture alu_arch;

--自定义 ALU_lib 程序包,程序文件 alu_lib.vhd
Library Ieee;
Use Ieee.Std_Logic_1164.All;
Package alu_lib Is                    --自定义程序包 alu_lib,这样说明可提高可读性
Subtype t_alu:Std_Logic_Vector(3 Downto 0);
    Constant alupass:Std_Logic_Vector(3 Downto 0):= "0000";
    Constant andop:Std_Logic_Vector(3 Downto 0) := "0001";
    Constant orop :Std_Logic_Vector(3 Downto 0):= "0010";
    Constant notop:Std_Logic_Vector(3 Downto 0) := "0011";
    Constant xorop:Std_Logic_Vector(3 Downto 0) := "0100";
    Constant plus:Std_Logic_Vector(3 Downto 0) := "0101";
    Constant alusub:Std_Logic_Vector(3 Downto 0) := "0110";
    Constant inc:Std_Logic_Vector(3 Downto 0) := "0111";
    Constant dec:Std_Logic_Vector(3 Downto 0) := "1000";
    Constant zero:Std_Logic_Vector(3 Downto 0) := "1001";
End Package alu_lib;
```

示例 3－10　ALU 的测试基准

```
--程序文件 alu_tb.vhd
Library Ieee;
Use Ieee.Std_Logic_1164.All;
Use Ieee.Std_Logic_Unsigned.All;
Use Work.alu_lib.All;                              --采用自定义程序包 alu_lib

Entity alu_tb  Is
End Entity alu_tb;

Architecture alu_tb_arch Of alu_tb Is
Signal sel:t_alu:=plus;
Signal a:Std_Logic_Vector(15 Downto 0):=(Others=>'1');
Signal b:Std_Logic_Vector(15 Downto 0):=(Others=>'0');
Signal c:Std_Logic_Vector(15 Downto 0);

    Component alu Is
        Port(sel:In Std_Logic_Vector(3 Downto 0);
```

```
            a,b:In Std_Logic_Vector(15 Downto 0);
            c:Out Std_Logic_Vector(15 Downto 0) );
    End Component alu;
Begin
DUT:alu Port Map( sel => sel, a => a,b => b, c => c);
    Process(a,b) Is
    Begin
        a<=a+1 After 50 ns;
        b<=b+1 After 50 ns;
    End Process;
    sel <= plus After 50 ns,                    --a+b
           alusub After 100 ns,                 --a - b
           andop After 150 ns,                  --a And b
           inc   After 200 ns,                  --a+1
           dec After 250 ns ;                   --a - 1
End Architecture alu_tb_arch;
```

3.8　工程文件管理与自定义库

Quartus Ⅱ提供工程档案压缩管理的功能。存档方法：Quartus Ⅱ→Project→Archive Project，生成文件 *.qar，同时会有日志文件。Qarlog 工程档案恢复：Quartus Ⅱ→Project→Restore Archive Project。

FPGA 开发环境如 Quartus Ⅱ，缺省将所有文件放在当前工作目录中，即 Work 库中，我们要利用 Files 设置将文件规划、分类和树形存放。

具体的做法：开发一个工程，工程名假设为 qtest，顶层工程文件目录名假设为 protop，源代码目录名假设为 srctop，仿真文件目录名假设为 simtop，上述三个目录还可以分别设子目录，库目录名为 mylib，存放结构如图 3-16 所示。

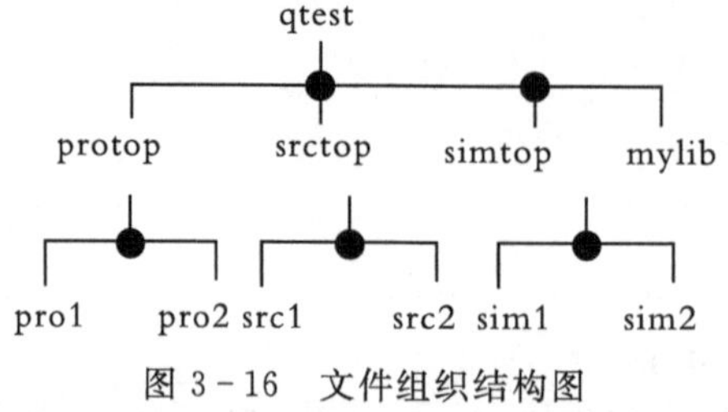

图 3-16　文件组织结构图

按照图 3-16 文件组织结构图，建立文件夹 mylib，将示例 3-9 自定义程序包 alu_lib.vhd 放在 mylib 目录中。

在 Quartus Ⅱ开发环境中，选择 Assignments→Settings，Libraries，找到库的路径文件夹 mylib，add 加入→OK，即指定了库的位置，如图 3-17 所示。

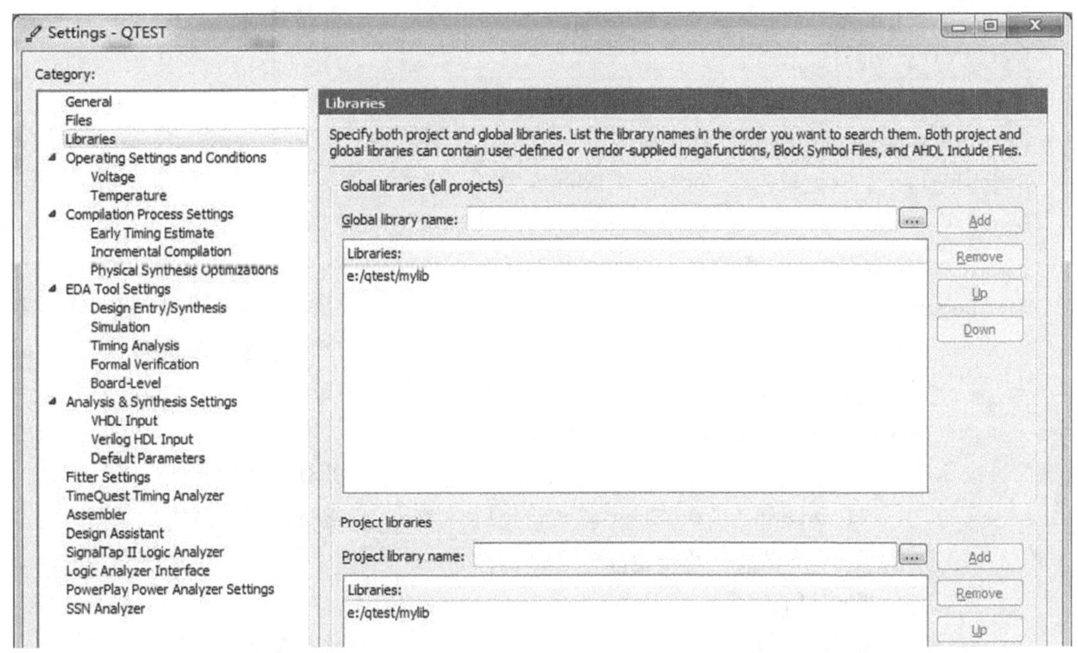

图 3-17 在 Quartus Ⅱ 中指定库的位置

示例 3-9 中 Library Ieee；改为 Library Ieee，mylib；或者增加一句 Library mylib。

示例 3-9 中，Use Work. alu_lib. All；改为 Use mylib. alu_lib. All。

这样就把缺省库 Work 库变成了自己指定库 mylib，实现了库按树形文件存档规范存放。

工程用到的程序、测试文件等可以通过 Assignments→Settings，Files，Add 加入，这样就可以实现按树形文件夹存放程序文件和测试文件进行开发。

指定库与文件的工作也可以利用 . tcl 来实现。

3.9 基于数据通道的加法设计

在系统设计中需要着重研究的问题是控制数据流传输与存储，在计算机系统设计中，数据、程序都通过总线传输，数据通道问题就是研究数据与程序在总线中的传输与存储的问题。

示例 3-11，3-12 分别给出了单输入通道与单双输入可选的数据通道程序。

示例 3-11 单输入数据通路行为级描述

```
--程序单输入数据通路 sdp. vhd
Library Ieee;
Use Ieee. Std_Logic_1164. All;
Use Ieee. Std_Logic_Unsigned. All;
Entity sdp Is
    Port (datain :In Std_Logic;                   --输入控制
          d1: In Std_Logic_Vector(7 Downto 0);    --数据输入
          clk,rst : In Std_Logic;                 --时钟,复位
```

```
        q : Out Std_Logic_Vector(7 Downto 0));     --数据输出
End Entity sdp;

Architecture sdp_arch Of sdp Is
Signal q1,q2: Std_Logic_Vector(7 Downto 0);
Begin
    Process(clk,rst) Is
    Begin
        If rst ='1'Then                             --中间寄存器与输出设初值
            q1<= "00000000";  q2<= "00000000";  q<= "00000000" ;
        Elsif clk'Event And clk='1'Then
            q1<= d1;                                --数据输入
            If datain ='0'Then q2<= q1;             --通道选择
            Else  q2<=q2+ q1;
            End If;
            q <=q2;                                 --数据输出
        End If;
    End Process;
End Architecture sdp_arch;
--在给第一个数时,选择信号 Datain 给低电平,其他时间给高电平
```

其 RTL Viewer 结果见图 3-18,仿真结果参看图 3-19。

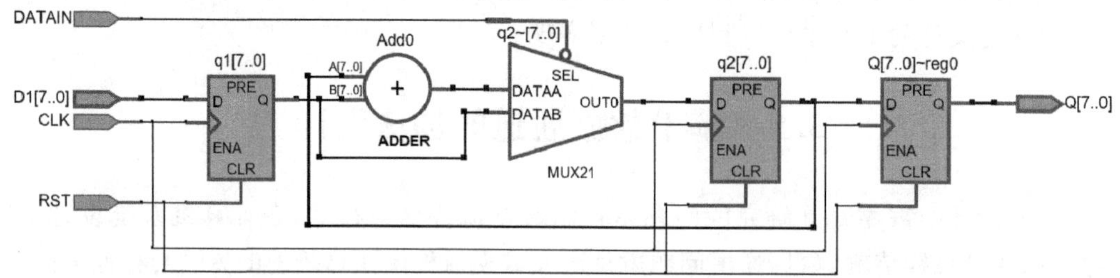

图 3-18　数据通道 1 个端口输入的串行加法 RTL Viewer 的结果

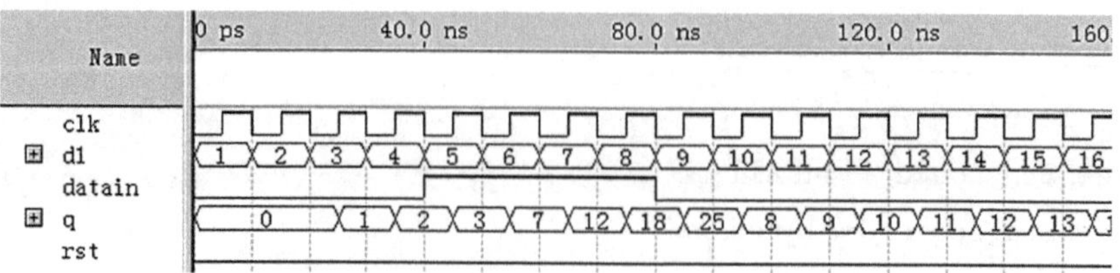

图 3-19　数据通道 1 个端口输入加法仿真结果

示例 3－12　单双可选输入数据通路行为级描述

```
--程序单双可选输入数据通路 sdpp.vhd
Library Ieee;
Use Ieee.Std_Logic_1164.All;
Use Ieee.Std_Logic_Unsigned.All;

Entity sdpp Is
    Port (sel1,sel2  :In Std_Logic;                  --数据选择信号
          d1: In Std_Logic_Vector(7 Downto 0);       --数据输入
          d2: In Std_Logic_Vector(7 Downto 0);       --数据输入
          clk,rst : In Std_Logic;                    --时钟与复位信号
          q : Out Std_Logic_Vector(7 Downto 0));
End Entity sdpp;

Architecture sdpp_arch Of sdpp Is
Signal q1,q2: Std_Logic_Vector(7 Downto 0);          --reg1,reg2 说明
Begin

    Process(clk,rst) Is
    Begin
        If rst ='1' Then                             --异步复位
            q1<= "00000000" ;q2<= "00000000" ;       --寄存器置初值
            q<= "00000000" ;
        Elsif clk'Event And clk='1' Then
            q1<= d1;                                 --数据传入 reg1
            If sel2 ='0' Then
                If sel1 ='0' Then q2<= q1;           --数据传入 reg2
                Else  q2<=q2+ q1;                    --数据传入 reg2
                End If;
                q<=q2;                   --前次加的结果与本次输入的数据相加
            Else
                q2<=d2;
                q<=q2+q1;
                                  --存入 reg1 中的 d1 与存入 reg2 中的 d2 相加
            End If;
        End If;
    End Process;
End Architecture sdpp_arch;
```

单双输入可选的数据通道程序仿真结果见图 3－20，RTL Viewer 结果见图 3－21。

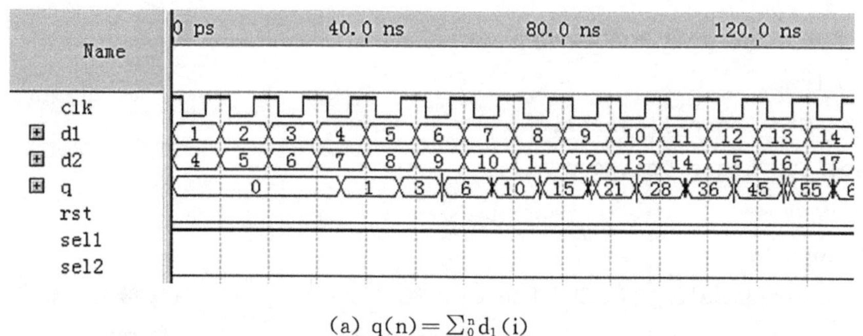

(a) $q(n)=\sum_0^n d_1(i)$

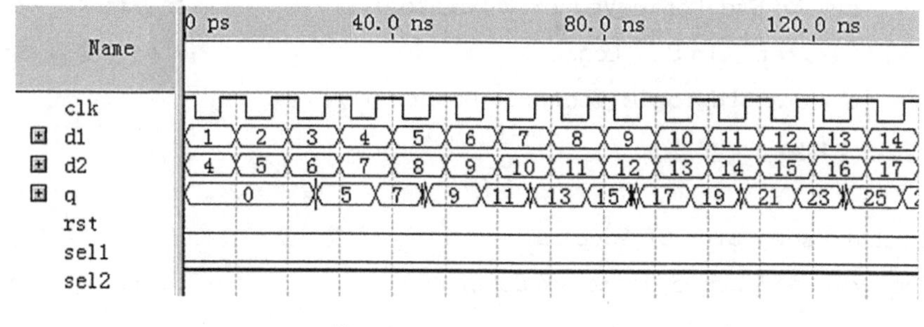

(b) $q(n)=d_1(n)+d_2(n)$

图 3－20　数据通道单双口输入可选的加法仿真

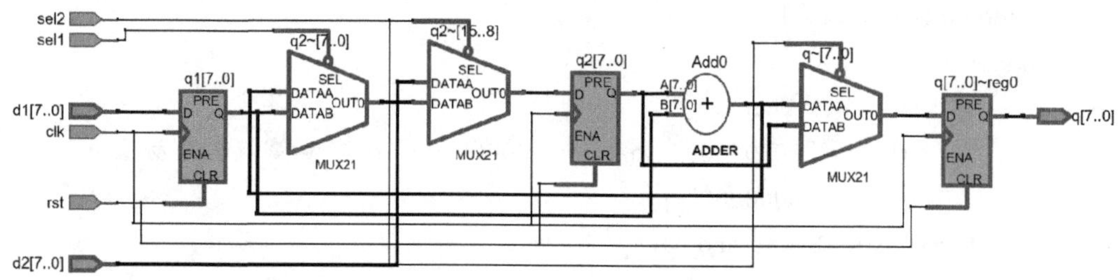

图 3－21　数据通道单双口输入可选的加法 RTL Viewer 的结果

利用上面程序通过 sel1 和 sel2 设置，可以实现从 1 个端口不断串行输入的数据进行两两相加的功能，也可以实现从两个端口同时送数的并行加法。

如果需给出结构化描述，则可以参考 RTL Viewer 来编写。这种改为结构化的方式是很有意义的，因为通常我们看到的 CPU 资料会给出数据通道图，结构化的描述更接近数据通道图。

习　题

3－1　用类属 Generic 指定位数和用循环语句编写一个 6 位的 BCD 码加法器。

3－2　用一个实体多个结构体的方式，编写 8 位加减法器电路程序。一个结构体采用 VHDL 结构化风格编写，一个采用 LPM 模块调用的方式编写，用 Testbeach 进行仿真，用 Configuration 分别对两个结构体进行综合比较。

3－3 调用 Altera 74 系列库中的 74LS181 以及 74LS182 实现 16 位具有先行进位功能的 ALU，编写 VHDL Testbeach 进行仿真；参照实现的功能用 VHDL 编写一个程序，进行仿真。采用 Quartus Ⅱ设计进行综合，比较综合结果。

3－4 采用包的形式来说明加法器的函数与过程，用类属 Generic 实现任意指定位的加法器，写出 3 种不同的 Testbeach 测试程序进行仿真。

3－5 设计一个数据通道，把控制数据通道的各种控制变化放到存储器中来控制，写出 Testbeach 测试程序进行仿真。

3－6 本章节有些案例只说明了加法，请给出相应的减法的设计与实现。

扩展学习与总结

1. 结构体的描述风格有几种？如何使用？比较它们的异同。
2. Generate 语句，If Generate 语句和 For 语句的作用和使用方法是什么？
3. 子程序有哪些，如何组成？他们的作用是什么？怎样使用？
4. 如何定义和使用过程与函数？如何传递参数和返回结果？
5. 了解块(Block)的卫式，块的功能和书写格式是什么？怎样使用？
6. 了解 VHDL 有哪些数值类属性(Attribute)和函数属性，这些属性如何说明，如何使用？
7. 进行寄存器的 RTL 描述时应该注意什么？
8. 如何自定义库？在 Quartus Ⅱ中如何指定自定义库？
9. 如何利用 Altera 的基本元件库以及 LPM 库？

4　Modelsim 仿真提高

本章的主要内容与方法：

(1)介绍 Modelsim 的仿真步骤。

(2)Model sim 操作方式。

(3)Modelsim 仿真方式。

(4)本章给出 Modelsim 的多种的测试激励方式的实例。

4.1　第三方仿真软件 Modelsim 仿真方式

Modelsim 操作方式主要有如下三种：

(1)用户图形界面(Graphical User Interface,GUI)；

(2)命令行(CMD)；

(3)批处理(.do,.tcl)。这三种可以混合运用。

4.2　四位加法器 Modelsim 仿真方式

层次化设计 4 位加法器 Modelsim 仿真要求。

(1)采用命令行法、Testbench 法、波形输入法给出仿真激励等。

(2)编写 Testbench 文件要求：

①信号赋值法。

②自定义类型数字序列法。

③自定义类型记录结构法。

④文件读取数据法。

(3)要给出不带门延迟的功能仿真和带门延迟的时序仿真。

4.3　Modelsim 仿真步骤

4.3.1　建工程(Project)与建立文件

建工程或文件时可从测试平台自身列表直接读取或者列表记录类型读取。

以四位加法器设计为例。采用例 3－1 实体与例 3－1b 结构体，将其中 Generic 中的正整数的值改为 4。实体名改为 adder4。将文件拷贝到 G:/shiyan1 目录下。

File→New→ Project,Add Existing File 或者 Creat New File,如图 4－1 所示。

在工程中加文件时选择 Add Existing File 或 Create New File,或者 Close 后,File→New→Source→VHDL,如图 4－2 所示。

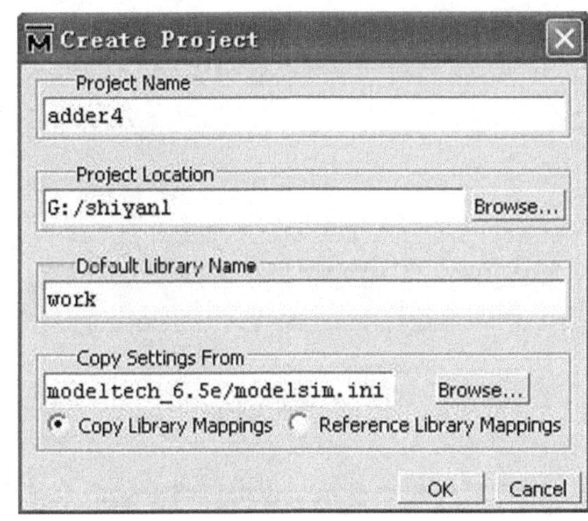

图 4 - 1　Modelsim 新建工程

图 4 - 2　在工程中加文件

4.3.2　编译(Compile)文件

(1)选 Add Existing File,添加已有文件,在我们所建的工程文件夹 G:/shiyan1 中选中 ：adder4. vhd,fulladder. vhd,单击 OK。

(2)选 Create New File,创建文件,打开编辑器,将例 3 - 1 加法程序输入,保存为adder 4. vhd 文件,再打开编辑器,将一位加法程序输入,保存为 fulladder. vhd 文件。进入 Project 窗口,右键选择 Add To Project→Existing File,选中 adder4. vhd,fulladder. vhd。

这时 Project 窗口存在了 adder4. vhd,fulladder. vhd 两个文件。在 Project 窗口,选中 adder4. vhd,fulladder. vhd 按右键,选择编译,Compile selected 或者 Compile all。编译成功后 Work 库中有了 adder4. vhd,fulladder. vhd。

4.4　Modelsim 仿真种类

Modelsim 有命令行式仿真、图形界面式(菜单)仿真、Testbench 仿真、宏命令 . do 文件仿真、. tcl 文件仿真,每种仿真都可以贯穿整个过程。

4.4.1　Modelsim 窗口开关与命令

Modelsim 可以进行多窗口协同工作,在主菜单 View 中,有 Project, Library, Object, Wave, List, Transcricpt 等开关选项,点击选项可以打开或关闭相应窗口。

如 Project 窗口,工程的 VHDL 及其测试文件都列在这个窗口。观察波形在 Wave 窗口,命令与命令行输入,以及程序信息在 Transcript 窗口中。

示例 4 - 1　Modelsim 命令行式仿真

Modelsim 命令行式仿真即所谓的特殊仿真器,它所使用的命令只在 Modelsim 这个环境中可以使用。下列命令逐行进入,可得到 4 位加法器的运算结果。

```
add list a b ci c co s
force a 1111
force b 0001
force ci 1
run 50 ns
force a 0101
force b 1110
force ci 0
run 50 ns
add wave *
run 200 ns
```

结果通常在Wave窗口可以看到波形或List窗口看到列表显示。也可以将上述命令行写进*.do文件，如testadd.do，运行testadd.do文件可直接看到仿真结果。运行testadd.do文件，方法是在Transcript窗口输入：

```
Vsim>do testadd.do
```

4.4.2 Testbench(设计基准文件)仿真

(1)建立设计基准文件。在设计顶层文件adder4编辑窗口按右键，在Show language templates中选择adder4.vhd生成测试基准文件的框架adder4_tb.vhd。

(2)打开adder4_tb.vhd，给输入信号激励初值，以及输入信号的变化情况。

①赋初值(在结构体开始之前)，例如：

```
Signal a Std_Logic_Vector (3 Downto 0) :="1111";
```

②给出所有输入激励的变化(在结构体中)，例如：

```
a<= "1010" After 50 ns, "1000" After 100 ns;
```

(3)保存及编译设计基准文件adder4_tb.vhd。

(4)仿真(Simulate)。

Simulate→Start Simulation→Work→adder4_tb.vhd

注意：不能优化，Enable optmization复选框中的勾去掉，如图4-3所示。

为了使Modelsim的仿真缺省为不优化，可以修改ModelSim安装目录下的modelsim.ini文件，在文件属性中取消read-only选项，然后打开modelsim.ini文件，修改VoptFlow=0。

(5)给波形窗口送入信号。

```
add wave *
```

通配符"*"指将所有信号送入波形窗口，系统默认是bin，二进制文件或者在Transcript窗口输入

```
add wave -dec *
```

上述代码指将所有信号以十进制送入波形窗口；命令行参数如果为hex，是信号以十六进制送入波形窗口；如果为oct，则为八进制。

(6)执行仿真(run)。

注意：在Transcript窗口输入

```
run 300 ns
```

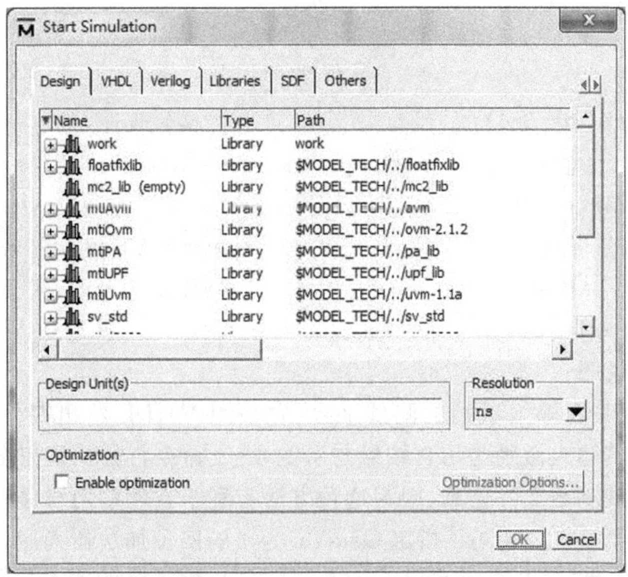

图 4-3 Modelsim 仿真开始界面

示例 4-2 四位加法器测试基准文件(信号赋值法)

```
Library Ieee;                                   --Ieee 库声明
Use Ieee.Std_Logic_1164.All;                    --引用程序包 Std_Logic_1164
Use Ieee.Std_Logic_Unsigned.All;                --引用程序包 Std_Logic_Unsigned

Entity adder4_tb Is
End Entity adder4_tb;

Architecture test1 Of adder4_tb Is
Signal a:Std_Logic_Vector(3 Downto 0) :="1111";              --信号说明与设初值
Signal b:Std_Logic_Vector(3 Downto 0) :="0001";
Signal ci:Std_Logic :='1';
Signal co:Std_Logic;
Signal s:Std_Logic_Vector(3 Downto 0);

Component adder4 Is
    Generic( N:=Positive);                       --Positive,正整数,Integers > 0
    Port(a,b:In Std_Logic_Vector(N-1 Downto 0);
         ci:In Std_Logic;
         co:Out Std_Logic;
```

```
        s:Out Std_Logic_Vector(N - 1 Downto 0) );
End Component adder4;

Begin
DUT: Generic Map( N=>4)                       --类属的传递
    adder4 Port Map(a => a,b => b, ci => ci,co => co,s => s );
    ci<='0' After 50 ns;
    a<="0101" After 50 ns;                    --加数 a 50 ns 后变为"0101"
    b<="1110" After 50 ns;                    --加数 b 50 ns 后变为"1110"
End Architecture test1;
```

示例 4-2 程序说明：

(1)示例 4-2 是只有输入激励的测试平台，它采用 VHDL 给出四位加法器的测试基准(信号赋值法验证)，其输入激励与仿真结果与示例 4-1 命令行方式是完全一致的。

(2)推荐采用测试基准文件仿真，因为这种方法有利于文件复用与移植等。

(3)adder4 即示例 3-1 addern 指定 Generic 为 4 的四位加法器程序，示例 3-1 addern N 为 8，指定的是 8 位加法，可以看到 Generic Map 具有参数重置的能力，这为设计与测试的复用(reuse)带来了很大的方便。

(4)a<="0101" After 50 ns；其中 After 50 ns 是不可综合句，是惯性延时(Inertial Delay)，是 VHDL 的默认延时，其描述器件本身固有的属性，信号的稳定值必须超过器件的固有延时才会生效；否则视为无效变化。

惯性(固有)延时是在仿真中使用最多的一种延时模型。

①对电路设计而言固有延时比较精确。

②它可以防止通过该电路毛刺的散布。

VHDL 还有传输延时概念，指器件间的连线延时，如 x 信号的传输延时：

```
a <= Transport x After 20 ns;
```

VHDL 还描述区间延时脉冲，如 x 信号的区间延时：

```
a<=Reject 4 ns Interial x After 10 ns;
```

示例 4-3　四位加法器测试基准文件(自定义类型数字序列法)

```
Entity testadder Is
End Entity testadder;

Architecture test2 Of testadder Is
Signal cin: Bit;
Signal cout: Bit;

    Component adder4 Is
        Port(a,b: In Bit_Vector( 3 Downto 0);
            ci: In Bit;
```

```
            co: Out Bit;
            s: Out Bit_Vector( 3 Downto 0) );
    End Component adder4;

Constant N: Integer:=11;                    --常数 N 初始化为整数 11
Type bv_arr Is Array(1 To N) Of Bit_Vector( 3 Downto 0);
                                            --自定义类型
Type bit_arr Is Array(1 To N) Of Bit;
Constant addend_array:
bv_arr: =("0111","1101","0101","1101", "0111", "1000",
         "0111","1000", "0000","1111", "0000");
                                            --加数序列
Constant augend_array:
bv_arr: =("0101","0101","1101","1101","0111","0111", "1000","1000",
         "1101","1111","0000");
                                            --被加数序列
Constant cin_array: bit_arr:=('0','0','0','0','1','0','0','0',
                              '1','1','0');
                                            --进位位序列
Constant sum_array: bv_arr:=("1100","0010","0010","1010","1111","1111",
                             "1111","0000","1110","1111", "0000");
                                            --和数序列
Constant cout_array: bit_arr :=('0','1','1','1','0','0','0','1',
                                '0','1','0');
                                            --进位输出序列
Signal addend,augend,sum: Bit_Vector(3 Downto 0);
Begin
    Process Is
    Begin
        For i In 1 To N Loop
            addend<=addend_array(i);
            augend<=augend_array(i);
            cin<=cin_array(i);
            Wait For 40ns;
            Assert(sum=sum_array(i) And cout=cout_array(i));
            Report "Wrong Answer"           --报告“错误答案”
            Severity error;                 --警告级别为“error”
        End Loop;
        Report "test finished";             --报告“测试完成”
```

```
    End Process;                    --test finished 后,重新开始进程,反复读数据表
    adder4: Generic Map( N=>4)
    Port Map(addend,augend,cin,sum,cout);
                                        --元件端口与信号位置对应
End Architecture test2;
```

程序说明：

(1)这是一个典型的快速测试平台。特点是激励列表中不仅有输入信息也有输出信息。这种方法的缺点是灵活性差些,输入复杂时表格会很大。

(2)快速测试平台往往可以在我们对程序做了小的变动或优化时,迅速了解设计是否依然正确。当设计进入样机生产及以后的环节时,这种思想也很好。

(3)程序中采用了断言 Assert,Assert 条件不满足时,报告“Wrong Answer”,Severity 的警告级别为 error。

(4)常量表测试完毕时,在 Transcript 窗口中可以看到按程序约定的报告“test finished”。

示例 4-4　一位加法器 Testbench(自定义类型记录结构法)

```
Library Ieee;
Use Ieee.Std_Logic_1164.All;
Use Ieee.Std_Logic_Unsigned.All;
Entity fulladder_tb  Is
End Entity fulladder_tb;

Architecture fulladder_tb_arch Of fulladder_tb Is
Signal x, y, cin   : Std_Logic;
Signal sum, cout:  Std_Logic;

    Component fulladder Is                  --全加器元件说明
        Port ( x,y, cin : In Std_Logic ;
               sum, cout : Out Std_Logic);
    End Component fulladder;
    Type test_rec Is Record                 --定义记录类型
        x   : Std_Logic ;
        y   : Std_Logic ;
        cin   : Std_Logic ;
        sum   : Std_Logic ;
        cout   : Std_Logic ;
    End Record test_rec;                    --结束记录类型说明
    Type test_array Is Array (Positive Range <>) Of test_rec;
    --Constant pattern:test_array:=(
```

```
    --(x=>'0',y=>'0',cin=>'0',sum=>'0',cout=>'0'), --名称对应方式
    --(x=>'0',y=>'0',cin=>'1',sum=>'1',cout=>'0'),
    --(x=>'0',y=>'1',cin=>'0',sum=>'1',cout=>'0'),
    --(x=>'0',y=>'1',cin=>'1',sum=>'0',cout=>'1'),
    --(x=>'1',y=>'0',cin=>'0',sum=>'1',cout=>'0'),
    --(x=>'1',y=>'0',cin=>'1',sum=>'0',cout=>'1'),
    --(x=>'1',y=>'1',cin=>'0',sum=>'0',cout=>'1'),
    --(x=>'1',y=>'1',cin=>'1',sum=>'1',cout=>'1'));
    Constant pattern:test_array:=(
    ('0','0','0','0','0'),      --位置对应方式
    ('0','0','1','1','0'),
    ('1','1','0','1','0'),      --有意给出一个错误数
    ('0','1','1','0','1'),
    ('1','0','0','1','0'),
    ('1','0','1','0','1'),
    ('1','1','0','0','1'),
    ('1','1','1','1','1'));
Begin
    DUT: fulladder                      --元件例化
    Port Map(x => x, y => y, cin => cin, sum => sum, cout => cout );

    Process Is                          --无敏感表进程,用于仿真,与 Wait 语句配合
    Variable vector:test_rec;
    Variable errors:Boolean:=False;  --变量 errors 为布尔类型,初值是 False

    Begin
        For i In pattern'Range Loop     --逐行读数据表 x,y,cin 列
            vector:= pattern(i);        --指定常数阵列第 i 行数据序列
            x<= vector.x;               --指定第 i 行数据序列中的被加数
            y<= vector.y;               --指定第 i 行数据序列中的加数
            cin<= vector.cin;           --指定第 i 行数据序列中的进位输入
            Wait For 100 ns;            --等待 100 ns
            Assert (sum= vector.sum and cout= vector.cout)
                                        --仿真数据与表格数据比较
            Report "Errors!"            --不同则报错
            Severity Note;              --错误级别为“注意”
        End Loop;
        Report "test finished";
        Wait;                   --test finished 后,程序始终等待,即只读一遍数据表
```

```
    End Process;
End Architecture fulladder_tb_arch;
```

示例 4 - 4 程序说明：

(1)这段测试程序也是快速测试平台，特点是采用了记录的方式。程序中，常数表格可以采用端口名称关联的方式(在程序中有注释)，也可以采用位置关联的方式。

(2)仿真选择 fulladder_tb. vhd 文件。表格数字第 3 行我们有意给出一个错误，使得第 3 行数据断言(Assert)不符合条件，这时在 Trancript 窗口，可以看到在运行到第三步时屏幕上按程序给出了 Note 级别的报告，Error(图 4 - 4)；如果我们将数据改回正确时，则没有信息(图 4 - 5)。两次运行在常量表测试完毕时，Transcript 窗口中可以看到按程序约定的报告 test finished。

```
Transcript
add wave -position end  sim:/fulladder_tb/sum
add wave -position end  sim:/fulladder_tb/cout
add wave -position end  sim:/fulladder_tb/pattern
VSIM 190> run
VSIM 190>
run
# ** Note: Errors!
#    Time: 300 ns  Iteration: 0  Instance: /fulladder_tb
run
run
run
run
VSIM 191> run
# ** Note: test finished
#    Time: 800 ns  Iteration: 0  Instance: /fulladder_tb
```

图 4 - 4　表格数字第 3 行有一个错误的信息面显示

```
Transcript
add wave -position end  sim:/fulladder_tb/y
add wave -position end  sim:/fulladder_tb/cin
add wave -position end  sim:/fulladder_tb/sum
add wave -position end  sim:/fulladder_tb/cout
add wave -position end  sim:/fulladder_tb/pattern
VSIM 198> run
VSIM 198>
run
run
run
run
run
VSIM 199> run
# ** Note: test finished
#    Time: 800 ns  Iteration: 0  Instance: /fulladder_tb
```

图 4 - 5　表格数字正确时的信息显示

(3)Wait 有关的语句在测试激励中很常用，它不需要先行定义敏感信号表。

Wait For 时间表达式--使进程暂停在表达式指定的时间。

Wait on 敏感表信号--进程暂停，直到某个信号发生变化。

Wait--表示永远暂停

Wait Until 条件表达式--使进程暂停，直到表达式成立时启动。

上面四个语句只有 Wait Until 语句可以综合。

4.5　Modelsim 连接器件库的仿真

第三方仿真软件 Modelsim 准备。

(1) 安装仿真环境:Modelsim se6.5。

(2) 建立工程。

建立 Modelsim Altera 库文件,进入 Modelsim 之后,在 Transcript 窗键入如下命令即可建立名为 cycloneⅡ 的 Modelsim Altera 仿真文件。

```
vlib cycloneii
vmap cycloneii cycloneii
vcom-work cycloneii c:/Altera/80/quartus/eda/sim_lib/cycloneii_atoms.vhd
vcom-work cycloneii c:/Altera/80/quartus/eda/sim_lib/cycloneii_components.vhd
vcom-work cycloneii c:/Altera/80/quartus/eda/sim_lib/Altera_mf_components.vhd
vcom-work cycloneii c:/Altera/80/quartus/eda/sim_lib/Altera_mf.vhd
```

注:(1)第一次在计算机上运行 Quartus Ⅱ与 Modelsim 时进行。

(2)220model.vhd 是 Work.lpm_Components。

(3)路径要根据实际安装目录来定,有的计算机可能是 d:/Altera/80/quartus/eda/sim_lib/……

4.6　Quartus Ⅱ+Modelsim VHDL 的功能仿真

在 Quartus Ⅱ新建 VHDL 文件,拷贝示例 3-1 代码,配置为结构体 one,Generic 为 8 位,实体改为 qadd,保存为 qadd.vhd。即设计 qadd.vhd 为结构化设计的 8 位加法器。

在 Assignmets→Settings→EDA Tools→Simulation→Tools name 中选择了用 Modelsim 仿真,生成了 qadd.vho。

用 Modelsim 仿真工具,采用 Textio 程序包的文件读取方式存取输入/输出数据,编写测试台文件进行仿真。

4.6.1　Textio 程序包

STD 标准库提供了 2 个程序包,Standard 和 Tcxtio 包。STD 库和 Standard 包在使用时都无需声明。Textio 程序包提供了 VHDL 仿真时与磁盘文件的交互。

深入了解 STD 标准库和 Textio 程序包可以参见 Quartus Ⅱ安装目录,如 10.1 版本,altera /10.1/quartus/libraries/vhdl/std,其中有 STD standard 和 Textio 87、93、2008 版的 VHDL 包说明。

在对 VHDL 源程序进行仿真时,有的输入/输出关系仅仅靠输入波形或编写 Testbench 中的信号输入是难以验证的。例如多位加法器,如果将所有的输入都全面验证几遍,很麻烦。若用 VHDL 设计一个处理器,需要读入指令和数据等,也很麻烦。这时候需要采用 Textio 包。

在验证加法器时,可以将所有输入保存在一个文本文件中,将其他软件计算出的结果保存在另外的文件中,以便事后判断是否正确及便于查找原因。

4.6.2 Textio 读写文件

文件类型的隐含声明定义了 File Open;File Close;Read;Write;Endfile;前四个为过程实现,最后一个用函数实现。

文本最好通过缓冲去进行读写,Textio 定义了 Readline,Writeline。

93 版的 Textio 程序包,不兼容 87 版。

```
File my_file:text Is In "my_input.vec";                  --87 版
File my_file:Open read_mode Is "my_input.vec";           --93 版
```

93 版对同一个文件既可以读也可以写,但不同时。

Textio 提供的用于访问文本文件的过程说明:

```
Procedure Readline(文件变量;行变量);
                                   --从指定文件读取一行数据到行变量中
Procedure Writeline(文件变量;行变量);
                                   --向指定文件写入行变量所包含的数据
Procedure Read(行变量;数据类型);
                                   --从行变量中读取相应数据类型的数据
Procedure Write(行变量; 数据变量; 写入方式;位宽);
                                   --将数据写入行变量
```

该过程写入方式表示写在行变量的左边还是右边,且其值只能为 left 或 right,位宽表示写入数据时占的位宽。例如,数据写入行变量的过程:

```
Write(Outline,Outdata,left,2);
```

表示将变量 Outdata 写入 Line 变量 OutLine 的左边,占 2 个字节。例如,读取整数的过程:

```
Procedure Read(L: Inout Line; Value: Out Integer; GOOD: Out Boolean);
```

其中,GOOD 用于返回过程是否正确执行,若正确执行,则返回 TRUE 。

查看 Textio 包,发现一些过程是重载过程,如何调用文件过程,可根据具体的应用,查看 Textio 包中的过程说明。

过程重载,类似于 C++,VHDL 提供了重载功能,即完成相近功能的不同过程可以有相同的过程名,但其参数列表不同,或参数类型、参数个数不同。

根据参数数据类型及参数个数的不同,有多种重载方式,Textio 提供了 Bit、Bit_Vector 、Boolean 、Character、Integer、Real、String、Time 数据类型的重载,同时,提供了返回过程是否正确执行的 Boolean 数据类型的重载。

在编译时选择语法标准为 VHDL - 93 标准。

示例 4 - 5 八位加法器测试基准文件(文件读取数据法)

```
Library Cyclone Ⅱ ;                        --Altera Cyclone Ⅱ 库
Library Ieee;
Use Cyclone Ⅱ . Cyclone Ⅱ _Components. All;
Use Ieee. Std_Logic_1164. All;
```

```
Use Std.Textio.All;                    --采用数据对象 File 时要引入 Textio 程序包

Entity qadd_tb Is
End Entity qadd_tb;
Architecture qadd_tb_arch Of qadd_tb Is
Signal a,b,s:Std_Logic_Vector(7 Downto 0) ;
Signal cin, cout:Std_Logic;
Signal ports: Std_Logic_Vector(26 Downto 1):=(Others=>'Z');
    Component qadd Is
        Generic(N:Positive);              --类属说明
        Port(a,b:In Std_Logic_Vector(N-1 Downto 0) ;
             cin:In Std_Logic;
             cout:Out Std_Logic;
             s:Out Std_Logic_Vector(N-1 Downto 0) );
    End Component qadd;
Begin
    DUT:qadd  Generic Map(N=>8);         --参数传递,元件 qadd 的 N 位数指定为 8
    Port Map( a => a,b => b, cin => cin,cout => cout, s => s) ;
             a<=ports(26 Downto 19);
             b<=ports(18 Downto 11);
             cin<=ports(10);
             s<= ports(9 Downto 2);
             cout<=ports(1);
    test: Process Is
    --File vector_file : text Is In "vectors";              --87 版标准
    File vector_file: Open Read_Mode Is "vectors.txt";      --93 版标准
    Variable  L : Line;                                      --变量 L 的类型是 L
    Variable  vector_time : Time;         --变量 vector_time 的类型是 Time
    Variable  R : Real;                   --变量 R 的类型是实数
    Variable  good_number : Boolean;      --变量 good_number 为布尔类型
    Variable  signo : Integer;
    Begin
        While Not Endfile(vector_file) Loop   --判断是否读到数据文件结尾
            Readline(vector_file, L);          --读取测试文件的行放入缓冲区 L 中
            If(L(1)='#') Then                  --L 中第一个字符为"#",读下一行
                Next;    --忽略本次循环的部分操作,调到 Loop 处,开始下一次循环
            End If;
            Read(L,R,GOOD => good_number);
            Next When Not good_number;         --表达式为真,执行 Next
```

```
    --开始下一次循环,否则继续
            vector_time := R * 1 ns;          --转换实数到时间变量
            If (now < vector_time) Then       --Wait until vector_time
                Wait For vector_time;         --now;给信号赋值等待的时间
            End If;
            signo := 26;                      --输入/输出的数据位数共有 26 个
            For i In L'Range Loop
                Case L(i) Is
                    When '0' => ports(signo) <= '0';              --Drive 0
                    When '1' => ports(signo) <= '1';              --Drive 1
                    When 'H' => Assert ports(signo) = '1';        --Test for 1
                    When 'L' => Assert ports(signo) = '0';        --Test for 0
                    When 'X' => Null;                             --Don't care
                    When '' | HT => Next;                         --跳过空格
                    When Others => Assert False; --断言有错,报告"不规则字符"
                                   Report "Illegal char In vector file: " & L(i);
                                   Exit;         --提前结束循环,直接跳出循环
                End Case;
                signo := signo - 1;
            End Loop;
        End Loop;
        Assert False Report "Test complete";
        Wait;                                                     --始终等待
    End Process;
End Architecture qadd_tb_arch;
```

测试基准文件调用的数据文件如下:

数据文件 vectors. txt

```
# Test vectors for 8 - bit adder
#  0 means force 0
#  1 means force 1
#  L means expect 0
#  H means expect 1
#  X means don't care
# time a b cin sum cout
# time 87654321 87654321 cin 87654321 cout
  0    11111111 00000000 0 XXXXXXXX X
  100 00000001 00000001 1 HHHHHHHH L
  200 00000010 00000001 0 LLLLLLHH L
```

```
300 10000000 00000001 0 LLLLLLHH L
400 11110000 00001111 0 HLLLLLLH L
500 00001111 11110000 1 HHHHHHHH L
600 10101010 10101010 1 LLLLLLLL H
700 00000000 00000000 0 LHLHLHLH H
800 00000000 00000000 0 LLLLLLLL L
```

qadd_tb 文件仿真结果如图 4-6 所示。

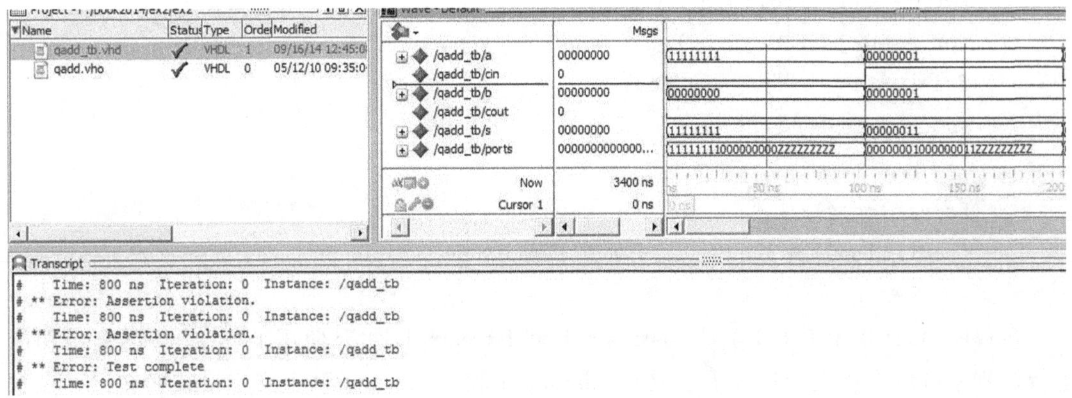

图 4-6 仿真 qadd_tb 文件

(1)仿真说明：

①这是典型的完全测试，在 VHDL 仿真时，可以直接读取输入文件作为设计的输入参数，并自动将结果与事先保存的文件相比较，给出一定的信息来确定结果的正确与否。程序在执行到 800 ns 时，停止，显示 Test complete。

②要注意 Wave 窗与 Transcript 窗在每一步执行时信息的对应。

(2)仿真程序说明：

①文件的 Open 语句默认是 Read_Mode，所以下面语句

```
File vector_file:Text Open Read_Mode Is "vectors.txt";
```

也可以写成：

```
File vector_file: Text Open Is "vectors.txt";
```

②程序中用到了 While Loop 条件循环，因为 1076.RTL 标准不支持 While，因而它只能在仿真程序中应用。For Loop 循环则更常用，For i in 0 to N-1 Loop，N 可以通过类属 Generic传递，或者通过对向量的属性操作，如 a'Range 来获得。

③在用 Textio 程序包文件时，一定要注意给读信号等待时间，文件操作比芯片内操作要慢得多，这个时间至少有几十纳秒。

④首先要仿真程序去适应设计，去适应环境，不要先动设计。如，有时对时序电路测试发生了仿真时钟触发沿与数据沿冲突，即建立时间不够，这时在仿真程序中将时钟反向一下就可以得到正确结果了。

⑤生成输入及预定结果文件的 C++程序。

可以使用 VC++、Matlab 等高级软件工具编写生成输入和预定结果文件的程序。由于设定输入为 8 位无符号数，因此，其范围为[0,127]。C++程序如下：

```
#include " iostream.h"
#include " stream.h"
    void main(void){
    int i,j;
    ofstreamfsin("d: \\shiyan1\\Modelsim\\addern\\testdata.dat");
    ofstreamfsout("d: \\shiyan1\\Modelsim\\addern\\result.dat");
    for(i=0;i<128;i++) {
    for(j=0;j<128;j++) {
    fsin<<i<<""<j<<Endl;
    fsout<<i+j<<Endl;
    }}
    fsin.close();
    fsout.close(); }
```

在程序中，使用了 C++类库 iostream. h 和 fstream. h，主要使用了“＜＜ ”的输出功能。运行该程序可以在规定的目录下生成 testdata. dat 和 result. dat 两个文本格式的文件。注意，一行输入多个数据时，之间以空格隔开即可。

4.7　Quartus Ⅱ ＋ Modelsim VHDL 的时序仿真

后仿真是时序仿真，其与前仿真的区别在于，测试文件所包含模型的结构不同。

前仿真是综合前逻辑模型，后仿真(指综合后)使用的是真实的门级结构模型，其中不但有逻辑关系，还包含实际门级电路和布线的延迟。综合后还有驱动能力等问题可借助软件工具分析。

4.7.1　四位加法器层次化设计时序仿真

(1)在 Quartus Ⅱ 中新建工程，如图 4－7 所示。

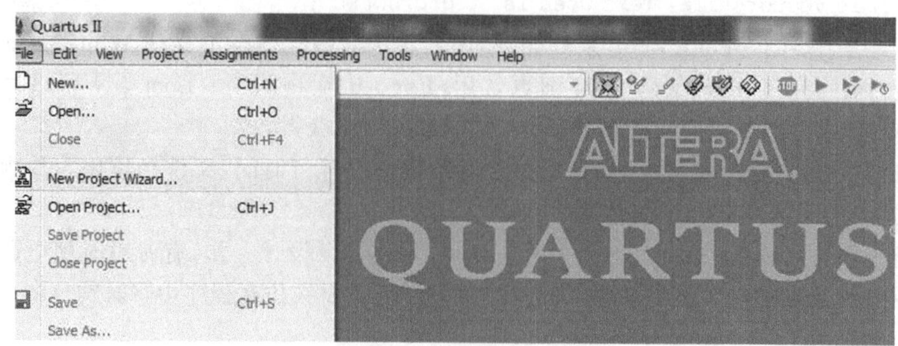

图 4－7　在 Quartus Ⅱ 中建工程

(2)设置工程目录 F:/add4,项目与实体名 adder4,如图 4－8 所示。

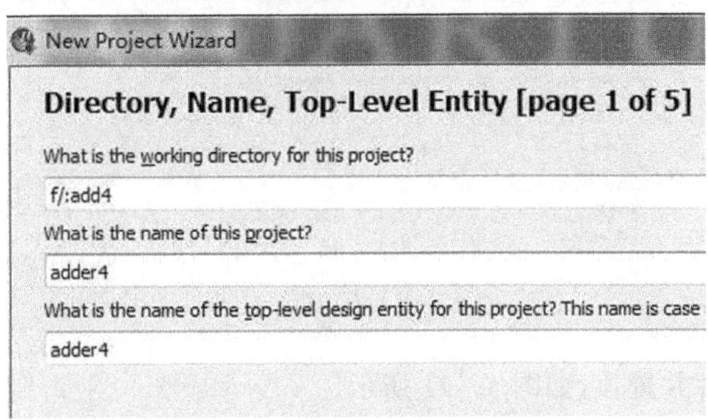

图 4－8 定义项目与实体名选器件

封装 Package:FBGA;引脚数目 Pin count:672;速度级别 Speed grade:8。选 EDA 工具仿真:Modelsim,VHDL 编译后自动启动门级仿真器如图 4－9、4－10 所示。

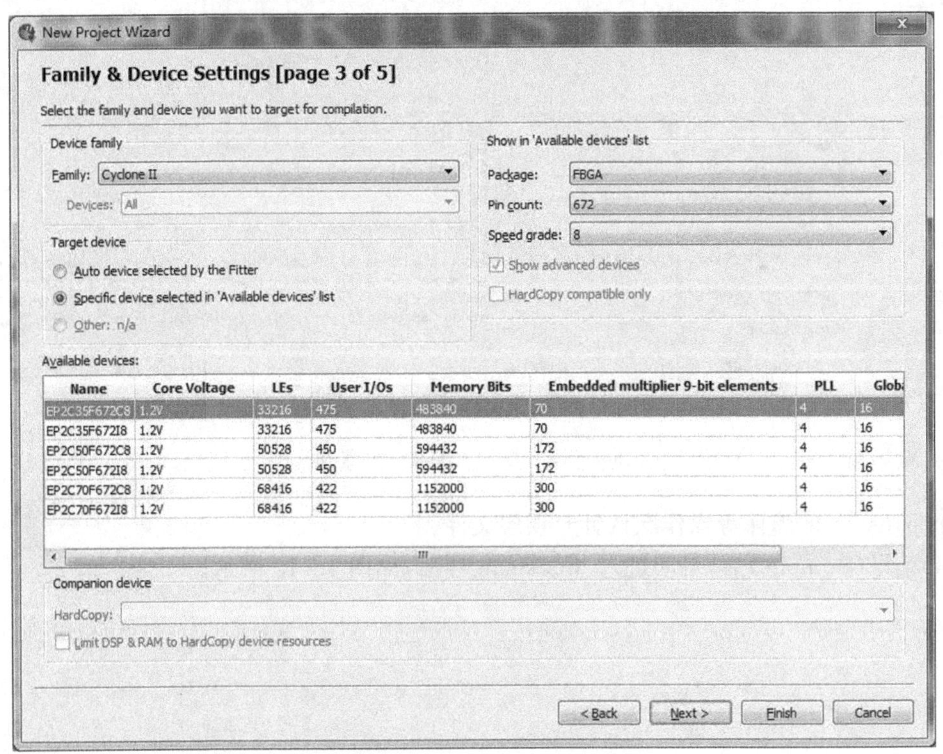

图 4－9 选器件 Cyclone Ⅱ系列 EP2C35F672C8 芯片

将写好的一位加法器文件和四位加法器顶层文件 fulladder. vhd, adder4,vhd 等工程所需 VHDL 文件拷贝到文件夹 F:/add4。

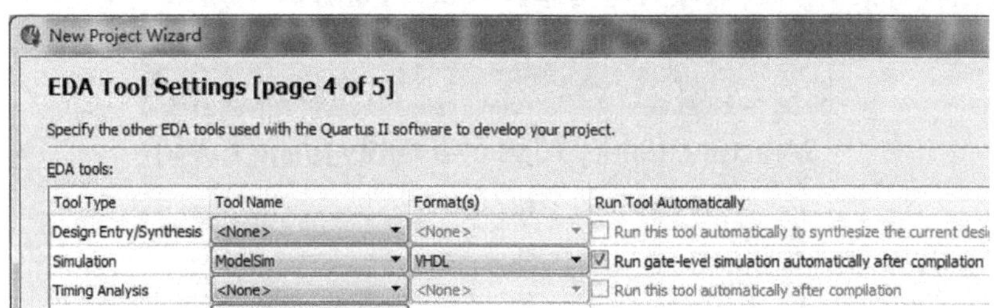

图 4－10　选 EDA 工具

(3) 打开顶层文件并置顶，如图 4－11 所示。

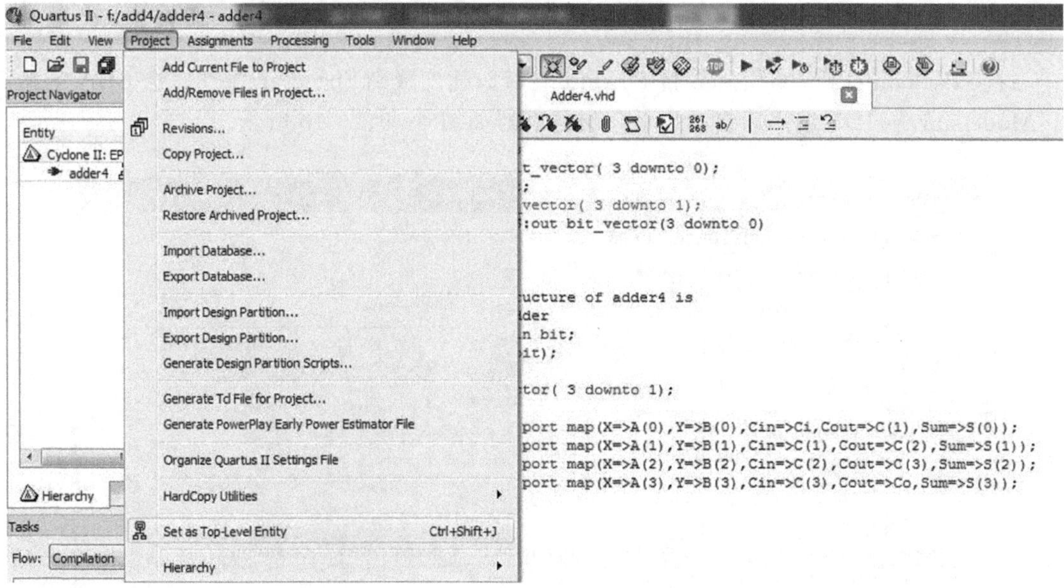

图 4－11　打开顶层文件并置顶

Quartus Ⅱ 的编译等操作默认针对顶层文件。

(4)编译(Compile)后，自动进入 Modelsim 界面，如图 4－12 所示。

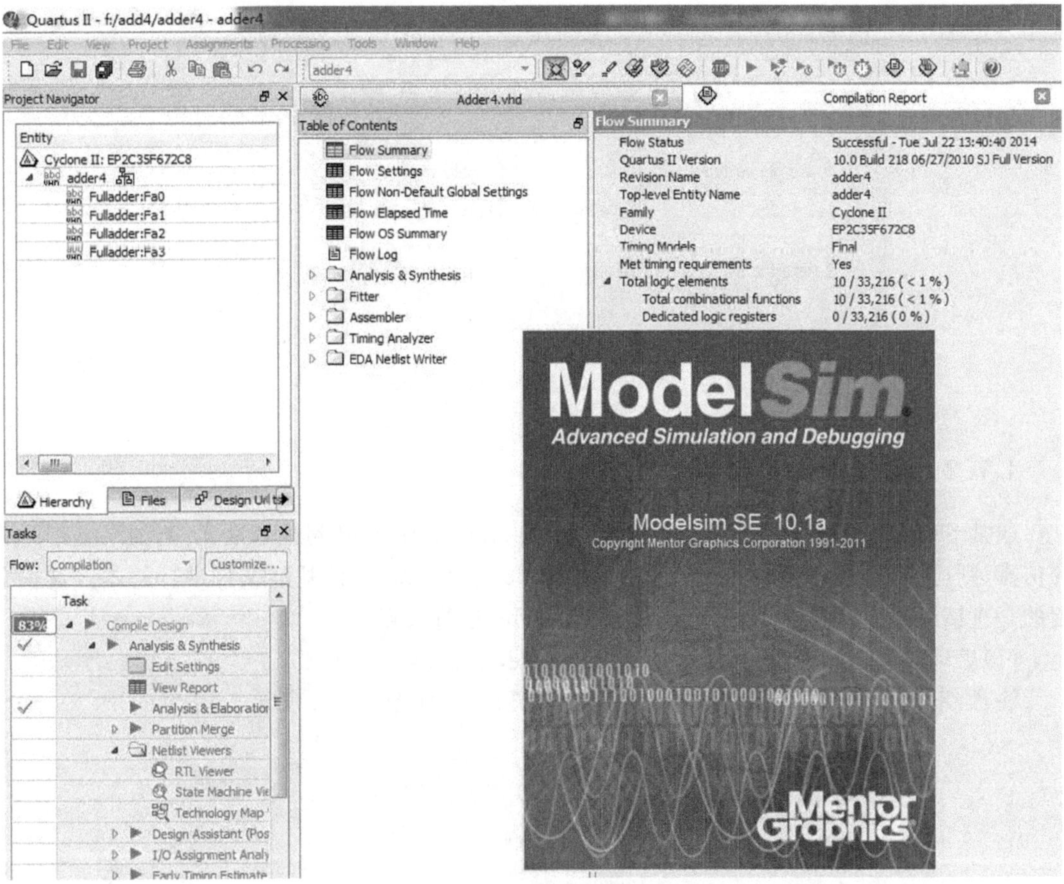

图 4-12 进入 Modelsim 界面

(5) 在 Modelsim 中新建工程，如图 4-13，4-14 所示。

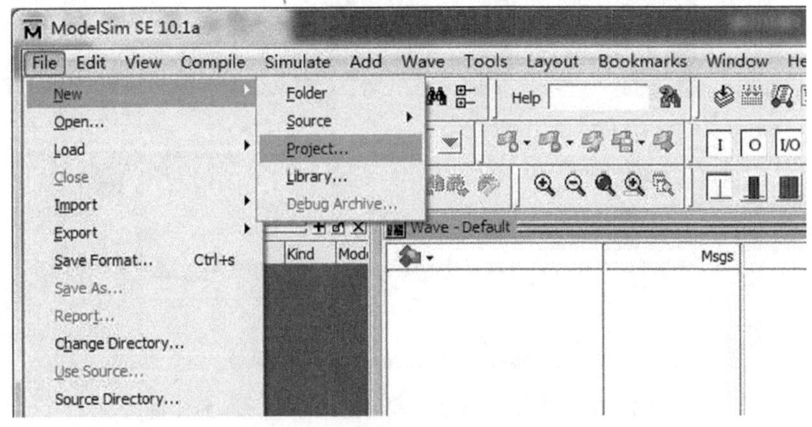

图 4-13 在 Modelsim 中新建工程

Add file to Project
File Name
F:/add4/adder4.vho
Browse...
Add file as type
default
Folder
Top Level
Reference from current location
Copy to project directory
OK
Cancel

图 4-14　选择 add4er4. vho

4.7.2　波形输入方式仿真

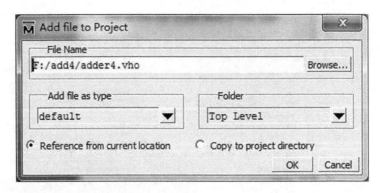

通常 Simulate 是针对由 adder4. vhd 产生的 adder4_tb. vhd 测试基准文件，如果是带时延的仿真，则应由采用 adder4. vho 文件产生的 adder4_tb. vhd 测试基准文件进行仿真。这里介绍的是直接给出波形对 adder4. vhd 或 adder4. vho 进行仿真。

(1)启动仿真程序，Simulate→Start Simulate，添加 . vho 文件。

注意：不能优化，即 Enable optimization 不要划钩，如图 4-15 所示。

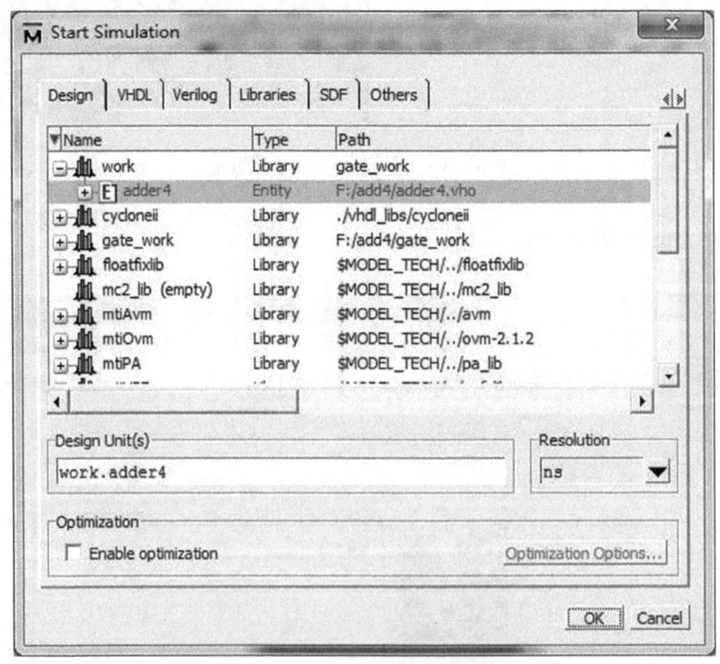

图 4-15　添加 . vho 文件和时延文件 * . sdo 的标签

(2)添加时延文件。

选 SDF 标签添加，时延文件 * . sdo，如图 4-16 所示。

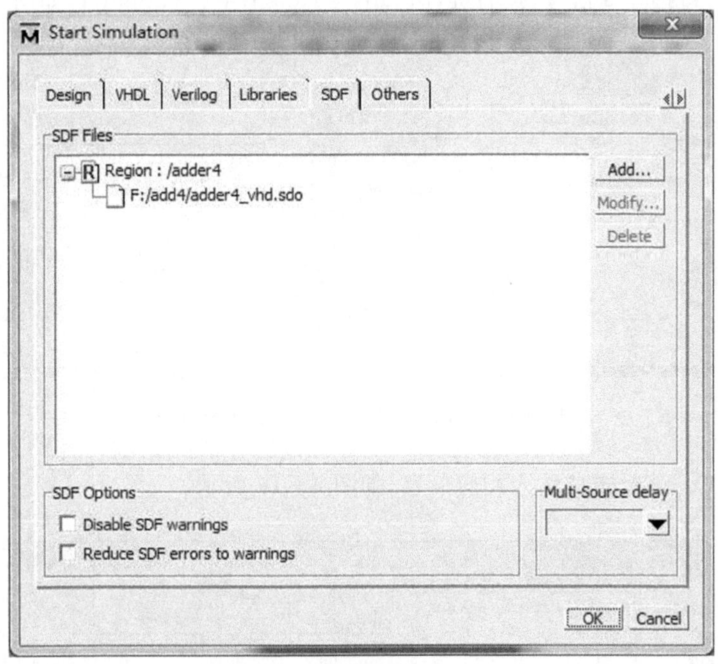

图 4-16 指定时延文件 . sdo

应用区对于波形激励仿真而言，即顶层实体名。

(3)用波形的方法给输入激励。

如果没有写测试基准文件，比较简单的设计也可以直接通过菜单中提供的波形输入功能进行仿真，如图 4-17 所示。

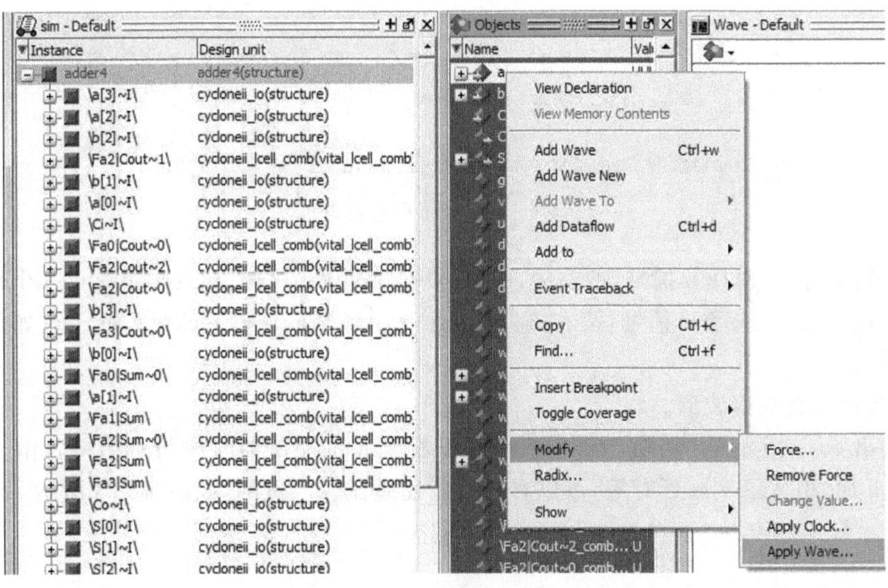

图 4-17 用波形的方法给输入激励

在 Object 窗口选择 a 信号，按右键，Modify→Add Wave→Counter，如图 4－18 所示。

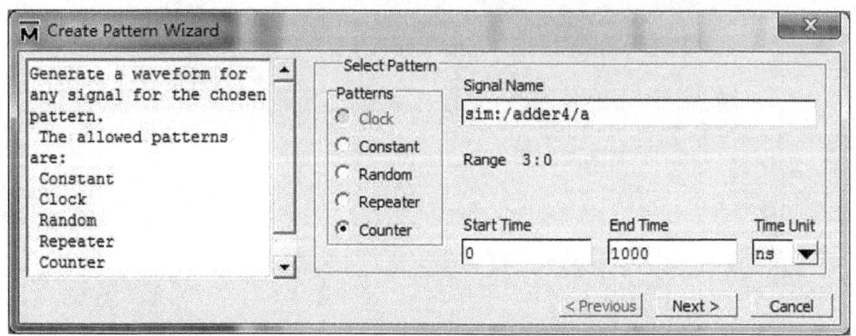

图 4－18 选择 a 信号用 Counter 给信号

在 Pattern：Counter 窗口输入相应信息，如图 4－19 所示。

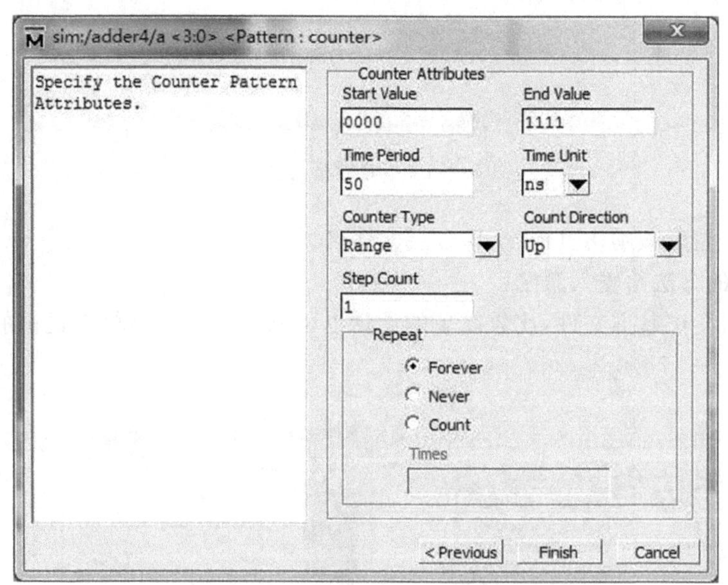

图 4－19 给 a 信号波形

在 Object 窗口选择 b 信号，按右键，Modify→Add Wave→Counter，如图 4－20 所示。

在 Object 窗口选择 ci 信号，按右键，Modify→Add Wave→Clock→Counter，如图4－21、4－22所示。

(4)运行仿真观察波形。

将输出 co，s 拖入 Wave 窗，点击 Run，可以看到 4 位加法器时序仿真结果。由于加进了 Cyclone Ⅱ库，引入了实际延时等，可以看到毛刺还是比较严重的，如图 4－23 所示。

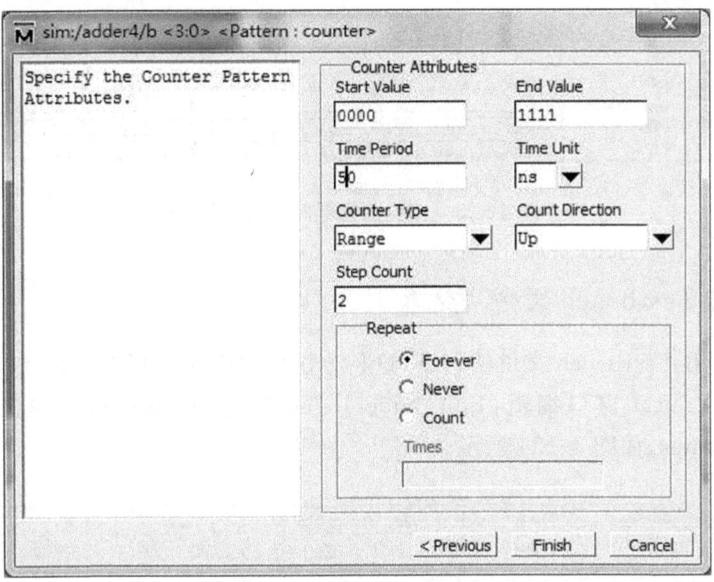

图 4-20　给 b 信号波形

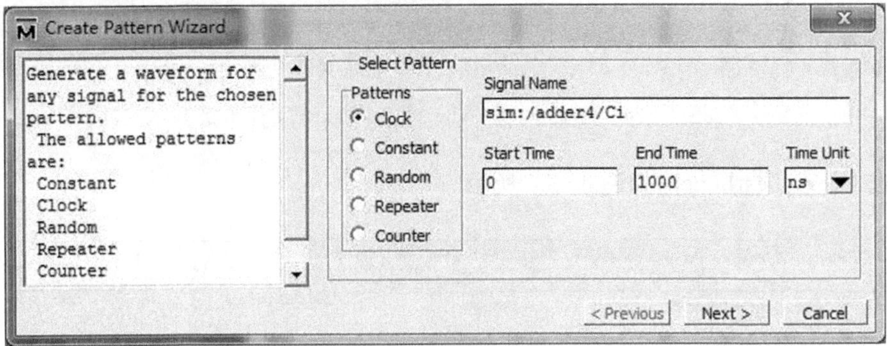

图 4-21　给 ci 信号 Clock 波形

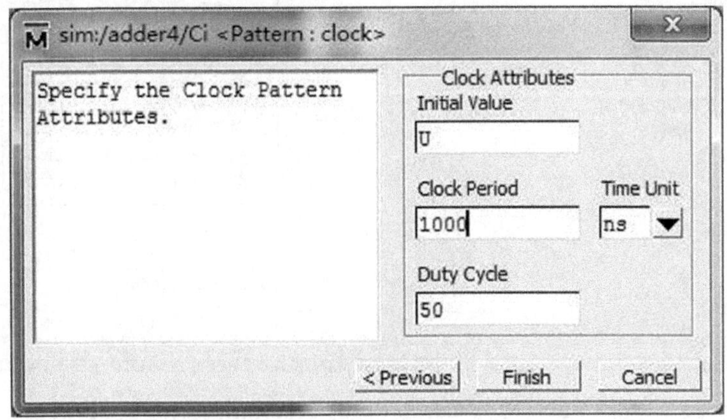

图 4-22　给 ci 信号波形的值

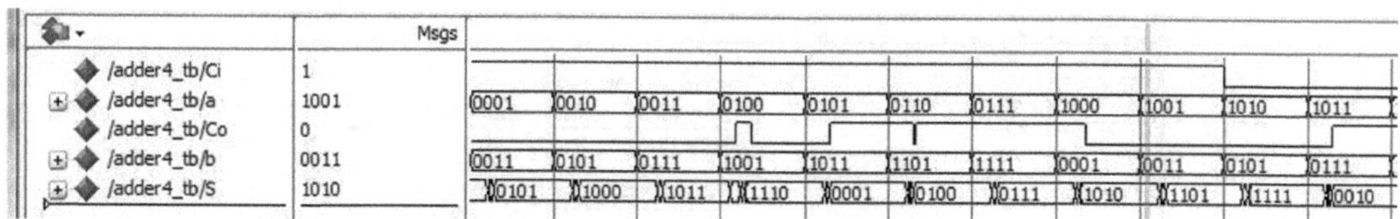

图 4 - 23　四位加法器时序仿真结果

4.7.3　用 Testbench 文件法仿真

设计通常会用 Testbench 文件法仿真，这种方法可以做比较复杂的设计验证。

(1)在平台 Project 窗口编辑(Edit) adder4. vho 文件，在编辑窗口单击 Source→Show Language Templates，如图 4 - 24 所示。

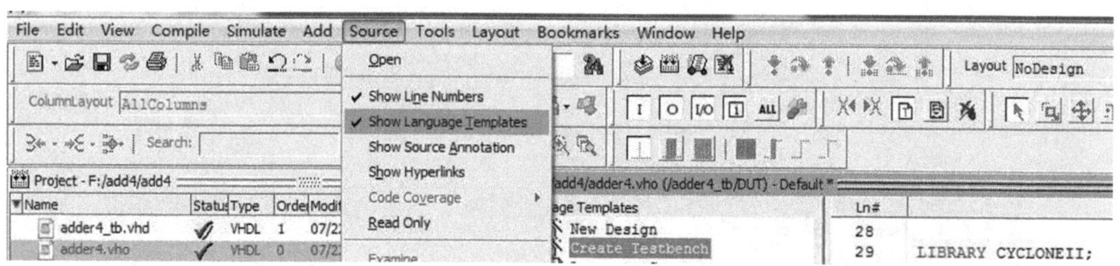

图 4 - 24　建立 adder4. vho 的 Testbench 文件步骤 1

(2)选 Creat Testbench，如图 4 - 25 所示。

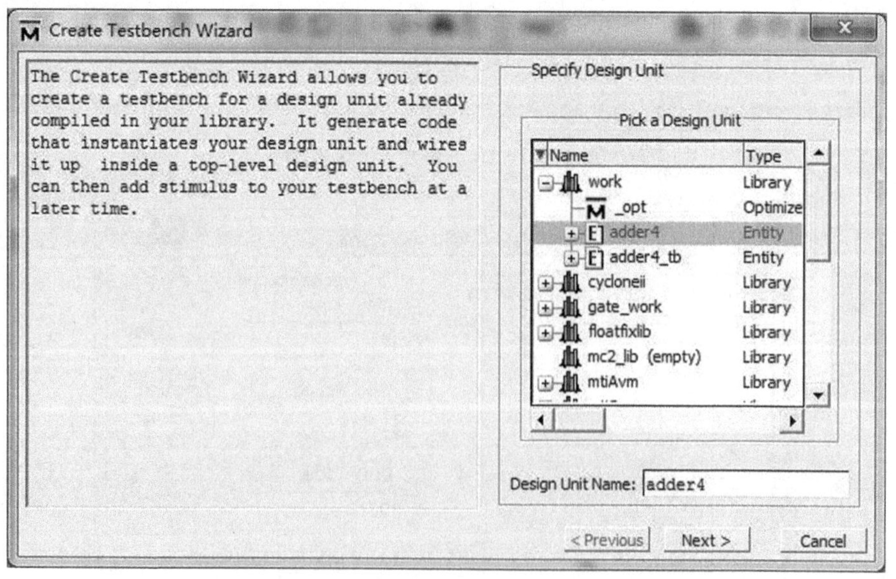

图 4 - 25　建立 adder4. vho 的 Testbench 文件步骤 2

(3)生成 Testbench 模板名字：adder4_tb. vhd，如图 4 - 26 所示。

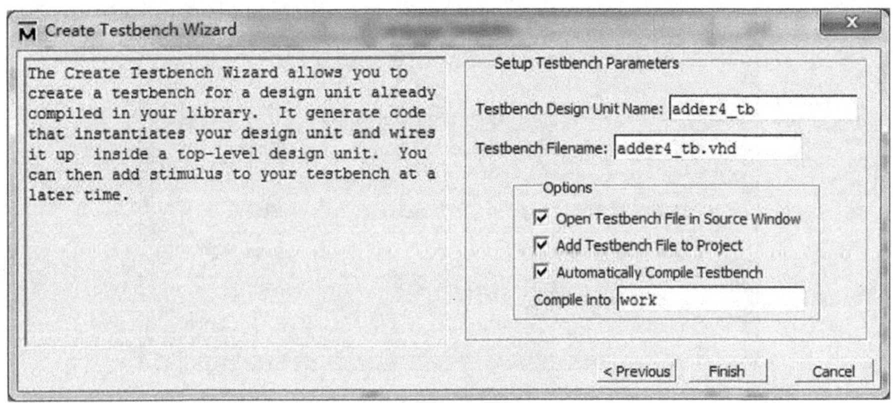

图 4-26　建立 adder4. vho 的 Testbench 文件步骤 3

(4)修改 adder4_tb. vhd,给输入端口赋初值,在结构体中给出端口的变化:Simulater→Strat Simulate,在 Work 库中选 adder4_tb. vhd,如图 4-27、4-28 所示。

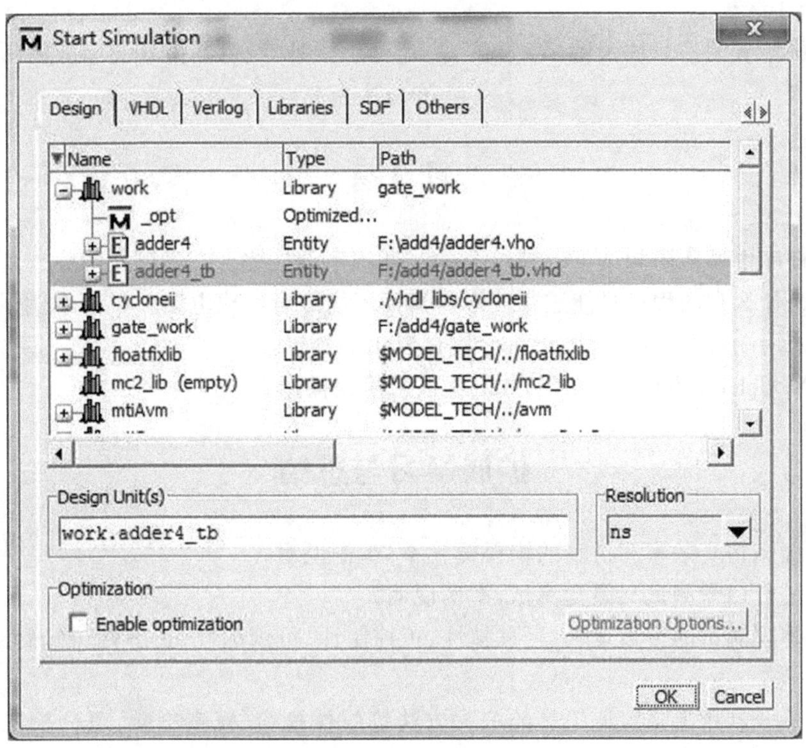

图 4-27　仿真 adder4. vho 的 Testbench

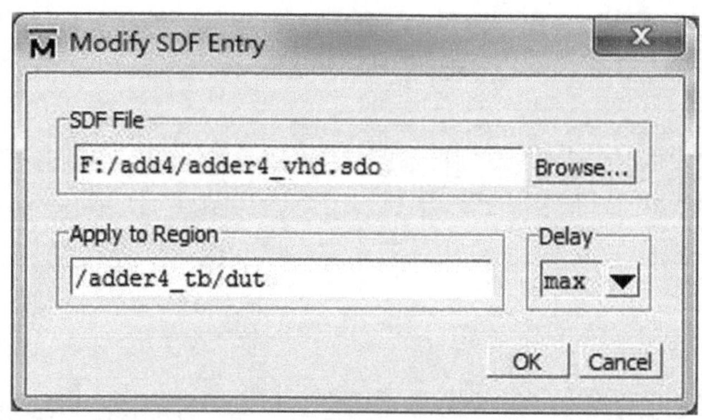

图 4-28 带时延仿真的设定

(5)选择时延文件 *.sdo,选择作用区,/testbench 实体名/测试单元,在 Transcript 窗口键入:

```
add wave *
run 1us
```

含义是将所有的信号加入到 Wave 窗口,运行 1 微秒,可得到仿真波形与波形输入法仿真波形相同。

习 题

4-1 程序中哪些语句是可以被综合的? 哪些不能被综合? 举例说明。

4-2 为什么有的程序只能进行波形仿真而不能综合生成电路? 举两个例子说明。

4-3 采用端口配置的方法,做一个测试复用的例子。

4-4 Modelsim 仿真中,采用.do 宏文件来简化操作。

扩展学习与总结

1. 如何提高仿真效率和仿真的错误覆盖率,举例说明。

2. 如何进行白盒测试与黑盒测试,举例说明。

3. 如何体现测试的可重复性、可移植性、可控制性、可观测性、正确性、精确性、预测性等,举例说明。

4. 如何综合利用形式验证工具的三大方法等效性检验、模型检验、理论验证进行设计验证,举例说明。

5. 查阅有关测试标准,说明阅读标准对测试开展的启发。

5 运算单元的设计提高

本章的主要内容与方法：

(1)常用程序包的介绍，状态机介绍。

(2)用不同的设计风格实现加减乘除。

(3)展现使用不同程序包的编程差异，了解数据转换的方法。

(4)增加寄存器进行模块间同步。

(5)针对系统级优化进行流水线设计(Pipeline)。

(6)采用状态机进行控制。

运算的问题非常重要，从 Quartus Ⅱ 版本以及器件与运算相关的改变之多，可以看到有关运算研究发展很快。如果我们希望了解这个变化，甚至跻身于这个变化研究之列，首先要能够不同风格、不同层次的去认识实践。

(1)本章运算用到的常用程序包介绍：

· Numeric_Std 程序包和 Arith 程序包。Numeric_Std 程序包的转换函数要少于 Arith 程序包的转换函数；Arith 程序包的算术操作少于 Numeric_Std 程序包。Numeric_Std 程序包是 Ieee 程序包，Arith 是 Synopsys 的程序包。

· Std_Logic_Unsigned 和 Signed 程序包。这两个包的区别是有无符号，这两个包 Overloading(重载)了 Integer，Std_Logic，Std_Logic_Vecoter 的运算符。Unsigned 和 Signed 程序包只能选取其中之一。同时处理有符号与无符号，要采用 Arith 程序包或 Numeric_Std 程序包。

(2)有关状态机的概念：

· 枚举类型(Enumerate)。

VHDL 语言枚举类型，是用文字符号来表示一组二进制数。在实际电路中，状态机的状态是一组枚举类型表示的。采用自定义类型方式。

· 自定义类型格式。

Type 数据类型名 Is　数据类型定义 Of 基本数据类型

如程序中状态说明：

```
Type state_type Is ( s1, s2, s3 ) ;
```

· 状态机时序同步、异步划分说明：

同步(Moore)：输入变化时，输出的变化要等待时钟的上升沿的到来，然后发生变化。输出仅为当前状态的函数。

异步(Mealy)：输入变化时，输出的变化立即发生。输出为当前状态和输入的函数。

· 状态机(State Machine)的软件设计组成：

状态机软件由说明部分、主控时序/组合 Process、辅助 Process 组成。

5.1 乘法器

示例 5 - 1,5 - 2 给出的并行乘法器是行为描述风格的。

示例 5 - 3a,b,c 给出的并行乘法器是结构化与数据流 RTL 描述风格,设计做了有流水线和无流水线的实现对比。

示例 5 - 4、5 - 5 探讨了串行乘法器的一般实现以及状态机实现。

示例 5 - 6 探讨了乘加法运算设计在不同级别器件的实现结果。

示例 5 - 1 有符号行为级描述乘法器

```
--有符号乘法器(Signed Multiplier)行为级描述
Library Ieee;
Use Ieee.Std_Logic_1164.All;
Use Ieee.Std_Logic_Arith.All;
Use Ieee.Std_Logic_Signed.All;

Entity Signed_mult Is
    Generic(N : Natural := 8);        --Natural,大于等于 0 的整数,用类属指定 8 位
    Port (a,b: In Std_Logic_Vector (N - 1 Downto 0);
          result: Out Std_Logic_Vector (2 * N - 1 Downto 0));
End Entity Signed_mult;

Architecture multiply_arch2 Of Signed_mult Is
Signal a_int, b_int: Signed (N - 1 Downto 0);
Signal pdt_int: Signed (2 * N - 1 Downto 0);
Begin
    a_int <= Signed (a);            --将 Std_Logic_Vector 转换为 Signed 有符号数
    b_int <= Signed (b);
    pdt_int <= a_int * b_int;       --用 Signed 类型运算
    result <= Std_Logic_Vector(pdt_int);
                                    --将 Signed 有符号数转换为 Std_Logic_Vector
End Architecture multiply_arch2;
```

示例 5 - 2 无符号乘法器行为级描述

```
/* 无符号乘法器(Unsigned Multiplier)与有符号乘法器的不同只是将 Signed 改
为 Unsigned */
Library Ieee;
Use Ieee.Std_Logic_1164.All;
Use Ieee.Std_Logic_Arith.All;
```

```
Use Ieee. Std_Logic_Unsigned. All;

Entity Unsigned_mult Is
    Generic(N : Natural := 8);
    Port (a,b: In Std_Logic_Vector (N-1 Downto 0);
          clk, aclr: In Std_Logic;
          result: Out Std_Logic_Vector (2 * N-1 Downto 0));
End Entity Unsigned_mult;

Architecture Unsigned_mult_arch Of Unsigned_mult Is
Signal a_reg, b_reg: Std_Logic_Vector (N-1 Downto 0);
Begin
    Process (clk, aclr) Is
    Begin
        If (aclr ='1') Then
            a_reg <= (Others => '0');  --等同于 a_reg <="00000000";
            b_reg <= (Others => '0');  --寄存器设初值
            result <= (Others => '0');
        Elsif (clk'Event And clk = '1') Then
            a_reg <= a;  b_reg <= b;                      --输入加寄存器
            result <= Unsigned(a_reg) * Unsigned(b_reg);   --输出加寄存器
                                    --将数据转换为无符号 Unsigned 类型运算
        End If;
    End Process;
End Architecture Unsigned_mult_arch;
```

5.1.1 乘法器非流水线与流水线研究

示例 5-3a,b 是组合电路无符号数乘法电路，这种算法将所有乘法项全计算出来，最后一并相加，这在硬件中称为阵列实现。这种方式速度快，资源消耗大，主要在时间要求较高的场合应用，在此基础上示例 5-3c 增加了流水线(Pipelinc)。

示例 5-3a 二进制四位乘法器的 RTL 级描述

```
--二进制四位乘法器顶层程序 multi4b.vhd
Library Ieee;
Use Ieee. Std_Logic_1164. All;
Use Ieee. Std_Logic_Unsigned. All;
Use Work. All;

Entity multi4b Is
```

```
    Port (clk:In Std_Logic;                              --流水线示例 multi_three 中要用
          x,y: In Std_Logic_Vector (3 Downto 0);         --被乘数、乘数
          p: Out Std_Logic_Vector (7 Downto 0) );        --积
End Entity multi4b;


Architecture multi one of multi4b Is
Signal c1,c2,c3, s1,s2,s3: Std_Llogic_Vector(3 Downto 0);
Signal xy0,xy1,xy2,xy3: Std_Llogic_Vector(3 Downto 0);

--|                        x3    x2    x1    x0  |        1 0 1 0  --被乘数
--|                        y3    y2    y1    y0  |      ×1 0 0 1   --乘数
--|--------------------------------------------|------------------------
--|               x3y0  x2y0  x1y0  x0y0       |        1 0 1 0
--|          x3y1  x2y1  x1y1  x0y1            |      0 0 0 0
--|          c12   c11   c10                   |
--|--------------------------------------------|------------------------
--|    c13   s13   s12   s11   s10             |                 --部分积之和
--|    x3y2  x2y2  x1y2  x0y2                  |    0 0 0 0
--|    c22   c21   c20                         |
--|--------------------------------------------|------------------------
--|c23   s23   s22   s21   s20                 |                 --部分积之和
--|x3y3  x2y3  x1y3  x0y3                      |  1 0 1 0
--|c32   c31   c30                             |
--|--------------------------------------------|------------------------
--|c33   s33   s32   s31   s30                 |
--|=p7   p6    p5    p4    p3    p2    p1   p0 | = 1 0 1 1 0 1 0  --积

Begin
    xy0(0) <= x(0) And y(0);  xy1(0)8 <=x(0) And y(1);
    xy0(1) <= x(1) And y(0);  xy1(1) <=x(1) And y(1);
    xy0(2) <= x(2) And y(0);  xy1(2) <=x(2) And y(1);
    xy0(3) <= x(3) And y(0);  xy1(3) <=x(3) And y(1);

    xy2(0) <= x(0) And y(2);  xy3(0) <=x(0) And y(3);
    xy2(1) <= x(1) And y(2);  xy3(1) <=x(1) And y(3);
    xy2(2) <= x(2) And y(2);  xy3(2) <=x(2) And y(3);
    xy2(3) <= x(3) And y(2);  xy3(3) <=x(3) And y(3);
```

```
    fa1: fulladder Port Map(xy0(2), xy0(1),c1(0), c1(1),s1(1));--对应示例 3-1k
    fa2: fulladder Port Map(xy0(3), xy0(2),c1(1), c1(2),s1(1));
    fa3: fulladder Port Map(s1(2), xy2(1),c2(0), c2(1),s2(1));
    fa4: fulladder Port Map(s1(3), xy2(2),c2(1), c2(2),s2(2));
    fa5: fulladder Port Map(c1(3), xy2(3),c2(2), c2(3),s2(3));
    fa6: fulladder Port Map(s2(2), xy3(1),c3(0), c3(1),s3(1));
    fa7: fulladder Port Map(s2(3), xy3(2),c3(1), c3(2),s3(2));
    fa8: fulladder Port Map(c2(3), xy3(3),c3(2), c3(3),s3(3));
    ha1:halfadder Port Map(xy0(1), xy1(0),c1(0), s1(0));        --对应示例 3-1o
    ha2:halfadder Port Map(xy1(3), c1(2),c1(3), s1(3));
    ha3:halfadder Port Map(s2(1), xy2(0),c2(0), s2(0));
    ha4:halfadder Port Map(s2(1), xy3(0),c3(0), s3(0));
        p(0) <= xy0(0); p(1) <= s1(0); p(2) <= s2(0);
        p(3) <= s3(0);  p(4) <= s3(1);  p(5) <= s3(2);
        p(6) <= s3(3);  p(7) <= c3(3);
End Architecture multi_one;
```

设计说明:RTL 级描述综合结果(图 5-1)与设计描述非常容易对应,如果做反标注后的调试,RTL 级结构化描述比较好做,同时有利于向 ASIC 移植,换言之,这个描述更低层,更有知识产权。

示例 5-3b 非流水线四位乘法器数据流描述

```
Architecture multi_two of multi4b Is
Signal x_mult_y0: Std_Logic_Vector (3 Downto 0);
Signal x_mult_y1: Std_Logic_Vector (3 Downto 0);
Signal x_mult_y2: Std_Logic_Vector (3 Downto 0);
Signal x_mult_y3: Std_Logic_Vector (3 Downto 0);

Begin
    Process(x,y) Is
    Begin                                                    --算法与 multi_one 相同

            x_mult_b0(0) <= x (0) And y (0);          --逻辑与
            x_mult_b0(1) <= x (1) And y (0);
            x_mult_b0(2) <= x (2) And y (0);
            x_mult_b0(3) <= x (3) And y (0);

            x_mult_b1(0) <= x (0) And y (1);
            x_mult_b1(1) <= x (1) And y (1);
            x_mult_b1(2) <= x (2) And y (1);
```

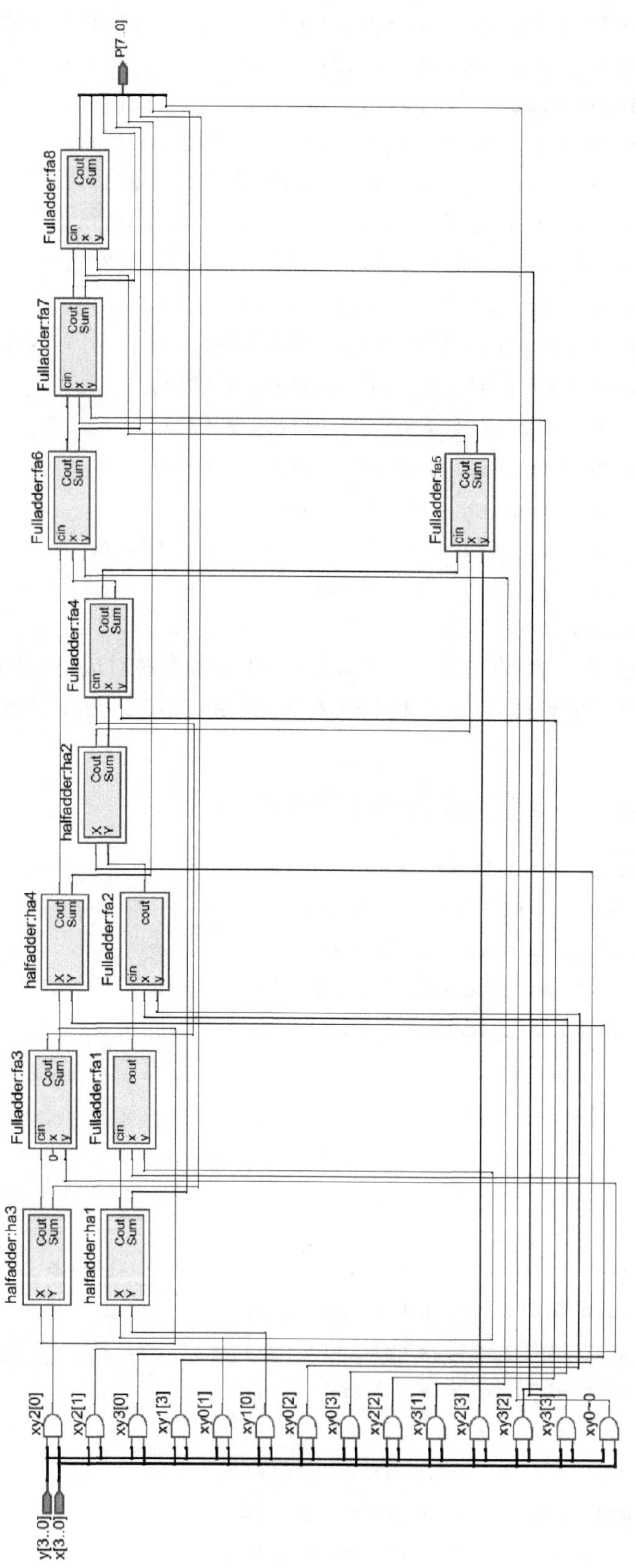

图 5-1　4 位乘法器的 RTL 级描述综合结果

```
            x_mult_b1(3) <= x (3) And y (1);

            x_mult_b2(0) <= x (0) And y (2);
            x_mult_b2(1) <= x (1) And y (2);
            x_mult_b2(2) <= x (2) And y (2);
            x_mult_b2(3) <= x (3) And y (2);

            x_mult_b3(0) <= x (0) And y (3);
            x_mult_b3(1) <= x (1) And y (3);
            x_mult_b3(2) <= x (2) And y (3);
            x_mult_b3(3) <= x (3) And y (3);
    End Process;

    p <=( "0000" & x_mult_y0 )                                --p 为乘积
         +( "000"& x_mult_y1 & '0' )                          --部分积之和
         +( "00" & x_mult_y2 & "00" )
         +( '0' & x_mult_y3 & "000" );
End Architecture multi_two;
```

其综合效果如图 5-2 所示。

示例 5-3c 流水线四位乘法器数据流描述

```
Architecture multi_three of multi4b Is
Signal x_mult_y0: Std_Logic_Vector(3 Downto 0);
Signal x_mult_y1: Std_Logic_Vector(3 Downto 0);
Signal x_mult_y2: Std_Logic_Vector(3 Downto 0);
Signal x_mult_y3: Std_Logic_Vector(3 Downto 0);
Signal ain,bin: Std_Logic_Vector(3 Downto 0);
Signal c_temp,tmp1,tmp2:Std_Logic_Vector(7 Downto 0);
Begin

    Process(clk,x,y) Is                                       --算法与 multi_one 相同
    Begin
        If(clk'Event And clk='1') Then                        --时钟上升沿触发
            xin <= x;                                         --输入端口 x 加触发器
            yin <= y;                                         --输入端口 y 加触发器
            L1:For i In 0 to 3 Loop
                                          --将结构体 multi_two 的相应语句改为循环语句
              x_mult_y0(i) <= xin (i) And yin (0);  --部分积
              x_mult_y1(i) <= xin (i) And yin (1);
```

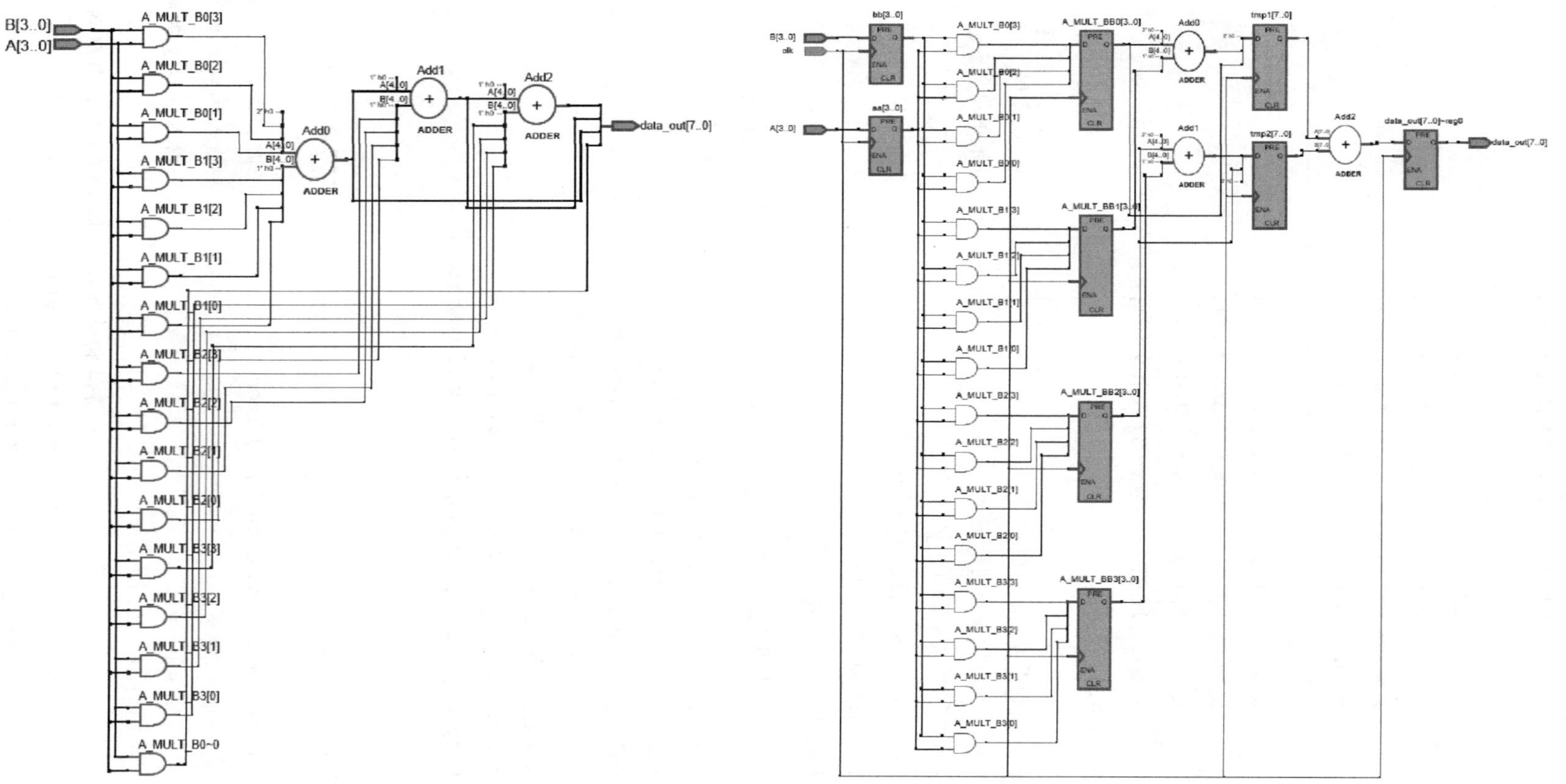

图 5-2　示例 5-3b 未加流水线与示例 5-3c 加流水线的综合结果

```
            x_mult_y2(i) <= xin (i) And yin (2);
            x_mult_y3(i) <= xin (i) And yin (3);
         End Loop L1;
         tmp1<=( "0000" & x_mult_y0 )+( "000"& x_mult_y1 & '0');
         tmp2<=( "00" & x_mult_y2 & "00" )+( '0' &x_mult_y3 & "000" );
         p <=tmp1+tmp2;                      --部分积之和,将连加拆分为 3 次加
      End If;
   End Process;
End Architecture multi_three;

Configuration multi4b_cfg Of multi4b Is
   For multi_two
   End For;
End Configuration multi4b_cfg;
```

程序说明:流水线的 Pipeline 结构本身不一定速度更快,但是毛刺更少,在系统流水级应用提高了系统性能,如图 5-3 所示,流水线指标主要可以通过吞吐率、加速比、效率、流水线的最佳段数来表示,这个例子特别有利于初学者了解 VHDL 的存储器的形成。

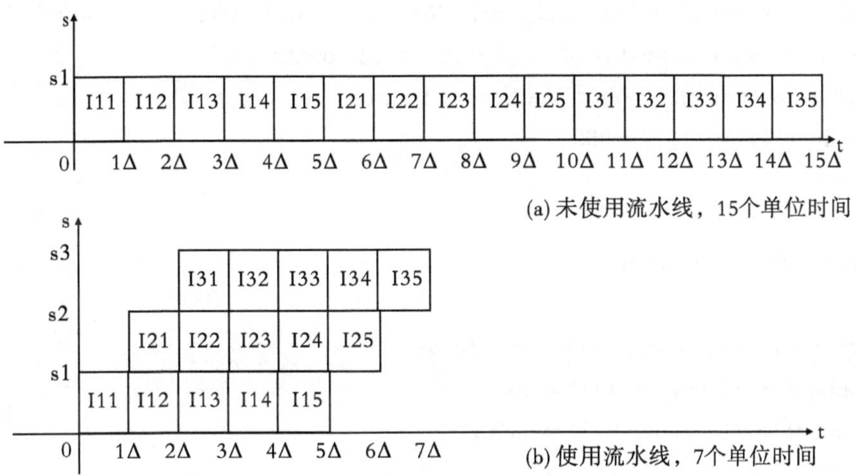

图 5-3 流水线系统级应用示意图

结构体 multi_three 是对结构体 multi_two 做的加流水线修改。上述 3 个结构体实现的有无流水线的 RTL Viewer(Quartus Ⅱ 主菜单 Tools→Netlist Viewer→RTL Viewer)比较结果如图 5-1,5-2 所示,其中图 5-1 的中的加法器是可以独立于综合器的,图 5-2 中的加法器则是综合器给出的有无流水线的对比。

示例 5－3d 乘法器的输出转换成 BCD

```
--输出控制模块,把乘法器的输出转换成 BCD 码在数码管上显示
--程序 bin2bcd.vhd

Library IEEE;
Use IEEE.Std_Logic_1164.All;
Use IEEE.Std_Logic_Arith.All;
Use IEEE.Std_Logic_Unsigned.All;

Entity bin2bcd Is
    Port (din: In  Std_Logic_Vector(7 Downto 0);                --输入 8 位 2 进制
          bcdout: Out  Std_Logic_Vector(11 Downto 0)  );
    --输出显示，已转换成 3 个字的 BCD 码

End Entity bin2bcd;

Architecture arch Of bin2bcd  Is
Signal data2,data3,data4 :Std_Logic_Vector(9 Downto 0);         --输出缓存
Signal hundred,ten,unit:Std_Logic_Vector(3 Downto 0);
Signal bcdbuffer:Std_Logic_Vector(11 Downto 0);
    --2'1111_1001_11=999

Begin
    bcdout<= bcdbuffer;

    bcdbuffer(11 Downto 8)<=hundred;
    bcdbuffer(7 Downto 4)<=ten;
    bcdbuffer(3 Downto 0)<=unit;

    get_hundred_value:Process(data2) Is                         --百位
    Begin
        data2<="00"&din;
        --get hundred value
        If data2>=900 Then
            hundred<="1001";                                    --9
            data3<=data2-900;
        Elsif data2>=800 Then
            hundred<="1000";                                    --8
```

```
        data3<=data2 - 800;
    Elsif data2>=700 Then
        hundred<="0111";                                --7
        data3<=data2 - 700;
    Elsif data2>=600 Then
        hundred<="0110";                                --6
        data3<=data2 - 600;
    Elsif data2>=500 Then
        hundred<="0101";                                --5
        data3<=data2 - 500;
    Elsif data2>=400 Then
        hundred<="0100";                                --4
        data3<=data2 - 400;
    Elsif data2>=300 Then
        hundred<="0011";                                --3
        data3<=data2 - 300;
    Elsif data2>=200 Then
        hundred<="0010";                                --2
        data3<=data2 - 200;
    Elsif data2>=100 Then
        hundred<="0001";                                --1
        data3<=data2 - 100;
    Else data3<=data2;
        hundred<="0000";
    End If;
End Process;

get_tens_value:Process(data3) Is                        --10 位
Begin
    If data3>=90 Then
        ten<="1001";                                    --9
        data4<=data3 - 90;
    Elsif data3>=80 Then
        ten<="1000";                                    --8
        data4<=data3 - 80;
    Elsif data3>=70 Then
        ten<="0111";                                    --7
        data4<=data3 - 70;
    Elsif data3>=60 Then
```

```
            ten<="0110";                                    --6
            data4<=data3 - 60;
        Elsif data3>=50 Then
            ten<="0101";                                    --5
            data4<=data3 - 50;
        Elsif data3>=40 Then
            ten<="0100";                                    --4
            data4<=data3 - 40;
        Elsif data3>=30 Then
            ten<="0011";                                    --3
            data4<=data3 - 30;
        Elsif data3>=20 Then
            ten<="0010";                                    --2
            data4<=data3 - 20;
        Elsif data3>=10 Then
            ten<="0001";                                    --1
            data4<=data3 - 10;
        Else data4<=data3;
            ten<="0000";
        End If;
    End Process;

    get_unit_value:Process(data4) Is                        --个位
    Begin
        If (data4>0) Then
            unit<=data4(3 Downto 0);
        Else unit<="0000";
        End If;
    End Process;
End Architecture arch;
--BCD 码显示很常用,方法也不是唯一的,也可以用去模和取余数的方法。
--这种行为级的程序在电路出问题时不好调试。
```

示例 5－4　串行乘法器

```
--本章乘法器主要是并行的,但是串行的实现方式也是需要关注的。
--本例 4 位串行乘法器的正确结果输出需要延时 4 个时钟之后得到。
--程序串行乘法器 smul.vhd
Library Ieee;
Use Ieee.Std_Logic_1164.All;
```

```
Use Ieee.Std_Logic_Unsigned.All;
Use Ieee.Numeric_Std.All;

Entity smul Is
    Port(a,b:In Std_Logic_Vector(3 Downto 0);
         q:Out Std_Logic_Vector(7 Downto 0) ;
         clk:In Std_Logic;
         load:In Std_Logic;
         ready:Out Std_Logic);
End Entity smul;
```

1010
×1001
累加器清零 00000 0000
第 1 部分积 1010
第 1 部分和 00101 0000
第 2 部分积 0000
第 2 部分和 00010 1000
第 3 部分积 0000
第 3 部分和 00001 0100
第 4 部分积 1010
第 4 部分和 001011010

```
Architecture behav Of smul Is                          --部分积累加右移乘法
Begin
    Process (clk) Is
        Variable count:Integer Range 0 To 4;
        Variable pa:Unsigned(8 Downto 0);
        Alias p:Unsigned(4 Downto 0) Is pa(8 Downto 4);  --别名
    Begin
        If Rising_Edge(clk) Then
            If load='1' Then
                p := (Others=>'0');
                pa(3 Downto 0 ):= Unsigned(a);     --a 转换为 Unsigned 数据类型
                count := 4;    --被乘数放在 9 位累加寄存器的高四位可节省寄存器
                ready<='0';
            ElsIf count > 0 Then
                Case Std_Logic'(pa(0)) Is      --pa(0)转换为 Std_Logic 数据类型
                    When '1' =>
                    p:= p + Unsigned(b);
                    When Others=>Null;
                End Case;
                pa := Shift_Right(pa,1);                --右移 1 位
                count := count - 1;
            End If;
            If count = 0 Then
                ready <= '1';
            End If;
            q<=Std_Logic_Vector(pa(7 Downto 0));        --乘积输出
        End If;
    End Process;
```

```
End Architecture behav ;
```

示例 5-5　状态机实现乘法器

```
Library Ieee;
Use Ieee.Std_Logic_1164.All;
Use Ieee.Std_Logic_Arith.All;

Entity mul8 Is
    Port(clk:In Std_Logic;                              --时钟
         a:In Std_Logic_Vector(7 Downto 0);
         x:in Integer Range -128 To 127;
         y:out Integer Range -32768 To 32767);
End Entity mul8;

Architecture mul8_arch Of mul8 Is
Type state_type Is(s1,s2,s3);                  --自定义类型进行状态的枚举说明
Signal state:state_type;         --声明信号 state 是自定义 state_type 类型的信号
Begin
    behav:Process Is                                    --行为描述
    Variable p,t: Integer Range -32768 To 32767;
    Variable count : Integer Range 0 To 7;
    Begin
        Wait Until clk='1';                             --clk 上升沿触发
        Case state Is
        When s1=>state<=s2;                             --状态 1 的转换
            count:=0;                                   --变量初始化
            p:=0;                                       --重置寄存器
            t:=x;
        When s2=>If count=7 Then state<=s3;             --状态 2 的转换
            Else
                If a(count)='1' Then
                    p:=p+t;
                End If;
                t:=t*2;                                 --右移位
                count:=count+1;
                state<=s2;
            End If;
            When s3=>y<=p; state<=s1;                   --状态 3,输出结果
        End Case;
```

```
    End Process behav;
End Architecture mul8_arch;
```

Quartus Ⅱ在 Netlist Viewer 中 State Machine Viewer 选项，可以查看具有状态机控制电路的综合结果，本例的查看结果如图 5-4 所示。

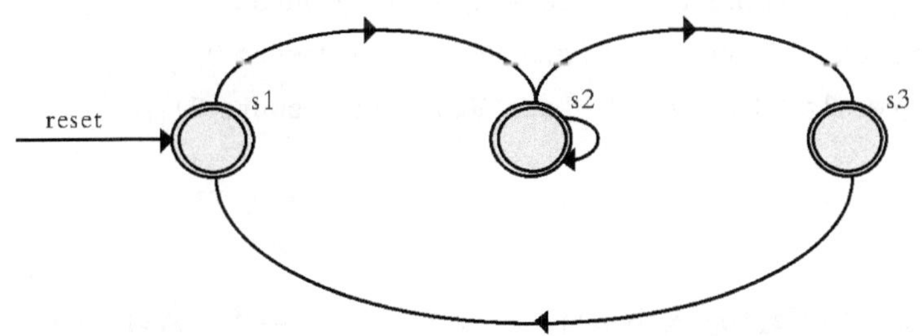

图 5-4 乘法器中状态机控制综合结果查看

示例 5-6 有符号乘加运算

程序特点：

(1)Quartus Ⅱ模板对有符号运算推荐了 Ieee. Numeric_Std 程序包，这个库更加开放。运算的输入/输出数据说明为 Signed 类型。

(2)程序采用了 Generic 参数指定位数的方式编写程序，使得程序的扩展性、灵活性得到提高。

(3)程序采用了同步设计，所有的运算的输入值都是寄存器同步后的数据。

(4)组合控制电路与时序电路描述分开在不同的进程中，这是一种很好的设计风格。

```
--Quartus Ⅱ VHDL Template 这是 Quartus Ⅱ 10.0 模板中的一个程序
--Signed Multiply-Accumulate 有符号的乘加累计运算

Library Ieee;
Use Ieee.Std_Logic_1164.All;
Use Ieee.Numeric_Std.All;

Entity Signed_multiply_accumulate Is
    Generic(Data_Width :Natural := 8);                 --指定 8 位乘加法运算
    Port (a : In Signed((Data_Width-1) Downto 0);      --8 位被乘数输入
          b : In Signed((Data_Width-1) Downto 0);      --8 位乘数输入
          clk : In Std_Logic;                          --时钟
          sload: In Std_Logic                          --加载数据信号
          accum_out: Out Signed((2*Data_Width-1) Downto 0));
End Entity Signed_multiply_accumulate;
```

```
Architecture rtl Of Signed_multiply_accumulate Is
--声明寄存器作为中间数据
Signal a_reg : Signed((Data_Width - 1) Downto 0);         --被乘数寄存器
Signal b_reg : Signed ((Data_Width - 1) Downto 0);        --乘数寄存器
Signal sload_reg : Std_Logic;
Signal mult_reg : Signed((2 * Data_Width - 1) Downto 0);
Signal adder_out : Signed((2 * Data_Width - 1) Downto 0);
Signal old_result : Signed((2 * Data_Width - 1) Downto 0);

Begin

    mult_reg <= a_reg * b_reg;                         --并行语句

    Process (adder_out, sload_reg) Is                  --控制进程
    Begin
        If (sload_reg = '1') Then
            old_result <= (Others => '0'); --累加数初值清零
        Else
            old_result <= adder_out;
        End If;
    End Process;

    Process (clk) Is                                   --数据通道进程
    Begin
        If (Rising_Edge(clk)) Then                     --时钟上升沿触发
            a_reg <= a;                                --输入加寄存器
            b_reg <= b;                                --输入加寄存器
            sload_reg <= sload;                        --输入加寄存器
            adder_out <= old_result + mult_reg;
                                                       --乘加累计结果通过寄存器后送出
        End If;
    End Process;
    accum_out <= adder_out;                            --输出累加的乘加结果
End Architecture rtl;
```

程序说明:示例 5-6 有符号乘加运算的 RTL Viewer 结果,如图 5-5 所示,乘法与加法并没有按照设计流水线的通常思维进行拆分,而是有意放在了一起。

分别观察 Stratix 器件以及 Cyclone Ⅱ 器件编译报告中的资源占用情况。从图 5-6,5-7 可以看到两种器件都有 9 位 DSP 模块。由于专用 DSP 模块的存在,拆分逻辑就显得没有必要了,综合器会将乘加运算自动地放到固定的单元。

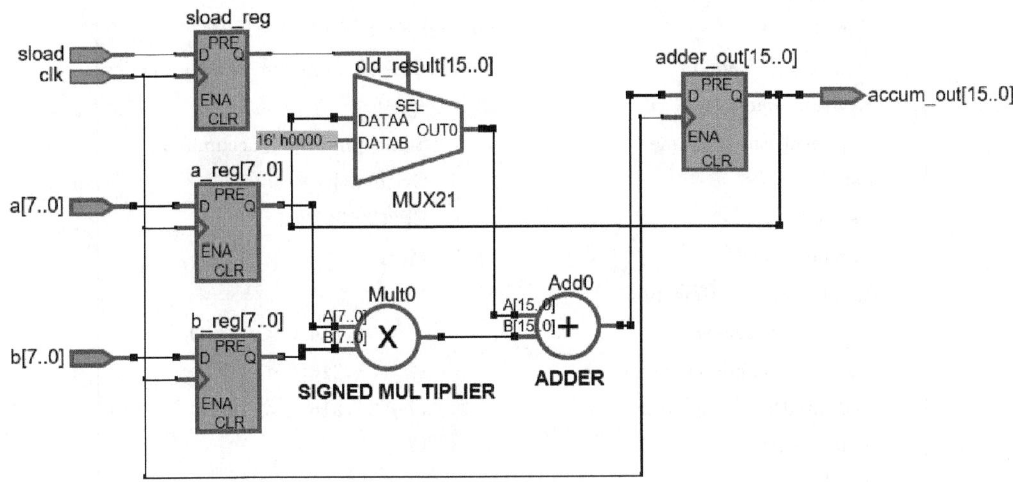

图 5-5　有符号乘加运算的 RTL Viewer 结果

Flow Status	Successful-Thu Sep 22 20:59:27 2016
Quartus Ⅱ Version 8.0	Build 215 05/29/2008 SJ Full Version
Revision Name	ex5_6
Top-level Entity Name	Signed_multiply_accumulate
Family	Stratix Ⅱ
Device	EP2S15F484C3
Timing Models	Final
Met timing requirements	Yes
Logic utilization	< 1 %
Combinational ALUTs	0 / 12,480 (0 %)
Dedicated logic registers	1 / 12,480 (< 1 %)
Total registers	1
Total pins	34 / 343 (10 %)
Total virtual pins	0
Total block memory bits	0 / 419,328 (0 %)
DSP block 9-bit elements	4 / 96 (4 %)
Total PLLs	0 / 6 (0 %)
Total DLLs	0 / 2 (0 %)

图 5-6　Stratix Ⅱ器件编译报告

从综合结果看，Stratix 器件和 Cyclone Ⅱ器件在实现这个电路是不同的。前者占用 4 个专用 DSP 乘法器组成，一个专用寄存器，没有占用组合逻辑单元；后者则占用 1 个嵌入 DSP 乘法单元及 32 个组合逻辑与 17 触发器完成，仅从资源占用这点看，就可以知道到 Stratix 器件在实现乘法乘加等有关 DSP 操作方面优于 Cyclone Ⅱ器件。

对于这个电路实现查看时序分析报告，Stratix Ⅱ系列速度可达 250 MHz 左右，而比 Cyclone Ⅱ系列则是 210 MHz 左右。

cyclone Ⅱ	
Flow Status	Successful-Fri Sep 16 18:01:24 2015
Quartus Ⅱ Version	8.0 Build 215 05/29/2008 SJ Full Version
Revision Name	ex5_6
Top-level Entity Name	Signed_multiply_accumulate
Family	Cyclone Ⅱ
Device	EP2C35F672C6
Timing Models	Final
Met timing requirements	Yes
Total logic elements	32 / 33,216 (< 1 %)
Total combinational functions	32 / 33,216 (< 1 %)
Dedicated logic registers	17 / 33,216 (< 1 %)
Total registers	17
Total pins	34 / 475 (7 %)
Total virtual pins	0
Total memory bits	0 / 483,840 (0 %)
Embedded Multiplier 9-bit elements	1 / 70 (1 %)
Total PLLs	0 / 4 (0 %)

图 5-7　Cyclone Ⅱ器件编译报告

就 FPGA 总体而言，有些芯片没有 DSP 模块，如，FLEX10K；有些芯片有 DSP 模块，但不同系列芯片的 DSP 的内涵是不同的，对于本例，Stratix Ⅱ芯片采用专用 DSP 模块，如乘法器、累加器和少量逻辑实现如图 5-8 所示，Cyclone Ⅱ芯片采用 2 个嵌入 DSP 乘法模块以及许多触发器和许多门电路实现。通过 Technology Map Viewer 来观察 2 个器件具体的实现技术和采用的元件的确是不同的，显然，Stratix Ⅱ性能更好。

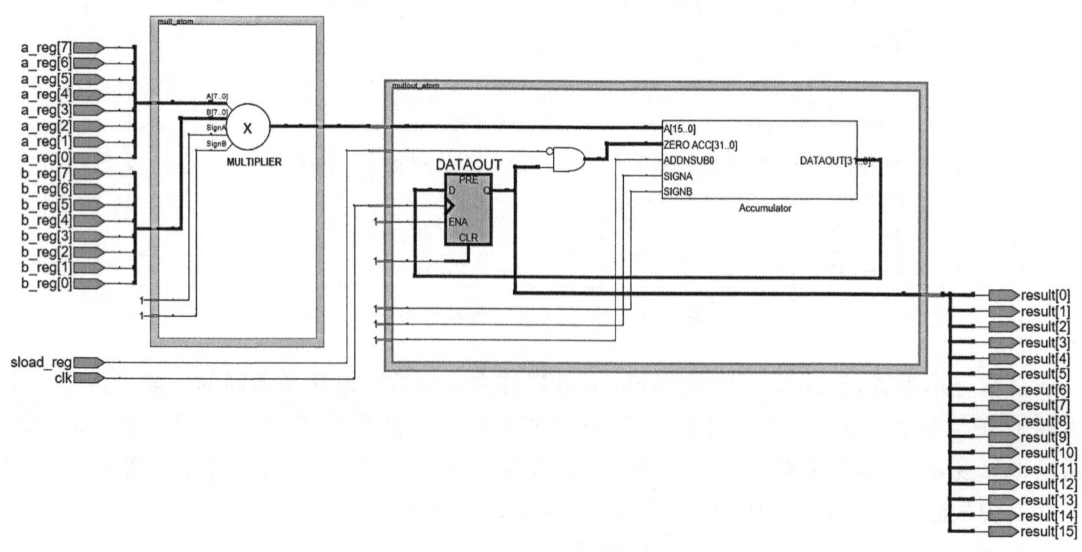

图 5-8　Stratix Technology Map Viewer 专用 DSP 模块底层

我们希望在设计中做到优化，因为优化与结构是如此的密切相关，同时一个非常重要的优化能力来自综合器，所以在设计中要关心不同级别的综合结果及其对比，加深对综合器的了解。

总的来说，一个好FPGA设计者要深入了解结构，了解综合器，在此基础上还要具有代码优化的能力。

5.1.2 硬件乘法器运算拓展

(1)移位相加。8位乘法器只需要占用16位移位寄存器和一个加法器的资源。大部分单片机与微处理都采用这种方法。该方法占用资源较少，速度较慢，8位乘法器需要8个时钟周期才能得到结果。

(2)查询表。查询表方法的速度等于存储器的速度，速度快，但占用存储空间较大。8位乘法器需要ROM的存储量为1 048 578位(1兆位)。

(3)逻辑树。输出数据的每一位都可以写成所有操作数的逻辑函数，这种方法的速度和查找表一样快。逻辑树也可以视为一个精简的查找表，8位数据的乘法器需要16个输入。

(4)加法树。采用加法树实现乘法，这种方法由移位相加组成。8位乘法器中，8×1可以用8与门实现，加法器树为操作数位数减1，如图5-9所示。如果采用流水线，乘法运算只需一个时钟周期。这种方法速度较快但占用资源较多。

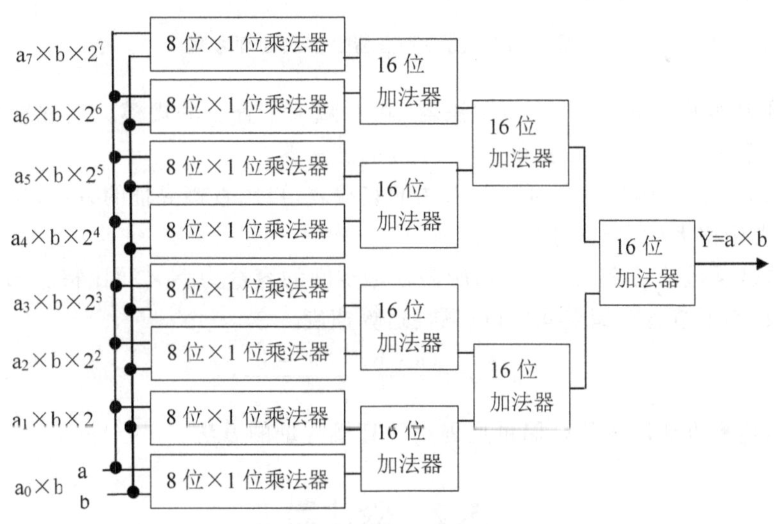

图5-9 加法树实现乘法

(5)混合乘法器。

①一次处理方案。

输入 $X=X_7:X_0$；$Y=Y_7:Y_0$

输出 $Z=Z_{15}:Z_0$

为了减少查表量，可以采用多次处理方案。

②多次处理方案。

$X=2^4(X_7:X_4)+X_3:X_0$;

$Y=2^4(Y_7:Y_4)+Y_3:Y_0$;

$$Z=(2^4(X_7:X_4)+X_3:X_0)\times(2^4\times(Y_7:Y_4)+Y_3:Y_0)$$
$$=2^8(X_7:X_4)\times(Y_7:Y_4)+2^4\times(X_7:X_4)\times Y_3:Y_0$$
$$+2^4X_3:X_0\times(Y_7:Y_4)+X_3:X_0\times Y_3:Y_0$$

混合乘法器是多次处理方案的应用,如图 5-10 所示。例如:8×8 位乘法器利用存储部分乘的小型查找表和加法器得到完整的乘积。

$X=2^4(X_7:X_4)+X_3:X_0$;

$$Z[15:0]=a_7:a_0\times X_7:X_0$$
$$=a_7:a_0\times(2^4\times(X_7:X_4)+X_3:X_0)$$
$$=a_7:a_0\times2^4\times(X_7:X_4)+a_7:a_0\times X_3:X_0$$

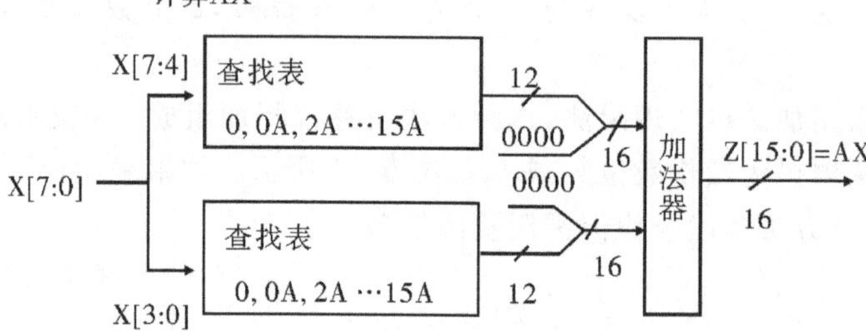

图 5-10 混合乘法器是多次处理方案

8×8 位乘法变成了两个 4×8 位乘法器,相对减少了查表的规模。实现这样的 8 位乘积运算只需要两个 $2^4\times12$ 的查找表和一个 16 位的加法器。

利用 ROM 宏模块构造一个可执行乘法的查找表,再用直接寻址的方法,这种方法运算速度快,但是要占用很大的存储空间。

当乘数与被乘数的总字长为 N 时,所需存储空间的容量为 $N\times2^N$ 比特。FPGA 内存储单元的容量有限,为了节省存储空间,可以牺牲运算速度。

$$a\times b=\frac{(a+b)^2-a^2-b^2}{2}$$

当 N>8,这种方法所需的存储量远远小于直接寻址的方法。

5.2 除法器

示例 5-7 八位除法器

```
Library Ieee;
Use Ieee.Std_Logic_1164.All;
Use Ieee.Std_Logic_Unsigned.All;
Use Ieee.Std_Logic_Arith.All;
```

```
Entity div8 Is
    Generic(Wn:Integer:=8;
            Wd:Integer:=6;
            Po2wnd:Integer:=8192;
            Po2wnl:Integer:=128;
            Po2wn:Integer:=255);
    Port(clk: In Std_Logic;
        n_in: In Std_Logic_Vector(Wn-1 Downto 0);            --被除数
        d_in: In Std_Logic_Vector(Wd-1 Downto 0);            --除数
        r_out: Out Std_Logic_Vector(Wn-1 Downto 0);          --余数
        q_out: Out Std_Logic_Vector(Wn-1 Downto 0));         --商
End Entity div8;

Architecture div8_arch Of div8 Is
Subtype twowords Is Integer Range -1 To po2wnd-1;            --自定义子类型说明
Subtype word Is Integer Range -1 To po2wnd-1;

Type state_type Is (s0,s1,s2,s3);
Signal state: state_type;
Begin

    states:Process Is                                         --无敏感表进程
    Variable r,d: twowords;                                   --双字变量
    Variable q:word;                                          --单字变量
    Variable count: Integer Range 0 To wn;
    Begin
        Wait Until clk='1';
        Case state Is
            When s0=>
                state<=s1;
                count:=0;
                q:=0;
                d:=Po2wnl * Conv_Integer(d_in);
                r:= Conv_Integer(n_in); --Std_Logic_Vector 类型转换为 Integer
            When s1=>
                r:=r-d;
                state<=s2;
            When s2=>
            If r<0 Then
```

```
                r:=r+d;
                q:=2*q;
            Else q:=2*q+1;
            End If;
            count:=count+1;
            d:=d/2;
            If count=Wn Then state<=s3;                       --division Ready?
            Else state<=s1;
            End If;
            When s3=>
            q_out<=Conv_Std_Logic_Vector(q,Wn);
                                          --Integer 转 Std_Logic_Vector 类型
            r_out<="00"&Conv_Std_Logic_Vector(r,Wd);
            state<=s0;
        End Case;
    End Process;
End Architecture div8_arch;
```

除法器仿真结果如图 5-11 所示。

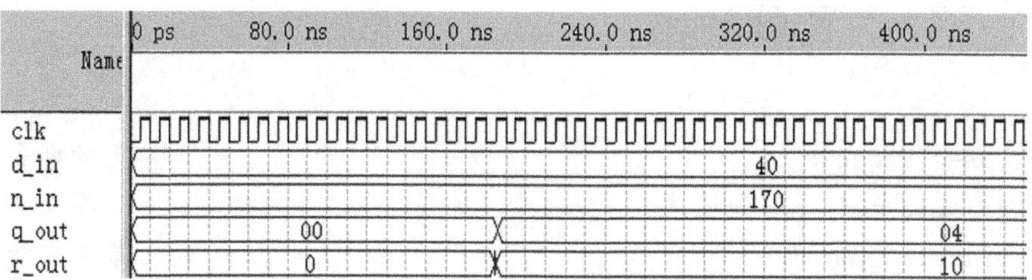

图 5-11　除法器仿真结果

示例 5-8　四位除法器

```
Library Ieee;
Use Ieee.Std_Logic_1164.All;
Use Ieee.Std_Logic_Arith.All;
Use Ieee.Std_Logic_Unsigned.All;

Entity divider4 Is
    Port(divident:In Std_Logic_Vector(3 Downto 0);          --被除数
         dividor:In Std_Logic_Vector(3 Downto 0);           --除数
         carrybit:Out Std_Logic;
         result:Out Std_Logic_Vector(3 Downto 0);           --商
         residual:Out Std_Logic_Vector(3 Downto 0));        --余数
End Entity divider4;
```

```
Architecture behav of divider4 Is
Begin
    Process(divident,dividor) Is
    Variable counter_1:Integer;
    Variable c,d,a,b,e,f,sig_1:Std_Logic_Vector(3 Downto 0);
    Begin
        a:=divident; b:=dividor;
        e:=a;        f:=b;
        counter_1:=0;
        If(b="0000") Then
            c:="1111"; d:="1111";
            carrybit<='1';
        Else
            If(a<b) Then                          --被除数小于除数不需要做除法
                c:="0000"; d:=a;
                carrybit<='0';
            Else
                If(a="0000") Then                 --0除以任何数都为0
                    c:="0000"; d:="0000";
                Else
                    For i In 3 Downto 0 Loop      --嵌套For循环结构,移位相除
                        If(f(3)='0')Then
                            For j In 3 Downto 1 Loop
                                                  --将除数移到第1位不为0的位与
                                f(j):=f(j-1);  --被除数第一位不为0的数对齐
                            End Loop;
                            f(0):='0';
                            counter_1:=counter_1+1;
                        End If;
                    End Loop;
                    For i In 3 Downto 0 Loop
                        If(i>counter_1) Then c(i):='0';
                        Elsif (e<f) Then          --当被除数小于除数
                            For j In 0 To 2 Loop
                                f(j):=f(j+1);
                            End Loop;
                            f(3):='0';
                            c(i):='0';            --商置"0"
                        Else                      --当被除数大于除数
```

```
                    e:=e-f;
                    c(i):='1';              --商置"1"
                    For j In 0 To 2 Loop
                        f(j):=f(j+1);       --移位
                    End Loop;
                    f(3):='0';
                End If;
            End Loop;
            d:=e;
        End If;
        carrybit<='0';
      End If;
    End If;
    result<=c;
    residual<=d;
  End Process;
End Architecture behav;
```

5.3 RTL级加减乘除运算整合

介绍这个案例的目的是了解一种整合的思路[7]。这个例子中利用了有限状态机FSM和数据通路等概念。整合是系统设计中的重要问题。区分系统与模块的重要线索之一就是看其有无控制。可以这样说,将原本各自为政的模块堆积在一起,增加控制协同工作,其过程叫整合,其结果叫系统。

5.3.1 乘除运算电路的控制

状态机方法是系统控制中常用的方法,状态图在系统设计中也很常见。本书有些状态机的程序没有给出状态图,因为给出了代码,状态图含在其中,不占用篇幅单列了。实际系统设计中,需要系统的状态图。系统的状态图的得来,不仅要想清楚整个系统的工作过程,还要有对每一个模块的工作过程分析的基础,根据系统要求可能要划分每一模块的状态,这个过程往往是自顶向下与自底向上互动的,在此基础上再做具体的设计。

本例乘除法采用相同的状态机来实现,状态机图常见有两种表示方法,如图5-12所示,图中左侧是早期资料中用的比较多的状态SM(State Machine)图表示方法。圆圈是状态框,带箭头的弧线上面的字表示状态的输入或者相应的输出;图中右侧是现在资料里普遍采用的表示方法,称为ASM(Algorithmic State Machine)图,方框表示状态框,菱形框是判断框,圆头矩形框是条件(输出)框。

图中s1是初始状态,s2是功能实现状态,s3是结果输出状态。这里状态图的含义是,初始状态为s1,在s1状态,当s=0时,下一个状态还是s1,当s=1时,转换到下一个状态s2。同理,在s2状态,当z=0时,状态在s2状态循环,当z=1时,状态转换到下一个状态s3。

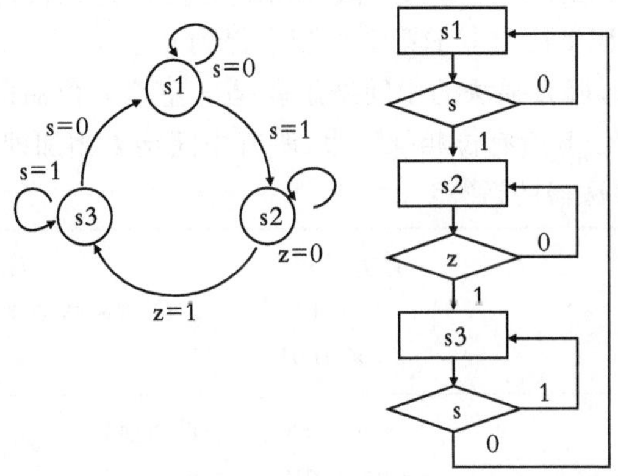

图 5-12 乘除法控制的状态图与 ASM 状态图

5.3.2 状态机控制的移位乘法

product:乘积;a:被乘数;b:乘数

算法:

```
product=0;
For k=0 To N-1 Do
     If b(k)=1 Then
          product= product+a;
     End If;
     Left-Shift a;
End For;
```

乘法设计的 ASM 图如图 5-13 所示。

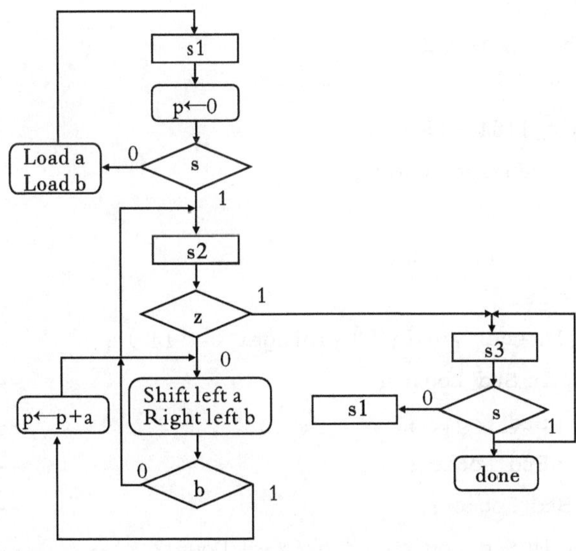

图 5-13 乘法设计的 ASM 图

被乘数 a 与乘数 b，首先用 a 与 b 的最低位相乘得到 s1，然后再把 a 左移 1 位与 b 的第 2 位相乘得到 s2，再将 a 左移 1 位与 b 的第 3 位相乘得到 s3。

如果 b 的相应位为 1，那么乘法的中间结果就是 a 左移 n 位后的结果，否则，如果 b 的相应位为 0，中间结果就是 0。b 所有位相乘结束，所有中间结果相加即得到 a 与 b 相乘的结果。

下面举例说明移位乘法器的运算：

步骤	运算表示	计算过程	注释
	p = a × b	1010 × 1001	load 被乘数与乘数
1	p	00000000	设初值，
		+ 00001010	a
	求和 p = p + a	00001010	部分积，b(0) = '1'
2	p	00010100	a 左移 1 位，b(0) = '0'，右移乘数 b 每次右移乘数 b
3	p	00101000	a 左移 1 位，b(0) = '0'，右移乘数 b
4	p	01010000	a 左移 1 位
		+ 00001010	a
	求和 p = p + a	01011010	结果，b(0) = '1'

乘法设计流程如图 5－13 所示，在此乘法计算中，每算出一个乘积项成为一个部分积，其得到要乘数 b 右移，运算位 b(0)为 1 时，通过被乘数 a 左移后与 a 对齐相加得到，b(0)为 0 时只移位不做运算，如此迭代，最后得到结果。这种方法耗费资源较少，但一个 4 位乘法需要 4 个周期才得到结果。

示例 5－9　RTL 级移位乘法器程序

```
--1 multiplier.vhd N 位乘法器
Library Ieee ;
Use Ieee.Std_Logic_1164.All ;
Use Ieee.Std_Logic_Unsigned.All ;
Use Work.All ;

Entity multiplier Is
    Generic ( N :Integer := 7; NN :Integer := 14 ) ;
    Port ( clock: In Std_Logic ;                                --时钟
           la : In Std_Logic ;                                  --加载被乘数
           lb : In Std_Logic ;                                  --加载乘数
           s: In Std_Logic ;                                    --开始
           dataa: In Std_Logic_Vector(N-1 Downto 0) ;           --multiplicand 被乘数
           datab: In Std_Logic_Vector(N-1 Downto 0) ;           --multiplier 乘数
```

```
        p: Buffer Std_Logic_Vector(NN - 1 Downto 0) ;         --porduct 乘积
        done : Out Std_Logic ) ;                               --运算完成标志
End Entity multiplier ;

Architecture multiplier_arch Of multiplier Is
Type state_type Is ( s1, s2, s3 ) ;                            --状态定义
Signal y : state_type ;                                        --状态说明
Signal ea, eb, ec, lc, ep, psel, z, zero: Std_Logic;
    --ea,eb 使能左移(multiplier)、右移(multiplicand)寄存器
    --ec,lc enable,load of downcounter
    --ep 使能乘积寄存器, psel select line of multiplexer
    --z detecter of zero, zero series input of shift
Signal a, ain, datap : Std_Logic_Vector(NN - 1 Downto 0) ;
    --a, ain 左移寄存器输入,输出
    --datap,output of multiplexer
Signal sum,nn_zeros :Std_Logic_Vector(NN - 1 Downto 0);
    --sum of product and multiplicand
    --2 * N - bit zero input to multiplicand
Signal b,n_zeros : Std_Logic_Vector(N - 1 Downto 0) ;
    --右移寄存器输出
    --N - bit zero load into register with multiplicand
Signal q:Integer Range 0 To n;                                 --count of downcounter
Begin
    fsm_transitions: Process ( clock ) Is
    Begin
        If (clock'Event And clock = '1') Then
            Case y Is
                When s1 =>
                If s = '0' Theny <= s1 ; Else y <= s2 ; End If ;
                When s2 =>
                If z = '0' Then y <= s2 ; Else y <= s3 ; End If ;
                When s3 =>
                If s = '1' Then y <= s3 ; Else y <= s1 ; End If ;
            End Case ;
        End If ;
    End Process fsm_transitions;

    fsm_outputs: Process ( y, s, la, lb, b(0) ) Is
    Begin
```

```
        ep <= '0'; ea <= '0'; eb <= '0'; done <= '0'; psel <= '0';
        Case y Is
            When s1 =>
            ep <= '1';
            If s = '0' and la = '1' Then ea <= '1';
            Else ea <= '0'; End If ;
            If s = '0' and lb = '1' Then eb <= '1';
            Else eb <= '0'; End If ;
            When s2 =>
            ea <= '1'; eb <= '1'; psel <= '1';
            If b(0) = '1' Then ep <= '1'; Else ep <= '0'; End If ;
            When s3 => done <= '1';
        End Case ;
    End Process fsm_outputs;

    --define the datapath circuit
    nn_zeros <= (Others => '0');                               --2 * N - bit zero
    n_zeros <= (Others => '0') ;                               --N - bit zero
    zero <= '0';
    ain <= n_zeros & dataa ;
    shifta: shiftlne Generic Map (N => NN)
        Port Map ( ain, la, ea, zero, clock, a ) ;
    shiftb: shiftrne Generic Map ( N => N )
        Port Map ( datab, lb, eb, zero, clock, b ) ;
    ec <= '1'; lc <= not s;
    count: downcnt Generic Map (N+1) Port Map(clock,ec,lc,q);
    z <= '1' When q = 0 Else '0';
    sum <= a + p ;
    --define the 2n 2 - to - 1 multiplexers for datap
    muxi: mux2to1 Generic Map ( N => NN )
        Port Map ( nn_zeros, sum, psel, datap ) ;
    regp: regne Generic map ( N => NN )
        Port Map ( datap, ep, clock, p ) ;
End Architecture multiplier_arch ;

--2 regne.vhd N 位带使能寄存器
Library Ieee ;
Use Ieee.Std_Logic_1164.All ;
```

```
Entity regne Is
    Generic ( N :Integer := 12 ) ;
    Port (r : In Std_Logic_Vector(N-1 Downto 0) ;              --寄存器输入
        e : In Std_Logic ;    --enable 1→enable 0→disable
        clock : In Std_Logic ;--clock Signal
        q : Out Std_Logic_Vector(N-1 Downto 0) ) ;             --寄存器输出
End Entity regnc ;

Architecture regne_arch Of regne Is
Begin
    Process ( clock ) Is
    Begin
        If clock'Event And clock = '1' Then                    --时钟上升沿触发
            If e = '1' Then
                q <= r ;                                       --数据存入寄存器
            End If ;
        End If ;
    End Process ;
End Architecture regne_arch ;

--3 downcnt. vhd n modules downcounter
Library Ieee ;
Use Ieee. Std_Logic_1164. All ;
Entity downcnt Is
    Generic ( modulus :Integer := 8 ) ;
    Port (clock : In Std_Logic ;
        e : In Std_Logic ;                        --使能 1→enable, 0→disable
        l : In Std_Logic ;                                     --加载 1→load
        q : Out Integer Range 0 To modulus-1 ) ;
End Entity downcnt ;

Architecture downcnt_arch Of downcnt Is
Signal count :Integer Range 0 To modulus-1 ;
Begin
    Process Is                                                 --无敏感表进程
    Begin
        Wait Until (clock'Event And clock = '1') ;
        If e = '1' Then
            If l = '1' Then
```

```
                count <= modulus - 1 ;                          --loading
            Else
                count <= count - 1 ;                            --counting
            End If ;
        End If ;
    End Process;
    q <= count ;                                                --计数输出
End Architecture downcnt_arch ;

--4 mux2tol. vhd n - bit 2 - to - 1 multiplexer
Library Ieee ;
Use Ieee. Std_Logic_1164. All ;
Entity mux2tol Is
    Generic ( N :Integer := 14 ) ;
    Port ( w0 : In Std_Logic_Vector(N - 1 Downto 0) ;          --input first term
           w1 : In Std_Logic_Vector(N - 1 Downto 0) ;          --input second term
           s: In Std_Logic ;                                    --select line
           f: Out Std_Logic_Vector(N - 1 Downto 0) ) ;         --选项输出
End Entity mux2tol ;

Architecture mux2tol_arch Of mux2tol Is                        --2 选 1 电路
Begin
    With s Select
    f <= w0 When '0',
    w1 When Others ;
End Architecture mux2tol_arch ;

--5 shiftlne. vhd n - bit right - to - left shift register
--with parallel load and enable
Library Ieee ;
Use Ieee. Std_Logic_1164. all ;
Entity shiftlne Is
    Generic ( N :Integer := 7 ) ;
    Port( r: In Std_Logic_Vector(N - 1 Downto 0) ;             --寄存器输入
          l: In Std_Logic ;                                     --加载
          e: In Std_Logic ;                                     --使能
          w: In Std_Logic ;                                     --串行输入
          clock: In Std_Logic ;                                 --时钟
          q: Buffer Std_Logic_Vector(N - 1 Downto 0) ) ;        --寄存器输出
```

```
End Entity shiftlne ;

Architecture shiftlne_arch Of shiftlne Is
Begin
    Process Is
    Begin
        Wait Until clock'Event And clock = '1' ;
        If e = '1' Then
            If l = '1' Then
                q <= r ;                            --parallel load 并行加载
            Else
                q(0) <= w ;                         --series input to lowest bit
                genbits: For i In 1 To N - 1 Loop
                    q(i) <= q(i - 1) ;              --左移
                End Loop genbits;
            End If ;
        End If ;
    End Process ;
End Architecture shiftlne_arch ;

--6 shiftrne.vhd n-bit left-to-right shift register
--with parallel load and enable
Library Ieee ;
Use Ieee.Std_Logic_1164.All ;
Entity shiftrne Is
    Generic ( N :Integer := 7 ) ;
    Port ( r : In Std_Logic_Vector(N - 1 Downto 0) ;            --寄存器输入
           l : In Std_Logic ;                                   --load 加载
           e : In Std_Logic ;                                   --enable 使能
           w : In Std_Logic ;                                   --串行输入
           clock: In Std_Logic ;                                --时钟
           q: Buffer Std_Logic_Vector(N - 1 Downto 0) ) ;       --寄存器输出
End Entity shiftrne ;

Architecture shiftrne_arch Of shiftrne Is
Begin
    Process Is
    Begin
        Wait Until clock'Event And clock = '1' ;
```

```
        If e = '1' Then
            If l = '1' Then
                q <= r ;                              --parallel load
            Else
                genbits: For i In 0 To N-2 Loop
                    q(i) <= q(i+1) ;            --shift high bit to low bit
                End Loop genbits ;
                q(n-1) <= w ;                   --series input to highest bit
            End If ;
        End If ;
    End Process ;
End Architecture shiftrne_arch ;
```

乘法器仿真结果如图 5-14 所示。

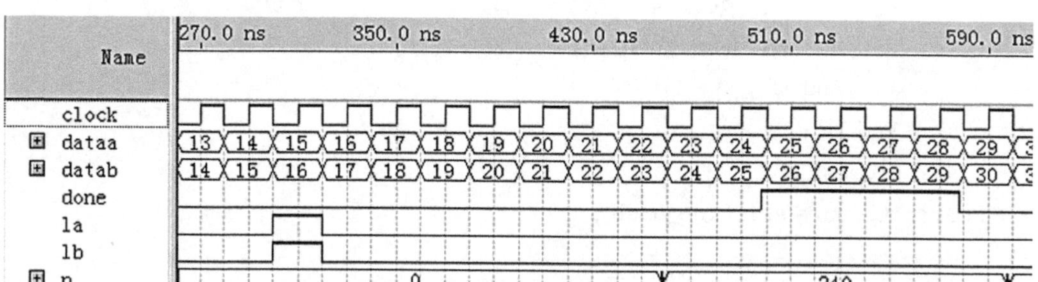

图 5-14　乘法器仿真结果

计算 p = 15×16=240,仿真输入加载被乘数 15,乘数 16,在 done 对应的位置显示结果为 p = 240,说明设计正确。

5.3.3　状态机控制的移位除法

算法:

```
r=0;
For i=0 To N-1 Do
    Left-Shift r||a;
    If r>=b Then
        q=1;
        r=r-b;
    Else
        q=0;
    End If;
End For;
```

其中,a 为被除数,b 为除数,q 为商,r 为余数。

除法设计的 ASM 流程如图 5－15 所示。从流程图可知：

程序运行开始则 s＝0，状态为 s1，加载 a，b；s＝1，状态转换到 s2。计算器 c 倒计数，c 的初值为 n＋1。

clock 触发时，余数和被除数左移，被除数将从最高位（MSB）移入余数 LSB，并检查余数是否大于除数，当条件成立时，商数项左移，移入 1 到被除数的寄存器的 LSB，并使余数减除数，结果返回余数寄存器。条件不成立时，商数项移入 0 到被除数的寄存器的 LSB。

伴随重复相同的判断和移位，在经过 n 个 clock 后，n 位的被除数全部移入了余数中，此时计数器计数到 0，令状态转换为 s3，信号 done 为 1，表示运算完成，直到 s 再度转为 0，done 为 0。

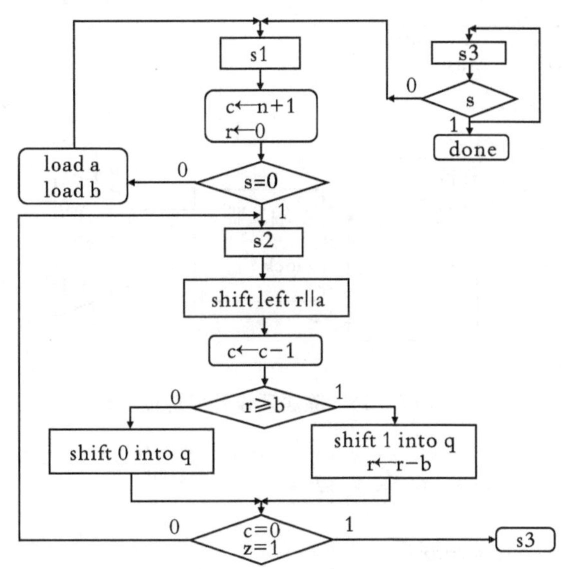

图 5－15　除法设计的 ASM 流程

除法器设计原理如图 5－16 所示。其中设计技巧归纳如下：

(1)减法器的设计：

将减数 b 先经过一个反向器取补码，且将加法器进位输入 cin 设定为 1 相加即可。

(2)实现 shift 1 or 0 into q：将减法器的进位输出端（cout）接到商数的移位输入。当 r≥b 时，减法器的输出为 1，反之输出为 0。

(3)为了节省存储寄存器，将商数输入存储被除数的寄存器。

(4)从流程图可见 shift r||a 和 r ← r－b 不可能在一个时钟内完成，改变设计，在余数和被除数之间插入一个 rr0 触发器，则 shift r||a 成为了 shift r||rr0||a，因此被除数的 MSB 将先移入此触发器，将余数寄存器的输出与触发器输出作为新的余数 r1 输入到减法器电路，因此当 r1＞＝b 时，减法器的输出将加载到余数寄存器，而不和移位相冲突。另外，触发器前多加了一个选择器用于清除功能，即只有在 s2 状态，在将被除数 MSB 移入此触发器，其他状态是选择器选择 0 输入触发器。

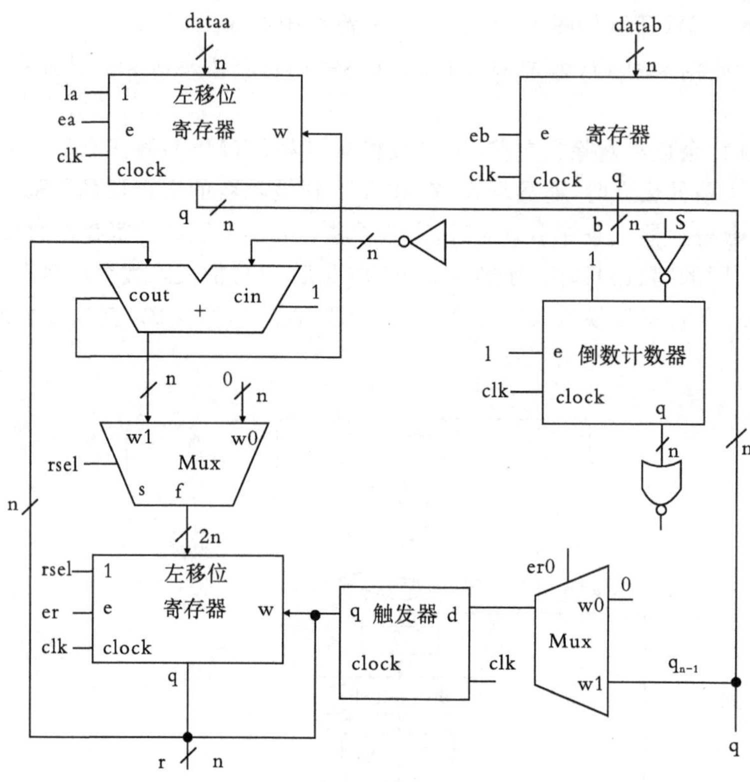

图 5-16　除法器设计原理图

示例 5-10　RTL 级移位除法器程序

```
--divider.vhd n-bit divider
Library Ieee;
Use Ieee.Std_Logic_1164.All;
Use Ieee.Std_Logic_Unsigned.All ;
Use Work.All ;

Entity divider Is
    Generic ( N : Integer := 7 ) ;
    Port ( clock: In Std_Logic ;              --时钟
           s : In Std_Logic ;                 --start operation 开始操作
           la: In Std_Logic ;                 --load of dividend 加载被除数
           eb: In Std_Logic ;                 --enable(load) of divisor 使能(加载)除数
           dataa, datab: In Std_Logic_Vector(N-1 Downto 0) ; --被除数、除数
           r : Buffer Std_Logic_Vector(N-1 Downto 0) ;        --remainder 余数
           q : Buffer Std_Logic_Vector(N-1 Downto 0) ;        --quotient 商
           done : Out Std_Logic ) ;                           --done operation
End Entity divider ;
```

```
Architecture divider_arch Of divider Is
Type state_type Is ( s1, s2, s3 ) ;                          --state declaration
Signal y: state_type ;                                       --状态信号定义
Signal zero, z, cout, lr, ec, ea, er0, er, lc, rsel, r0: Std_Logic ;
    --z detecter of zero, zero series input of shift
    --cout subtractor carry—out
    --lr, er load, enable of quotient
    --ec, lc load of downcounter, enable of downcounter,
    --ea,er0 enable of dividend, enable of rr0 register,
    --rsel selected line of register r's multiplexer
    --r0 output of register a's multiplexer
Signal a, b, datar : Std_Logic_Vector(N—1 Downto 0) ;
    --output of register a, b (dividend, divisor)
    --parallel load of register r(remainder)
Signal sum : Std_Logic_Vector(N Downto 0) ;                  --减法器的和
Signal count : Integer Range 0 To N—1 ;                      --倒计数的范围
Begin

    fsm_transitions: Process ( clock ) Is                    --状态机转换进程
    Begin
        If (clock 'Event And clock = '1') Then
            Case y Is
                When s1 => If s = '0' Then y <= s1 ; Else y <= s2 ; End If;
                When s2 => If z = '0' Then y <= s2 ; Else y <= s3 ; End If;
                When s3 => If s = '1' Then y <= s3 ; Else y <= s1 ; End If;
            End Case ;
        End If ;
    End Process fsm_transitions;

    fsm_outputs: Process ( s, y, cout, z ) Is
    Begin
        lr <= '0' ; er <= '0' ; er0 <= '0' ;                 --设初值
        ea <= '0' ; done <= '0' ;rsel <= '0' ;
        Case y Is
            When s1 => er <= '1' ;
            If s = '0' Then lr <= '1' ;
                If la = '1' Then ea <= '1' ; Else ea <= '0' ; End If ;
            Else ea <= '1' ; er0 <= '1' ; End If ;
```

```
            When s2 => rsel <= '1'; er <= '1'; er0 <= '1'; ea <= '1';
            If cout = '1' Then lr <= '1'; Else lr <= '0'; End If;
            When s3 => done <= '1';
        End Case;
    End Process fsm_outputs;

    --定义数据通道电路
    zero <= '0';
    --divisor
    regb:regne Generic Map(N => N)Port Map(datab, eb, clock, b);
    --remainder
    shiftr: shiftlne Generic Map(N => N )
            Port Map (datar, lr, er, r0, clock, r );
    --flip-flop with multiplexer
    ff_r0: muxdff1 Port Map(zero, a(N-1), er0, clock, r0 );
    --dividend
    shifta:shiftlne Generic Map(N => N)
              Port Map ( dataa, la, ea, cout, clock, a );
    q <= a;  ec <= '1'; lc <= Not s;
                                                  --downcounter
    counter:downcnt Generic Map(N+1 ) Port Map(clock, ec, lc, count);
                                                  --nor gate zero detector
    z <= '1' When count = 0 Else '0';
    sum <= r & r0 + (Not b +1);          --补码相加,减法器 subtractor
    cout <= sum(N);
                                         --multiplexer of register r(remainder)
    datar <= (Others => '0') When rsel = '0' Else sum(N-1 Downto 0);
End Architecture divider_arch;

--调用元件:
--downcnt.vhd
--regne.vhd
--shiftlne.vhd
--muxdff1.vhd
--前 3 个元件共享示例 5-9 的调用元件文件 downcnt.vhd,regne.vhd,shiftlne.vhd

--程序文件 muxdff1.vhd
Library Ieee;
Use Ieee.Std_Logic_1164.All;
```

```
Entity muxdff1 Is
    Port ( w0 : In Std_Logic ;                    --input first term
           w1 : In Std_Logic;                     --input second term
           s: In Std_Logic ;                      --select line
           clock: In Std_Logic;
           r0: Out Std_Logic ) ;                  --output seleted term
End Entity muxdff1;

Architecture muxdff1_arch Of muxdff1  Is
Signal f1:Std_Logic;
Begin
    With s Select                                 --2 选 1
    f1 <= w0 When '0',
           w1 When Others ;

    Process(clock) Is                   --加 D 触发器是电路调试中常用的方法
    Begin
        If(clock'Event And clock='1') Then
            r0<=f1;                               --触发器
        End If;
    End Process;
End Architecture muxdff1_arch;
```

除法器程序仿真如图 5-17 所示。

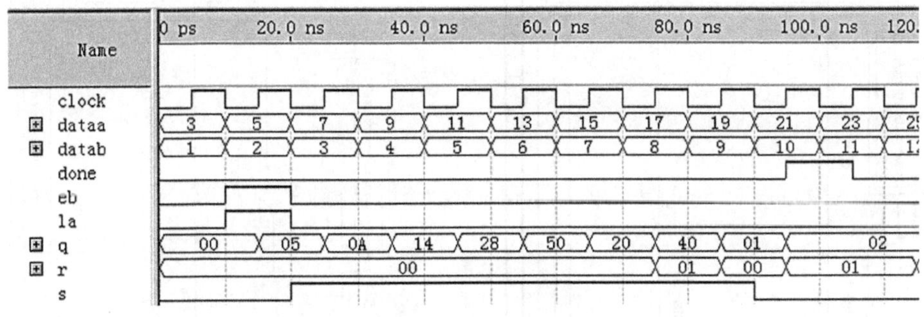

图 5-17　除法器仿真结果

仿真说明:计算,q = 5÷2,输入加载被除数 5,除数 2,在 done 对应的位置显示结果商为 q=2,余数为 r=1;仿真验证说明了设计正确。

乘法器与除法器的 RTL Viewer 结果如图 5-18 所示。除法程序中的状态机、寄存器、左移寄存器、右移寄存器、倒计数程序与乘法器相同。两者不仅代码设计中可以共享复用,在代码综合实现方面也有共享复用的可能。

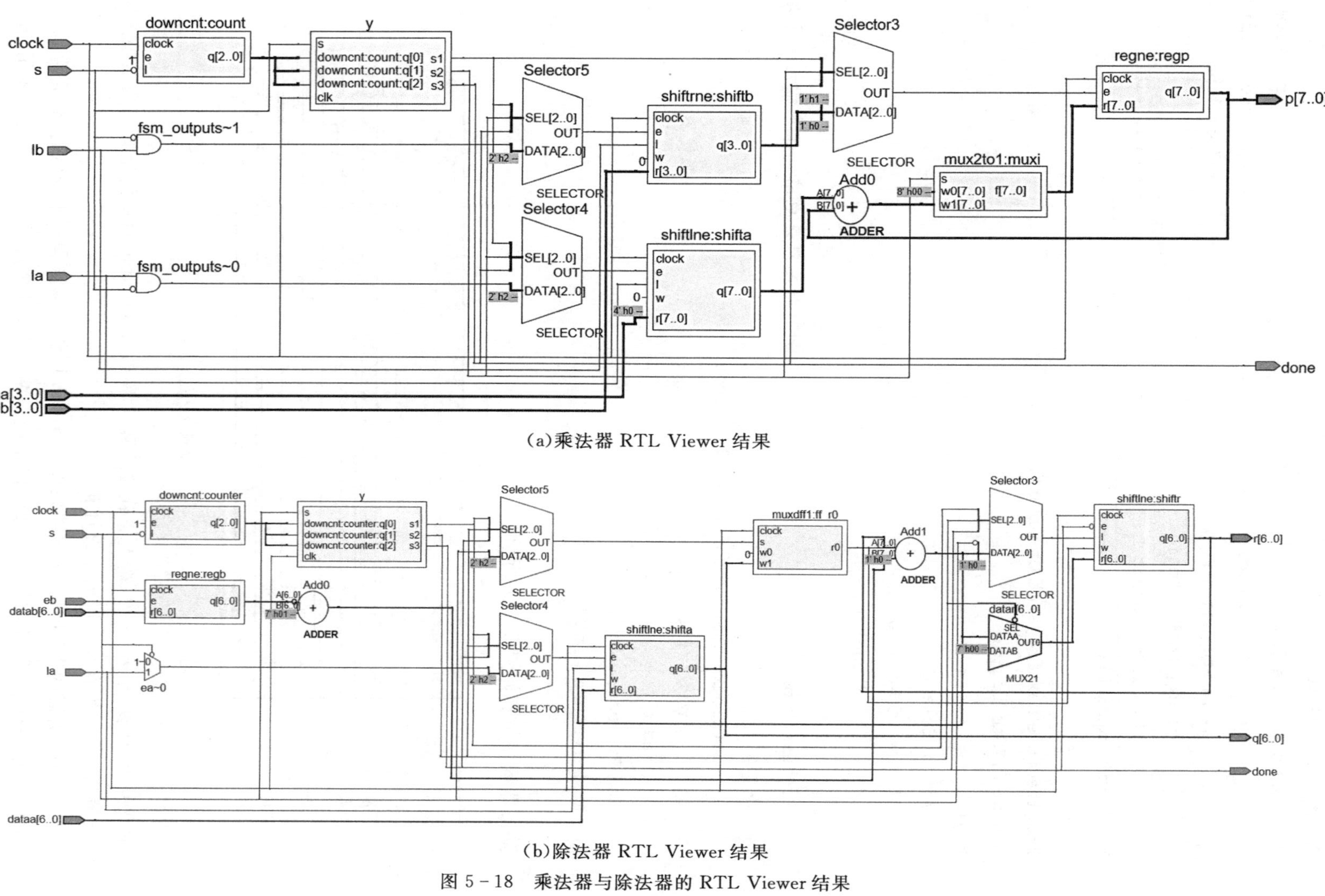

(a)乘法器 RTL Viewer 结果

(b)除法器 RTL Viewer 结果

图 5-18 乘法器与除法器的 RTL Viewer 结果

5.3.4 加减乘除整合与 BCD 加减电路的控制

加减法器控制通道状态图与 ASM 图，如图 5－19 所示：
s＝0 状态，等待输入数据；s＝1，使能加减法工作。
z＝0，倒计数为 0；z＝1，s2 状态，运算完成状态。

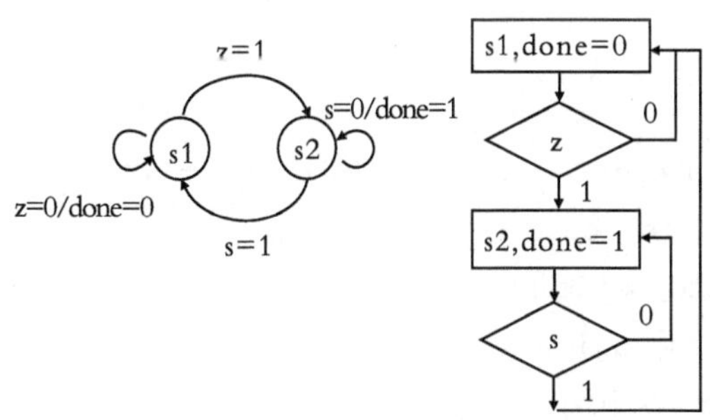

图 5－19 加减电路整合状态图与 ASM 图

示例 5－11 加减乘除电路顶层

通过将示例 5－11a 示例 5－11b 主程序执行 File→Create/Update→Create Symbol Files for Current File 成为 Symbol，在新建图形中作为 2 个元件插入，选中元件按右键，执行弹出菜单命令 Generate Pins For Symbol 加入引脚，再对图稍加修改，即可得到如图 5－20 所示的加减电路整合顶层原理图。乘除法的输出是 16 进制，加减法是 BCD 码，如果希望乘除法的输出是 BCD 码，可以参考示例 5－3d 修改。

加减乘除电路整合顶层原理如图 5－20 所示。

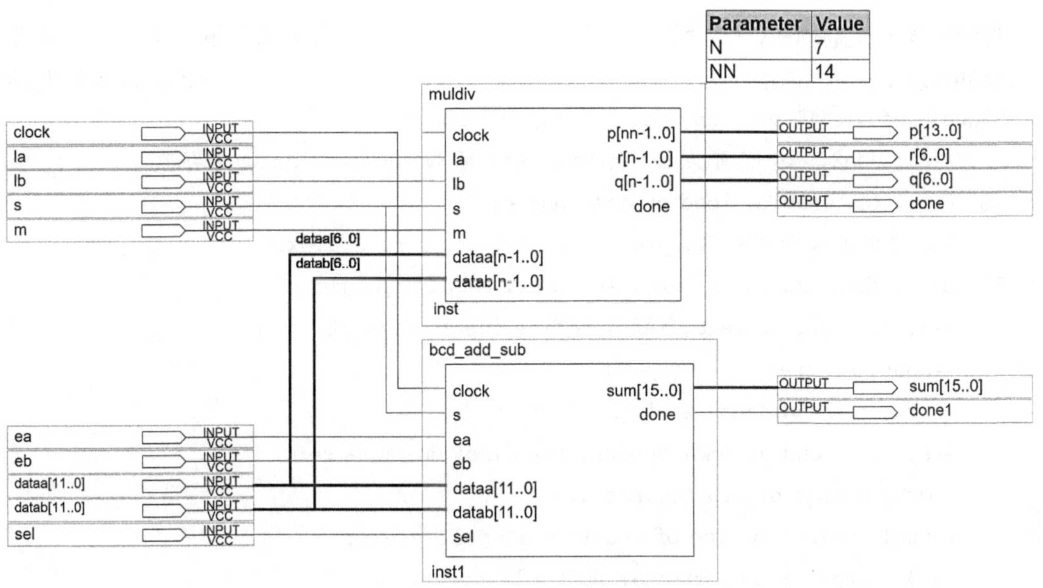

图 5－20 加减乘除电路整合顶层原理图

示例 5-11a 乘除电路整合

```
Library Ieee;
Use Ieee.Std_Logic_1164.All;
Use Ieee.Std_Logic_Unsigned.All ;
Use Work.All ;                                               --引用当前工程文件

Entity muldiv1 Is
    Generic ( N : Integer := 7 ; NN : Integer := 14 ) ;
    Port( clock:In Std_Logic ;                               --时钟
          la : In Std_Logic ;                                --加载 dataa
          lb : In Std_Logic ;                                --加载 datab
          s: In Std_Logic ;                                  --开始信号
          dataa: In Std_Logic_Vector(N-1 Downto 0) ;         --被乘数或被除数
          datab: In Std_Logic_Vector(N-1 Downto 0) ;         --乘数或除数
          m: In Std_Logic;                                   --乘除选择
          p: Buffer Std_Logic_Vector(NN-1 Downto 0) ;        --积 porduct
          r : Buffer Std_Logic_Vector(N-1 Downto 0) ;        --余数 remainder
          q : Buffer Std_Logic_Vector(N-1 Downto 0) ;        --商 quotient
          done : Out Std_Logic ) ;                           --操作完成信号
End Entity muldiv1;

Architecture muldiv1_arch Of muldiv1 Is
Type state_type Is ( s1, s2, s3 ) ;                  --自定义枚举类型状态机说明
Signal y : state_type ;                                      --状态机信号说明
Signal psel, ea1, ea2, eb, ec, lc, ep: Std_Logic;
    --ea1, ea2, eb, 使能左移(multiplier)、右移(multiplicand)寄存器
    --ec, lc, enable, load of downcounter
    --ep, 使能乘积寄存器, psel, select line of multiplexer
Signal z, zero, cout, lr, er0, er, rsel, r0 : Std_Logic ;
    --z, detecter of zero, zero, series input of shift
    --cout, subtractor carry-out
    --lr, er, load, enable of quotient
    --ec, lc, load of downcounter, enable of downcounter,
    --er0, enable of rr0 register,
    --rsel, selected line of register r's multiplexer
    --r0, output of register a's multiplexer
Signal a, ain, datap, sum, nn_zeros: Std_Logic_Vector(NN-1 Downto 0) ;
    --左移移位寄存器输出 a
```

```
    --input of shift—left regster ain
    --output of multiplexer datap
    --sum of product and multiplicand
    --2 * n—bit zero input to multiplicand nn_zeros
Signal ad, b, bd, datar, n_zeros: Std_Logic_Vector(N—1 Downto 0) ;
    --ad, b 的寄存器输出(dividend)
    --parallel load of register r(remainder)
    --N-bit zero load into register with multiplicand
Signal sumd: Std_Logic_Vector(N Downto 0) ;          --sum of subtractor
Signal count: Integer Range 0 To N—1 ;               --range of downcounter
Begin

    fsm_transitions: Process ( clock ) Is             --状态机转换,控制通道
    Begin
        If (clock'Event And clock = '1') Then
            Case y Is
                When s1 => If s = '0' Then y <= s1 ; Else y <= s2; End If;
                When s2 => If z = '0' Then y <= s2 ; Else y <= s3; End If;
                When s3 => If s = '1' Then y <= s3 ; Else y <= s1; End If;
                When Others=>NULL;
            End Case ;
        End If ;
    End Process fsm_transitions;

    fsm_outputs1: Process ( y, s, la, lb, b(0),cout, z) Is
    Begin
        done <= '0' ;
        If m='1' Then                                 --initialize value
            ep <= '0' ; ea1 <= '0' ; eb <= '0'  ; psel <= '0' ;
        Else
            lr <= '0' ; er <= '0' ; er0 <= '0' ;
            ea2 <= '0' ; rsel <= '0' ;
        End If;
        Case y Is
            When s1 =>
            If m='1'  Then
                ep <= '1' ;
                If s = '0' And la = '1' Then ea1 <= '1' ;
                Else ea1 <= '0' ;End If ;
```

```
            If s = '0' And lb = '1' Then eb <= '1';
            Else eb <= '0';End If ;
        Else
            er <= '1';
            If s = '0'  Then lr <= '1';
                If la = '1'  Then ea2 <= '1';
                Else ea2 <= '0'; End If ;
            Else ea2 <= '1'; er0 <= '1';End If ;
        End If ;
        When s2 =>
        If m='1'  Then
            ea1 <= '1'; eb <= '1'; psel <= '1';
            If b(0) = '1' Then ep <= '1';
            Else ep <= '0'; End If ;
        Else
            rsel <= '1'; er <= '1'; er0 <= '1'; ea2 <= '1';
            If cout = '1' Then lr <= '1';
            Else lr <= '0'; End If ;
        End If ;
        When s3 => done <= '1';
    End Case ;
End Process fsm_outputs1;

--定义数据通道电路
nn_zeros <= (Others => '0');                     --2*N位零
n_zeros <= (Others => '0') ;                     --N位零
zero<= '0';
ain <= n_zeros & dataa ;
sum <= a + p ;
--define the datapath circuit(divider)
q <= ad ;
sumd <= r & r0 + (Not bd +1) ;                   --减法器
cout <= sumd(N) ;
datar <= (Others => '0') When rsel = '0' Else sumd(N-1 Downto 0) ;
                                     --multiplexer of register r(remainder)

ec <= '1'; lc <= Not s;
z <= '1' When count = 0 Else '0';                --nor gate zero detector
--数据通道--
```

```
    shiftam:shiftlne Generic Map(N=>NN) Port Map(ain,la,ea1,zero,clock, a) ;
                                            --左移移位寄存器
    shiftb:shiftrne Generic Map(N=>N) Port Map(datab,lb,eb,zero,clock, b) ;
                                            --右移移位寄存器
    regp:regne Generic Map(N=>NN) Port Map(datap, ep, clock,p ) ;
                               --带使能端的 D 触发器,datap 输入,p 输出
    muxi:mux2to1 Generic Map(N=>NN) Port Map (nn_zeros, sum, psel, datap) ;
                        --二选一 define the 2N 2-to-1 multiplexers for datap
    shifta:shiftlne Generic Map(N=>N) Port Map(dataa,la,ea2,cout,clock, ad) ;
                                            --左移移位寄存器,dividend
    shiftr: shiftlne Generic Map(N=>N) Port Map(datar,lr,er,r0,clock,r);
                                            --左移移位寄存器,remainder
    ff_r0: muxdff1 Port Map(zero, ad(N-1), er0, clock, r0 ) ;
                                            --二选一带输出缓存触发器
    regb: regne Generic Map(N=>N) Port Map(datab, lb, clock, bd ) ;
                               --带使能端的 D 触发器,datab 输入,bd 输出
    counter: downcnt Generic Map(N+1) Port Map(clock, ec, lc, count);
                                        --N+1 带使能与加载的递减计数
End Architecture muldiv1_arch ;
```

乘除整合电路仿真结果如图 5-21 所示。

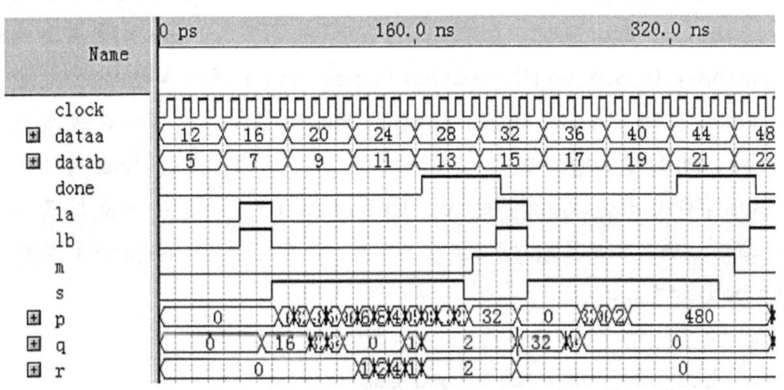

图 5-21 乘除整合电路仿真结果

计算了两组数,16÷7;32×15;仿真结果:q=2,r=2;p=480;说明设计正确。

乘除法整合 RTL Viewer 的结果如图 5-22 所示。可以看到这是比较简易的整合,其中乘除法共享了控制与倒计数模块,由于本例不同时做加减乘除,所以还可以进一步共享单元做面积优化。

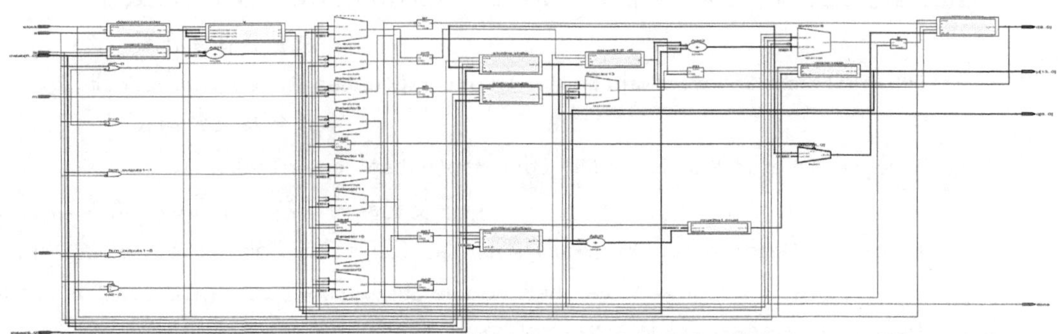

图 5-22　乘除法整合 RTL Viewer

示例 5-11b　BCD 加减电路整合

```
--bcd_add_sub.vhd 状态机实现带启动结束标志的 3 位 BCD 加减法
Library Ieee ;
Use Ieee.Std_Logic_1164.All;
Use Work.All;

Entity bcd_add_sub Is

    Port(clock : In Std_Logic ;
        s : In Std_Logic ;                                  --操作开始信号
        ea,eb: In Std_Logic ;                               --加载 a,b 的使能
        dataa : In Std_Logic_Vector(11 Downto 0) ;          --被加数或被减数
        datab : In Std_Logic_Vector(11 Downto 0) ;          --加数或减数
        sel: In Std_Logic ;                                 --加减选择
        sum : Out Std_Logic_Vector(15 Downto 0) ;           --加减结果的 BCD 码表示
        done : Out std_logic) ;                             --操作完成标志
End Entity bcd_add_sub;

Architecture bcd_add_sub_arch Of bcd_add_sub Is
Type state_type Is ( s1, s2 ) ;                             --有限状态机状态定义

Signal y: state_type ;                                      --有限状态机信号的定义
Signal ec : Std_Logic ;                                     --减计数使能
Signal lc : Std_Logic ;                                     --减计数加载 0 检测
Signal a,b: Std_Logic_Vector(11 Downto 0) ;
Signal count: Integer Range 0 To 7 ;                        --计数
```

```
Signal sumc: Std_Logic_Vector(11 Downto 0) ;
                                        --bcd.sum -> negative.a interconnection
Signal co : Std_Logic;                              --bcd3.co
Signal negative_com,z : Std_Logic;
                                            --complement enable of negative sum
Begin

    fsm_transition: Process ( clock ) Is            --状态机转换进程
    Begin
        If clock'Event And clock = '1' Then
            Case y Is
                When s1 => If z = '0' Then y <= s1 ; Else y <= s2; End If;
                When s2 => If s = '1' Then y <= s2 ; Else y <= s1; End If;
            End Case ;
        End If ;
    End Process fsm_transition;

    fsm_output: Process (y) Is                      --状态机输出进程
    Begin
        Case y Is
            When s1 => done <= '0' ;
            When s2 => done <= '0' ;
            When Others=>NULL;
        End Case;
    End Process fsm_output;
    bcd: bcd3 Port Map(a, b, sel, sel, co, sumc) ;  --3 个字的 bcd 加/减电路

    negative_com <= sel And (Not co) ;
    --bcd.sumc 9's only if sel='1' for selecting sub mode and co='0'
    --represent negative sum
        sum(15 Downto 13) <= "000" When negative_com ='0'Else "111" ;
        sum(12)<=((Not sel) And co);                --根据表 5-1 得到的逻辑
    complement: negative Port Map(sumc, negative_com, sum(11 Downto 0)) ;
    --数据通道--

    rega: regne Generic Map(N => 12 )Port Map(dataa, ea, clock, a ) ;
    regb: regne Generic Map(N => 12 )Port Map(datab, eb, clock, b ) ;
    --数据寄存器 datatb,datatb register
```

```
    ec <= '1'; lc <= Not s;

    counter: downcnt Generic Map(modulus => 8)
    Port Map(clock, ec, lc, count);             --N+1 带使能与加载的递减计数

    z <= '1' When count = 0 Else '0';               --检测 0, done
End Architecture bcd_add_sub_arch;

--调用层次的程序--
--bcd3.vhd3 digits bcd adder/subtractor                3 个字的 bcd 加减法
Library Ieee;
Use Ieee.Std_Logic_1164.All;
Use Work.All;

Entity bcd3 Is
    Port(a,b: In Std_Logic_Vector(11 Downto 0);
         ci,sel: In Std_Logic;
         co: Out Std_Logic;
         sum: Out Std_Logic_Vector(11 Downto 0));
End Entity bcd3;

Architecture behavior Of bcd3 Is
Signal cc: Std_Logic_Vector(1 Downto 0);
Begin
    bcd1:bcd
    Port Map(a(3 Downto 0), b(3 Downto 0), ci, sel, cc(0), sum(3 Downto 0));
    bcd2:bcd
    Port Map(a(7 Downto 4), b(7 Downto 4), cc(0), sel, cc(1), sum(7 Downto 4));
    bcd3:bcd
    Port Map(a(11 Downto 8), b(11 downto 8), cc(1), sel, co, sum(11 Downto 8));
End Architecture behavior;

--bcd.vhd 1 digits bcd adder/subtractor                1 个字的 bcd 加减运算
Library Ieee;
Use Ieee.Std_Logic_1164.All;
Use Work.All;

Entity bcd Is
    Port(a,b: In Std_Logic_Vector(3 Downto 0);
```

```
        ci,sel: In Std_Logic;                       --sel,0 加法,1 减法
        co: Out Std_Logic;
        sum: Out Std_Logic_Vector(3 Downto 0));
End Entity bcd;

Architecture behavior Of bcd Is
Signal b2: Std_Logic_Vector(3 Downto 0);            --取 9 补的结果
Begin
    U0: com9s Port Map (b,sel,b2);
    U1:bcdadd Port Map (a,b2,ci,co,sum);
End Architecture behavior;

--com9s.vhd9's generator  9 的补码电路
Library Ieee ;
Use Ieee.Std_Logic_1164.All;

Entity com9s Is
    Port( a: In Std_Logic_Vector(3 Downto 0);
          sel: In Std_Logic;
          z: Out Std_Logic_Vector(3 Downto 0));
End Entity com9s;
```

表 5-1 BCD 码与对应的 9 补码

BCD 码				9 补码			
a3	a2	a1	a0	z3	z2	z1	z0
0	0	0	0	1	0	0	1
0	0	0	1	1	0	0	0
0	0	1	0	0	1	1	1
0	0	1	1	0	1	1	0
0	1	0	0	0	1	0	1
0	1	0	1	0	1	0	0
0	1	1	0	0	0	1	1
0	1	1	1	0	0	1	0
1	0	0	0	0	0	0	1
1	0	0	1	0	0	0	0

```
Architecture behavior Of com9s Is                  --表 5-1 逻辑
Begin
    Process (sel,a) Is
    Begin
        If sel='1' Then
            z(3)<=(Not a(3)) And (Not a(2)) And (Not a(1));
            z(2)<=a(2) Xor a(1);
            z(1)<=a(1);
            z(0)<=Not a(0);
        Else z<=a;
        End If;
    End Process;
End Architecture behavior;

--bcdadd.vhd 1 digit bcd adder                     一个字的 bcd 加法
Library Ieee ;
Use Ieee.Std_Logic_1164.All;
```

```
Use Work.All;

Entity bcdadd Is
    Port( a, b: In Std_Logic_Vector(3 Downto 0);
          ci: In Std_Logic;
          co: Out Std_Logic;
          sum: Out Std_Logic_Vector(3 Downto 0));
End Entity bcdadd;

Architecture behavior Of bcdadd Is
Signal s, a2: Std_Logic_Vector(3 Downto 0);
Signal y, c4, zero, nouse: Std_Logic;
Begin
    U0: fadd4 Port Map (a,b,ci,c4,s);                        --参见示例 3-1
    y<=c4 Or (s(3) And s(2)) Or (s(3) And s(1));             --s 大于 9 的逻辑
    a2<='0' & y & y & '0';                                   --y=1,a2=6
    co<=y; zero<='0';
    U1: fadd4 Port Map (a2,s,zero,nouse,sum);                --大于 9 加 6
End Architecture behavior;

--negative.vhd correct negative number circuit              负数修正电路
Library Ieee ;
Use Ieee.Std_Logic_1164.All;
Use Work.All;

Entity negative Is
    Port(a: In Std_Logic_Vector(11 Downto 0);
         sel: In Std_Logic;
         z: Out Std_Logic_Vector(11 Downto 0));
End Entity negative;

Architecture behavior Of negative Is
Signal a2, zero: Std_Logic_Vector(3 Downto 0) ;
Signal unuse: Std_Logic ;
Begin
    zero <= "0000" ;
    com1: com9s Port Map (a(3 Downto 0), sel, a2);
    bcd1: bcdadd Port Map (zero, a2, sel, unuse, z(3 Downto 0));
    com2: com9s Port Map (a(7 Downto 4), sel, z(7 Downto 4));
```

```
    com3: com9s Port Map (a(11 Downto 8), sel, z(11 Downto 8));
End Architecture behavior;
```

3 位 BCD 加减电路整合仿真结果如图 5－23 所示。从图上可以看出：Sel 为低时，做加法，计算 103 ＋ 6，ea，eb 加载了被加数和加数 103 和 6，仿真结果和 sum 为 109，加法结果正确。Sel 为高时，做减法，计算 110－20，ea，eb 加载了被减数和减数 110 和 20，仿真结果和 sum 为 90，减法结果正确。要注意 BCD 仿真数字要给 16 进制数据。

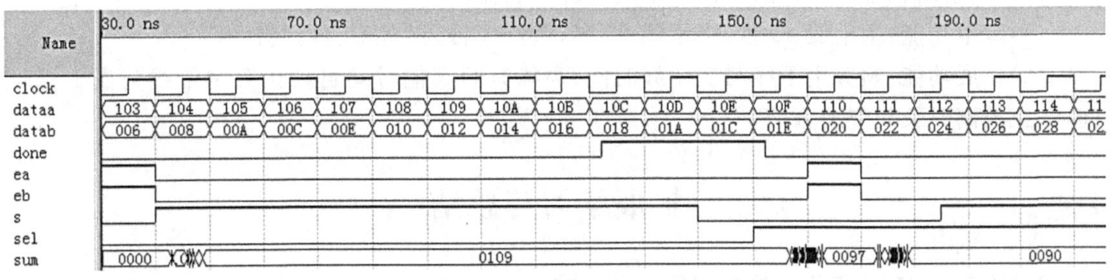

图 5－23 BCD 加减电路整合仿真结果

程序说明：

(1)乘除法的符号问题很简单，正常运算，最后的符号为运算的 2 个数字符号位的异或非。减法运算的符号位则与参与运算 2 个数的绝对值大小有关，电路实现时，这个程序对减数求 9 的补码，但若发生负数输出会出现尚未求补的数，因而设计了负数修补电路，negative. vhd。

(2)对 BCD 求 10 的补码与对 9 求补码加 1 的结果一样，根据表 5－1 BCD 码与对应的 9 补码，进行卡诺图化简后，编写了 9 的补码电路 com9s. vhd。

(3)为了在第 4 位可以显示最大值 1 或者负号，加减法程序约定了第 4 个字符的编码表，如表 5－2 所示。

表 5－2 第 4 个字符编码的逻辑函数

sel	co	sum(15～12)
0	0	0000 加法模式下没有进位
0	1	0001 加法模式下有进位
1	0	0010 减法模式下和为负，七段显示时在编码
1	1	0000 减法模式下和为正

习　题

5-1　设计实现 RTL 级的 BCD 加减乘除计算，从占用资源、性能实现、通用性等方面，比较各运算的综合结果。

5-2　讨论如何利用好 FPGA 结构与特点对本章的运算做优化，如采用一些原语会有什么改进？

5-3　采用状态机的方法设计一个交通灯控制器。

(1)基本功能：主干线 45 秒，支干线 25 秒，黄灯 5 秒，拐弯 15 秒，倒计时。

(2)扩展功能：预置主干线与支干线时间；高峰时间其他时间自动预置主干线与支干线时间，人工控制时长与交通方向。

扩展学习与总结

1. VHDL 有哪些运算？这些运算有什么特点？如何使用？
2. VHDL 顺序描述语句有哪些？并行描述语句有哪些？如何使用？
3. 查阅资料学习和总结同步与异步状态机程序的编写方法。
4. FPGA 在开发数量为多少时转为 ASIC？

6　系统的计数分频与定时设计

本章的主要内容与方法：

(1)介绍了各种计数与分频。

(2)比较计数的同步与异步级联。

(3)比较二进制技术与 LFSR 计数器，查看报告了解性能与占用资源，介绍设计约束。

(4)比较了不同标准的描述与示范了设计规范。

(5)介绍芯片的锁相环(PLL)。

6.1　可变模计数器

在复杂数字系统中，为了保证系统各部分时钟与基准时钟的同步，需要微调各部分的时钟频率。这一工作可由可变模计数器完成。首先把本地时钟利用锁相环进行倍频，得到一个高建时钟，然后利用可变模计数器对此高倍频时钟进行 2^N 分频，并把计数器的最高位的输出作为调整后的同步时钟。如果本地时钟与基准时钟的相位不同步，当超过一定的门限时，就通过改变计数的模值来对其位进行修正，使其始终保持与基准时钟的同频同相。

可变模计数器起到了对本地时钟频率进行闭环微调的作用。

如果本地时钟的快慢在允许的变化范围之内，计数器正常工作，完成对高频时钟的 2^N 分频，得到一个与基准时钟同步的输出时钟，此时计数器的模值为 16。

如果本地时钟的慢于基准时钟，计数器的值多加 1，此时计数器的模值为 15，本地时钟频率提高，调整后，计数器的模值恢复为 16。

如果本地时钟的快于基准时钟，计数器停止计数，此时计数器的模值为 17，本地时钟频率降低，调整后，计数器的模值恢复为 16。

Local 时钟为外部时钟源引入的时钟，clk 为倍频时钟，counter 为工作在 clk 频率的计数器，counter[1]为调整后的一个时钟脉冲。

对外部时钟的调整并非是插入或扣除 clk 的一个时钟周期，而是根据本地时钟的快慢把时钟脉冲提前或延迟一个 clk 时钟周期。

本地时钟的调整精度为倍频后时钟的一个周期，倍频值越大，每次调整量越小，精度越高。

示例 6－1a　可变模计数器

```
Library Ieee;
Use Ieee.Std_Logic_1164.All;
Use Ieee.Std_Logic_Unsigned.All;

Entity modcounter Is                                   --可变模计数器实体描述
    Port(clk,clk_en,reset:In Std_Logic;
```

```
        flag:In Std_Logic_Vector( 1 Downto 0);
        counter: Out Std_Logic_Vector( 1 Downto 0));
End Entity modcounter;

Architecture modcounter_arch Of modcounter Is
Signal score:Std_Logic_Vector( 1 Downto 0);
Begin
    Process(clk, reset) Is
    Begin
        If(reset= '1') Then score<=(Others=>'1');
        ElsIf(clk'Event And clk='1') Then
            If(clk_en= '1') Then
                If(flag="01") Then score<=score+2;
                                                    --本地时钟的慢于基准时钟
                ElsIf( flag="11") Then score<=score;
                                                    --本地时钟的快于基准时钟
                Else score<=score+1;
                                                --本地时钟的快慢在允许的变化范围之内
                End If;
            End If;
        End If;
    End Process;
    counter<=score;
End Architecture modcounter_arch;
```

示例 6-1b 可预置模计数器

```
--输入"00","01","10"和"11"可以得到模 3,5,7,8 计数
Library Ieee;
Use Ieee.Std_Logic_1164.All;
Use Ieee.Std_Logic_Unsigned.All;
Entity dcounter Is                                  --可预置模计数器实体描述
    Port(clk,clr:In Std_Logic;
        s:In Std_Logic_Vector(1 Downto 0);
        q :Out Std_Logic_Vector(3 Downto 0));
End Entity dcounter;

Architecture dcounter_arch Of dcounter Is
Signal count:Std_Logic_Vector(3 Downto 0);
Begin
```

```
    Process(clk,clr,s) Is
    Begin
        q<= count;
        If (clr = '0') Then  count <= "0000";
        Elsif(clk'Event And clk='1') Then
            If(s="00") Then                                          --3 计数
                If(count >= "0010") Then  count <= "0000";
                Else count <= count+1;
                End If;
            ElsIf(s="01") Then                                       --5 计数
                If(count >= "0100") Then  count <= "0000";
                Else count <= count+1;
                End If;
            ElsIf(s="10") Then                                       --7 计数
                If(count >= "0110") Then  count <= "0000";
                Else count <=  count+1;
                End If;
            ElsIf(s="11") Then                                       --8 计数
                If(count >= "0111") Then  count <= "0000";
                Else count <=  count+1;
                End If;
            End If;
        End If;
    End Process;
End Architecture dcounter_arch;
```

示例 6-2　四位环形计数器

四位环形计数器有 4 种计数值，其原理及仿真图见图 6-1、6-2。

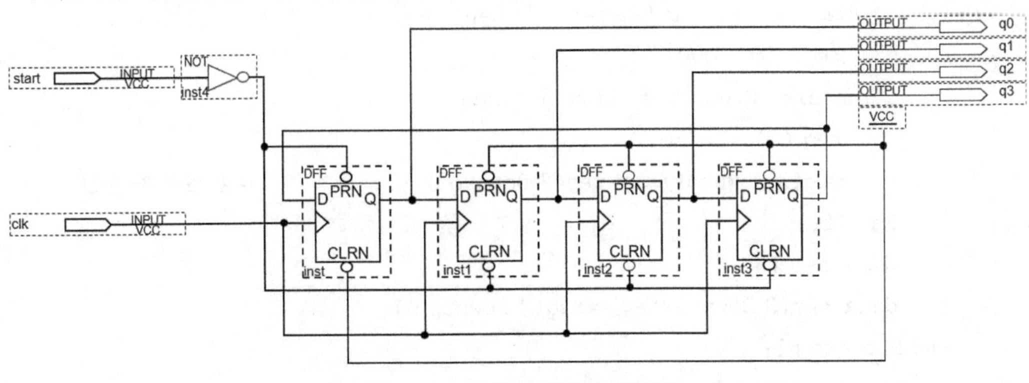

图 6-1　四位环形计数原理图

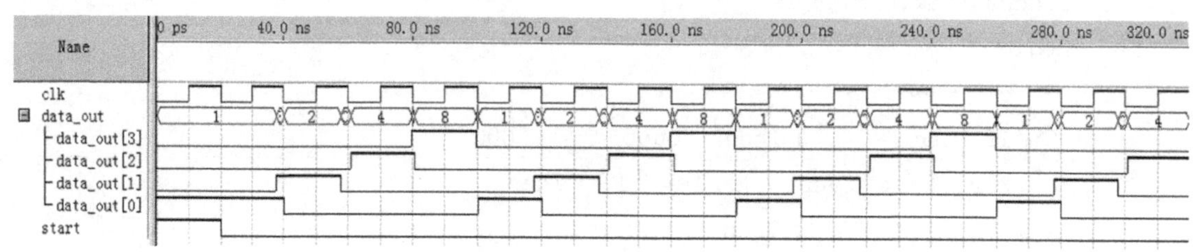

图 6-2　四位环形计数仿真

图 6-2 所示计数器也可以是编码电路，每个编码中只有一个'1'存在，因而又称独热码(One Hot Code)，由于 FPGA 中有丰富的寄存器资源，所以这种编码在 FPGA 设计中最为常用。n 个寄存器可以有 2^{n-1} 个计数，这类计数器冗余计数多，应用时往往要做处理。

```
--四位环形计数 VHDL 程序代码,fourbitohc.vhd
Library Ieee;
Use Ieee.Std_Logic_1164.All;

Entity fourbitohc Is                                          --四位环形计数
    Port(clk,start: In Std_Logic;
         data_out: Out Std_Logic_Vector(3 Downto 0));
End Entity fourbitohc;

Architecture fourohc_arch Of fourbitohc Is
Signal reg:Std_Logic_Vector(3 Downto 0):="0001";              --计数寄存器置初值
Signal clr,d:Std_Logic;
Begin

    p1:  Process(clk) Is
    Begin
        clr<=Not start;
        If start='1'  And clr='0' Then
            reg <= "0001";
        Elsif clk'Event And clk='1' Then
            reg (0) <= d;
            reg(3 Downto1)<=reg(2 Downto 0);                  --左移位寄存器
        End If;
        d<=reg (3);
        data_out(3 Downto 0)<=reg(3 Downto 0);
    End Process p1;
End Architecture fourohc_arch;
```

示例 6-3 四位与 N 位扭环形计数器

四位扭环形计数器又称约翰逊(Johnson)计数器,它非常节省资源,可以有 2n 种计数,其中 n 为寄存器的个数,如图 6-3 所示。

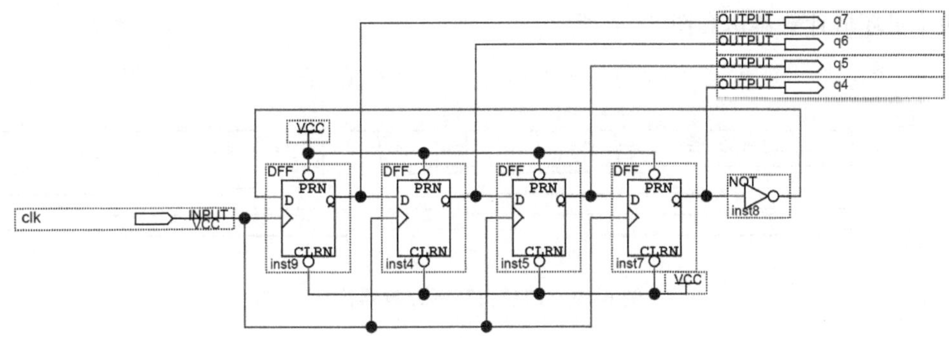

图 6-3 四位扭环形计数

应用:四位扭环形计数器可以构成一种简单的循环彩灯图案设计。

```
--N 位扭环形计数程序, nbitrc.vhd
Library Ieee;
Use Ieee.Std_Logic_1164.All;

Entity nbitrc Is                  --n 位扭环计数,库与程序包的说明与示例 6-2 程序相同
    Generic(n : Natural := 4);                              --默认为 4 位扭环计数
    Port(clk,reset: In Std_Logic;
        data_out: Out Std_Logic_Vector(n-1 Downto 0));
End Entity nbitrc;

Architecture nbitrc_arch Of nbitrc Is
Begin
    Process(clk,reset) Is
    Variable reg:Std_Logic_Vector(n-1 Downto 0);
    Begin
        If reset = '1' Then reg := (Others => '0');  --计数初值
        Elsif (clk'Event And clk='1') Then
            If reg(n-1) = '0' And reg(0) = '0' Then
                                                --计数进入无效状态时(2^(n-1)-2n)
                reg(n-1):= '1';reg := (Others => '0');--回到第 2 个顺序计数值
            Else
                reg := Not reg(0) & reg(n-1 Downto 1);
            End If;
```

```
        End If;
        data_out <= reg;
    End Process;
End Architecture nbitrc_arch;
```

四位扭环形计数器仿真图 6-4 所示。

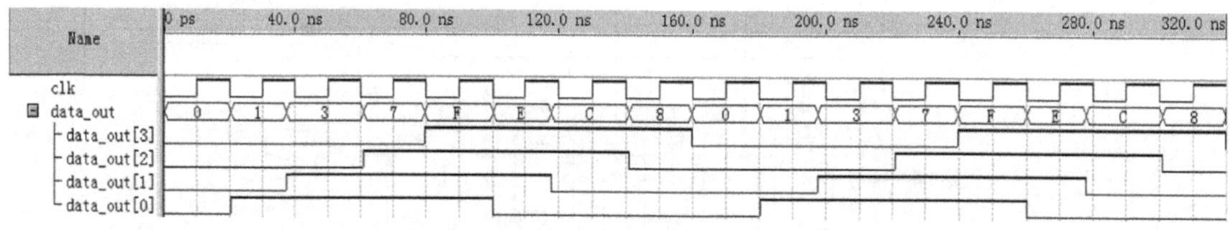

图 6-4　四位扭环形计数器仿真

示例 6-4　计数器的应用——节拍器

```
Library Ieee;
Use Ieee.Std_Logic_1164.All;
Use Ieee.Std_Logic_Unsigned.All;

Entity jiepai Is
    Port( clk,en,reset: In Std_Logic;                  --时钟、使能、复位
          stop,start: In Std_Logic;                     --停止,开始位
          q: Out Std_Logic_Vector(3 Downto 0));         --4 节拍输出
End Entity jiepai;

Architecture jiepai_arch1 Of jiepai Is
Signal qs: Std_Logic_Vector(1 Downto 0);
Signal qt: Std_Logic_Vector(3 Downto 0);
Begin

    count:Process(clk) Is                               --4 循环计数
    Begin
        If(reset='1') Then
            qs<="00";
        Elsif(clk'Event And clk='1') Then
            If(en='1') Then
                If(qs="11") Then qs<="00";
                Elsif(qs<"11") Then
                    qs<=qs+1;                           --4 计数
                End If;
```

```
                q<=qt;                              --同步输出
            End If;
        End If;
    End Process count;

    ecode:Process(start,stop,qs) Is                 --计数值编码
    Bcgin
        If(start='1' And stop='0') Then
            Case qs Is
                When "00"=>qt<="0001";
                When "01"=>qt<="0010";
                When "10"=>qt<="0100";
                When "11"=>qt<="1000";
                When Others=>Null;
            End Case;
        Else qt<="0000";
        End If;
    End Process ecode;
End Architecture jiepai_arch1;

Architecture jiepai_arch2 Of jiepai Is
Signal qs: Std_Logic_Vector(1 Downto 0);
Signal qt: Std_Logic_Vector(3 Downto 0);

Begin

    count1:Process(clk) Is                          --4 非循环计数
    Begin
        If(reset='1') Then
            qs<="00";
        Elsif(clk'Event And clk='1') Then
            If(en='1') Then
                If(qs="11") Then qs<="11"; q<="1000" ;
                Elsif(qs<"11") Then
                    qs<=qs+1; q<=qt ;
                End If;
            End If;
        End If;
    End Process count1;
```

```
    ecode1:Process(start,stop,qs) Is                           --计数值编码
    Begin
        If(start='1'And stop='0') Then
            Case qs Is
                When "00"=>qt<="0001";
                When "01"=>qt<="0010";
                When "10"=>qt<="0100";
                When "11"=>qt<="1000";
                When Others=>qt<="1000";
            End Case;
        Else qt<="0000";
        End If;
    End Process ecode1;
End Architecture jiepai_arch2;

Configuration jiepai_cfg Of jiepai Is                          --配置语句
    For jiepai_arch1
    End For;
End Configuration jiepai_cfg;
```

节拍器循环计数仿真结果如图 6－5 所示，非循环计数仿真结果如图 6－6 所示。

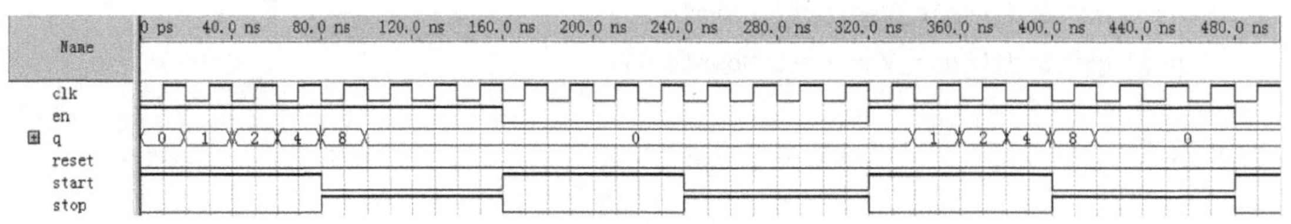

图 6－5　节拍器循环计数仿真结果

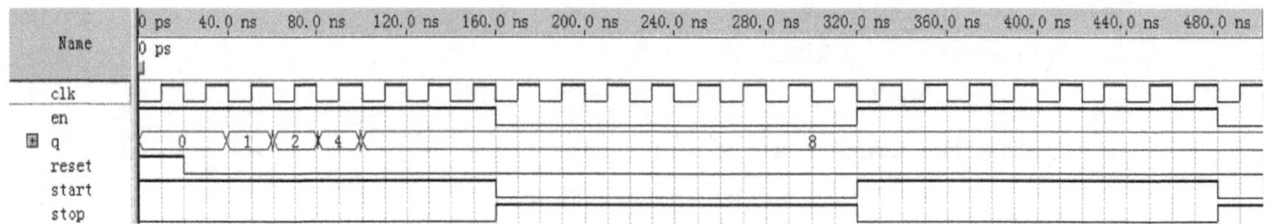

图 6－6　节拍器非循环计数仿真结果

6.2　异步与同步计数器设计比较

10 计数器级联 BCD10 的倍数计数器，即所谓模 10 计数器在电子产品设计中极为常用，模 10 计数器以 10 计数器为基础，可以采用异步与同步设计来连接。

6.2.1 模10计数与级联

示例6-5 模10计数器

```
Library Ieee;
Use Ieee.Std_Logic_1164.All;
Use Ieee.Std_Logic_Unsigned.All;                          --声明重载函数包

Entity count10a Is
    Port(clk,clr,en: In Std_Logic;
         q: Out Std_Logic_Vector(3 Downto 0);             --计数输出
         count: Out Std_Logic);                           --进位输出
End Entity count10a;

Architecture one Of count10a Is                           --进位超前一位的模10计数
Signal coun: Std_Logic_Vector(3 Downto 0);
Begin
    q<=coun;
    Process(clk,clr,en) Is
    Begin
        If(clr='1') Then  coun<="0000";                   --异步清零
        Elsif(clk'Event And clk='1') Then                 --上升沿触发
            If(en='1') Then                               --同步使能
                If(coun=9) Then  coun<="0000";
                Else coun<=coun+'1';                      --0到9计数
                End If;
            End If;
        End If;
    End Process;
    count<='1' When en='1' And (coun=9) Else '0';        --进位采用并行语句
End Architecture one;

Architecture newone Of count10a Is
Signal coun: Std_Logic_Vector(3 Downto 0);
Signal count1: Std_Logic;
Begin
    q<=coun;

    Process(clk,clr,en) Is
    Begin
```

```
        If(clr='1') Then  coun<="0000";                      --异步清零
        Elsif(clk'Event And clk='1') Then                    --上升沿触发
            count<= count1;                                  --将进位延迟1个时钟周期输出
            If(en='1') Then                                  --同步使能
                If(coun=9) Then  coun<="0000";
                Else coun<=coun+'1';
                End If;
            End If;
        End If;
    End Process;
    count1<='1' When en='1' And (coun=9) Else '0';           --进位采用并行语句
End Architecture newone;

Architecture two Of count10a Is                              --进位对齐的模10计数
Signal coun: Std_Logic_Vector(3 Downto 0);                   --信号法
Begin
    q<=coun;

    p1:Process(clk,clr,en) Is
    Begin
        If(clr='1') Then coun<="0000";                       --异步清零
        Elsif(clk'Event And clk='1') Then                    --上升沿触发
            If(en='1') Then                                  --同步使能
                If(coun=9) Then  coun<="0000";
                Else coun<=coun+'1';
                End If;
            End If;
        End If;
    End Process p1;

    p2:Process(clk) Is
    Begin
        If(clk'Event And clk='1') Then
            If (en='1' And (coun=9)) Then count<='1'; --进位采用顺序语句
            Else count<='0';
            End If;
        End If;
    End Process p2;
End Architecture two;
```

```
Architecture three Of count10a Is                              --进位对齐的模10计数
Begin

    Process(clk,clr,en) Is
    Variable coun: Std_Logic_Vector(3 Downto 0);               --变量法
    Begin
        If(clr='1') Then
            coun:="0000";
        Elsif(clk'Event And clk='1') Then
            If(en='1') Then
                If(coun=9) Then
                    coun:="0000";
                    count<='1';
                Else
                    coun:=coun+'1';
                    count<='0';
                End If;
            End If;
        End If;
        q<=coun;
    End Process;
End Architecture three;

Configuration c10_tb_cfg of count10a Is                        --配置语句
    For two
    End For;
End Configuration c10_tb_cfg ;
```

模 10 计数器结构 one 仿真结果如图 6－7 所示。结构体 newone,two,three 仿真结果如图6－8所示。

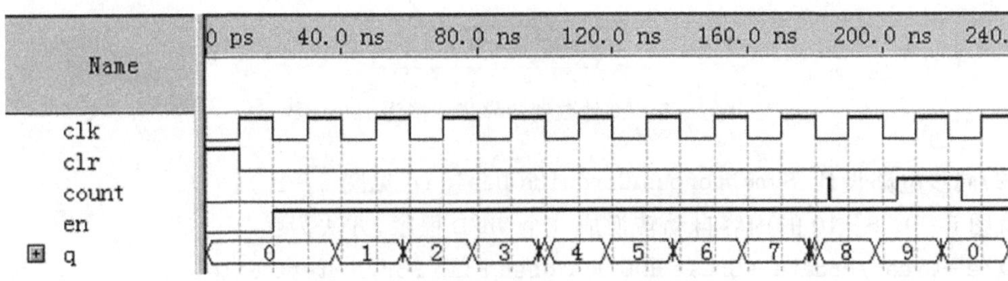

图 6－7　模 10 计数器结构 one 仿真结果

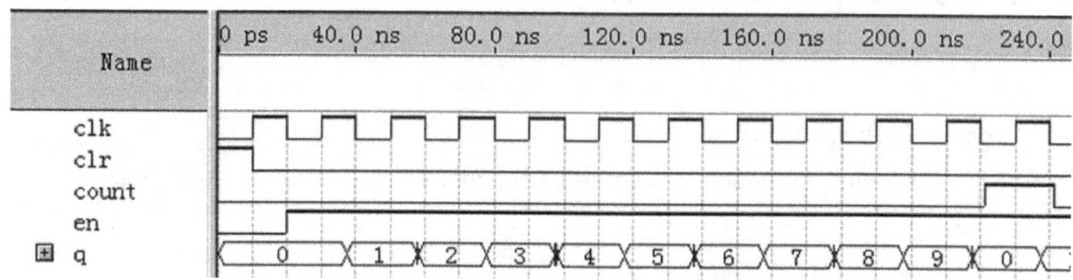

图 6-8　模 10 计数器结构体 newone,two,three 仿真结果

程序说明：

(1)结构体 one 与结构体 two 进位输出都是占空比为 1∶10 的 10 分频。

(2)计数结构体 one 的进位提前了 1 位。修改方法是使结构体 newone 进位正确,即将进位带回到进程中,直接对进位位进行了延时。

(3)修改方法是将原来并行语句改为了由时钟上升沿触发的顺序语句。

(4)结构体 three 采用了变量计数。仿真结果与结构体 newone,two 相同。

(5)计数与运算在进位时最容易出错的,仿真时要重点关注。

示例 6-6　级联 BCD100 计数器讨论

(1)异步电路设计(Asynchronous Circuit Design)(见图 6-9)。

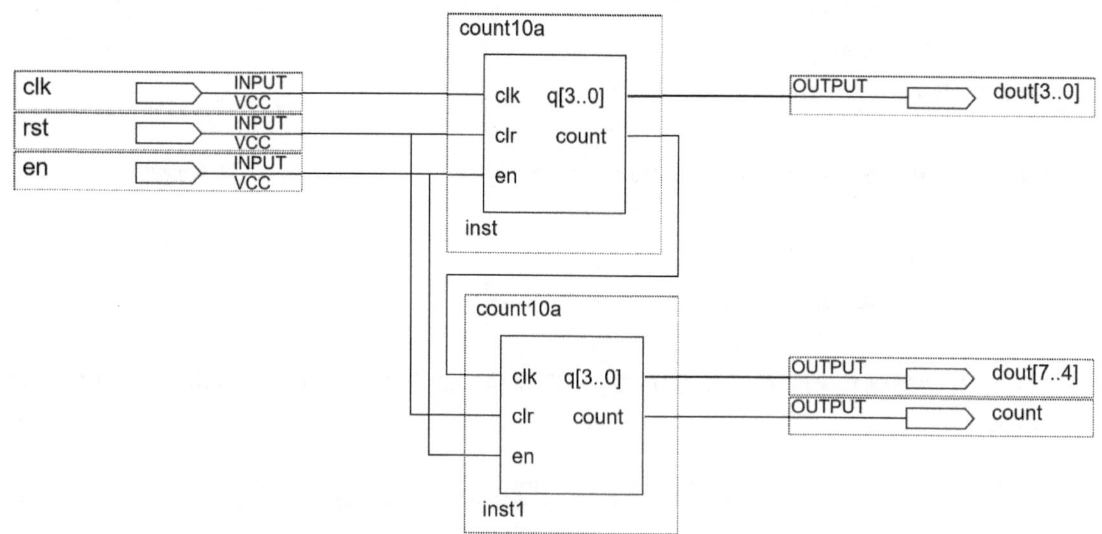

图 6-9　10 计数器的异步连接为 100 计数

(2)同步电路设计(Synchronous Circuits Design)(见图 6-10)。

将图 6-9，6-10 的内容自动转换成 2 个 VHDL 程序,方法为：

File→Creat/Updata→Creat HDL Designer File For Current File

将其合成为便于比较分析的 1 个实体 2 个结构体的形式。

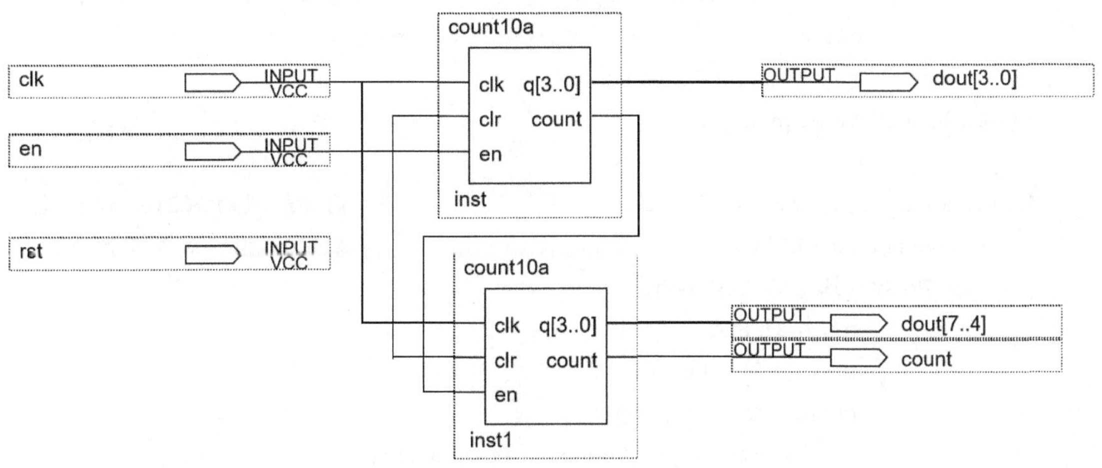

图 6-10 10 计数器的同步连接为 100 计数

```
Library Ieee;
Use Ieee.Std_Logic_1164.All;
Use Ieee.Std_Logic_Unsigned.All;

Entity c100 Is                                        --BCD100 计数器实体声明
    Port( clk, rst, en:  In  Std_Logic;               --时钟,异步复位,同步使能端
          count: Out Std_Logic;
          dout : Out Std_Logic_Vector(7 Downto 0));
End Entity c100;

Architecture asyn_type Of c100 Is                     --异步级联的 BCD100 计数器
    Component count10a Is
        Port( clk, clr, en: In Std_Logic;
              count : Out Std_Logic;
              q : Out Std_Logic_Vector(3 Downto 0));
    End Component count10a;

Signal SYNTHESIZED_WIRE_0 :  Std_Logic;
Begin

    b2v_inst : count10a
    Port Map( clk => clk, clr => rst, en => en,
              count => SYNTHESIZED_WIRE_0,
              q => dout(3 Downto 0));

    b2v_inst1 : count10a
```

```
    Port Map( clk => SYNTHESIZED_WIRE_0,          --进位输出连到了时钟端
              clr => rst, en => en,count => count,
              q => dout(7 Downto 4));
End Architecture asyn_type;

Architecture syn_type Of c100 Is                  --同步级联的 BCD100 计数器
    Component count10a Is                         --采用 count10a 结构体 one
        Port( clk : In Std_Logic;
              clr : In Std_Logic;
              en : In Std_Logic;
              count : Out Std_Logic;
              q : Out Std_Logic_Vector(3 Downto 0));
    End Component count10a;
Signal dout0,dout1:Std_Logic_Vector(3 Downto 0);
Signal SYNTHESIZED_WIRE_0 : Std_Logic;

Begin

    b2v_inst : count10a
    Port Map(clk => clk, clr => rst, en => en,
             count => SYNTHESIZED_WIRE_0,
             q => dout(3 Downto 0));
    b2v_inst1 : count10a
    Port Map( clk => clk, clr => rst,
              en => SYNTHESIZED_WIRE_0,           --进位输出连到了使能端
              count => count,
              q => dout(7 Downto 4));
End Architecture syn_type;

Architecture syn_typex Of c100 Is                 --同步级联的 BCD100 计数器

    Component count10a Is                 --采用 count10a 对齐结构 newone,不推荐
        Port( clk : In Std_Logic;
              clr : In Std_Logic;
              en : In Std_Logic;
              count : Out Std_Logic;              --进位输出
              q : Out Std_Logic_Vector(3 Downto 0));
    End Component count10a;
```

```
Signal dout0,dout1:Std_Logic_Vector(3 Downto 0):="0000";
Signal SYNTHESIZED_WIRE_0 :  Std_Logic;

Begin
    b2v_inst : count10a
        Port Map( clk => clk, clr => rst, en => en,
                  count => SYNTHESIZED_WIRE_0,
                  q => dout0(3 Downto 0));

    b2v_inst1 : count10a
        Port Map( clk => clk, clr => rst,
                  en => SYNTHESIZED_WIRE_0,
                  count => count,
                  q => dout1(3 Downto 0));

    Process(clk) Is
    Variable dout2,dout3:Std_Logic_Vector(3 Downto 0);
    --同步级联的BCD100计数器采用对齐的10计数需要修正才能正确
    Begin
        If clk'Event And clk='1' Then
            dout2:=dout1;
            dout(3 Downto 0)<=dout0;
            dout3:=dout2(3 Downto 0)+SYNTHESIZED_WIRE_0;
            If (dout3="1010") Then
                dout3:="0000";
            Else
                dout3:= dout2(3 Downto 0)+SYNTHESIZED_WIRE_0;
            End If;
            dout(7 Downto 4)<=dout3;
        End If;
    End Process;
End Architecture syn_typex;

Configuration  c100_tb_cfg of c100 Is                  --配置语句
--For asyn_type                                        --异步级联指定
--For syn_typex
    For syn_type                                       --同步级联指定
        For b2v_inst :count10a Use Entity Work.count10a(one);
        End For;
```

```
        For b2v_inst1 :count10a Use Entity Work.count10a(one);
        End For;
    End For;
End Configuration c100_tb_cfg ;
```

同步级联 100 计数 Syn type(one)的仿真和 syn typex(newone)的仿真如图 6－11 所示。

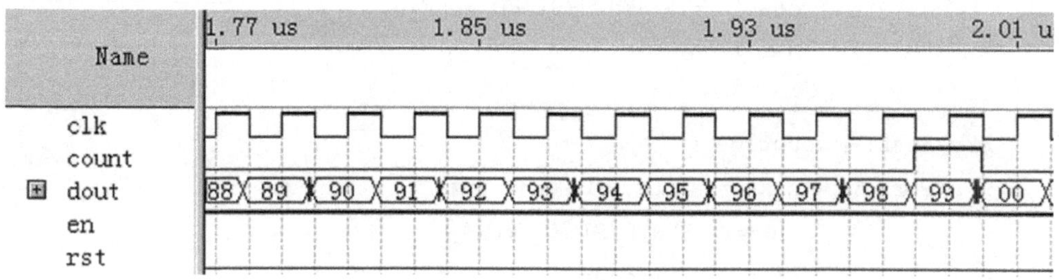

图 6－11 同步级联 100 计数 syn type(one)的仿真和 syn typex(newone)的仿真

6.2.2 级联中的设计原则

(1)级联遵循 FPGA 同步设计原则。

根据 FPGA 结构特点，在 FPGA 设计中最好采用同步设计，可以看到采用超前位 10 计数结构体 syn_type(one)描述的同步级联很容易得到正确结果，syn_type(one)指 100 计数结构体 syn_type 中调用模 10 计数器结构体 one。

事实上，同步级联，由于使能信号要等待时钟的到来，反倒 10 计数采用结构体 one 的进位提前的形式是正确的 syn type(one)。采用对齐位 syn typex(newone)，asyn_type(newone) 结果都有问题，如图 6－12 和图 6－13 所示。结构体 syn typex(newone)修正了图 6－13 的问题。

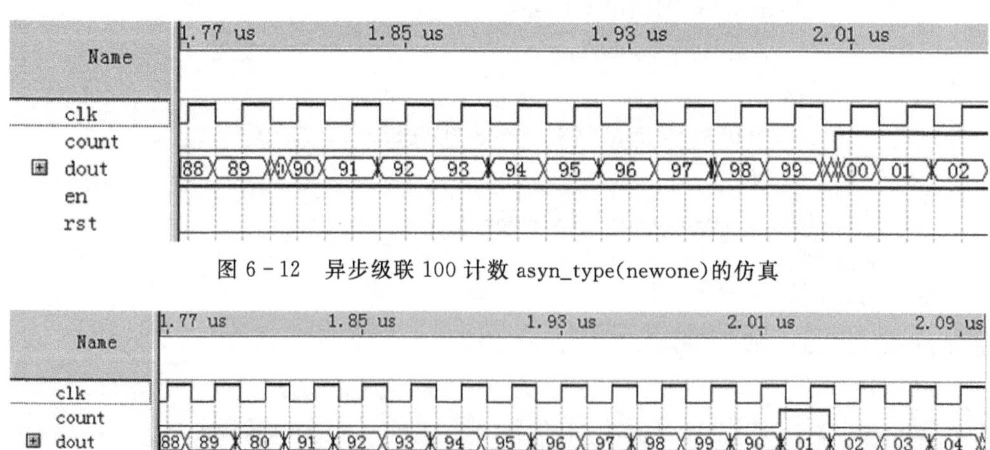

图 6－12 异步级联 100 计数 asyn_type(newone)的仿真

图 6－13 同步级联 100 计数 syn typex(newone)修正前仿真

同步超前计数 syn_type(one)级联，电路的毛刺比异步级联更少，其实即使在运算过程中我们也是不希望有许多的毛刺的，因为毛刺还会带来功耗的增加。

(2)硬件原则与模块的易用性原则。

示例 6-5 和示例 6-6 采用 newone 结构体，在底层 10 计数的进位看是对了，级联却必须通过修正才能得到正确计数，显然不符合模块的易用性原则，不推荐在级联时运用。

syn_typex(newone)结构体中生硬的修改方式是我们不推荐的，硬件原则要本着电路运行的规律来体现。

asyn_type(newone)级联 100 进位的宽度不是我们需要的 1 个周期，而是 10 个周期。我们可以设计一个产生单脉冲的电路来解决。参见示例 6-15。

(3)面向应用及其背景的原则，即系统原则。

对应用而言，一个设计原型的对与错和好与坏都不是绝对的，甚至对立的特性可互相转换，所以要把设计置于应用的背景，明确设计的应用条件，把握住各种转换。如利用模 10 计数器(结构体 one)同步级联方式继续级联可以构造 10、100、1000、10000 计数。如果需要把 10 的整数倍的数读出，则可以借鉴示例 6-5 将 one 结构体改为 newone，two，three 的方法。

(4)设计的多元化选择与应用探索原则，即工程原则。如利用模 10 计数器可以进行分频。把每一级的进位输出或计数最高位引出，即为相应的不等占空比的分频，100 计数进位对应 100 分频。

示例 6-6a 数据流方式的 BCD 计数

示例 6-6 讨论的是结构化描述风格的 BCD 计数，结构化风格是复杂设计调试的主要的风格。

事实上，一些小设计采用数据流方式是很便捷的。比如 BCD 计数就可以采用如下同步设计程序来实现，综合器能够给出很好的结果。如果希望更多位的计数，在此基础上也很容易实现扩展。

RTL 级的优化设计对于硬件工程师来说吸引力很大，但是设计之初，设计原型的迅速实现无疑是很重要的。

```
Library Ieee;
Use Ieee.Std_Logic_1164.All;
Use Ieee.Std_Logic_Unsigned.All;

Entity bcdcount1 Is
    Port ( rst:In Std_Logic;                              --复位信号
           clk :In Std_Logic;                             --系统时钟
           bcd1,bcd2,bcd3:Out Std_Logic_Vector(3 Downto 0));
End Entity bcdcount1;

Architecture behav Of bcdcount1 Is
Signal b1,b2,b3:Std_Logic_Vector(3 Downto 0);        --BCD 计数器
Begin
```

```
    com:Process(rst,clk) Is
    Begin
        If rst ='1' Then                                    --复位
            b1<="0000";b2<="0000";b3<="0000";
        Elsif clk'Event And clk='1' Then
            If b1="1001" Then b1<="0000";         --此 If 语句完成个位 BCD 计数
                If b2="1001" Then b2<="0000";     --此 If 语句完成十位 BCD 计数
                    If b3="1001" Then b3<="0000";
                                                  --此 If 语句完成百位 BCD 计数
                    Else b3<=b3+1; End If;
                Else b2<=b2+1; End If;
            Else b1<=b1+1; End If;
        End If;
    End Process com;
    bcd1<=b1; bcd2<=b2; bcd3<=b3;
End Architecture behav;
```

6.3 查看设计报告与 TimeQuest 时序分析

在进行 FPGA 开发综合后，一定要查看设计报告以及利用时序分析工具深入了解设计。

知识点：

1. 动态时序分析(Dynamic Timing Analysis)

传统仿真是从仿真起始至结束由许多事件发生来校验功能。采用逻辑仿真器在验证功能的同时验证时序，它以逻辑模拟方式运行，需要输入向量作为激励。

随着规模增大，所需要的向量数量以指数增长验证所需时间长，占到整个设计周期的 50%，而最大的问题是难以保证足够的覆盖率。

2. 静态时序分析(Static Timing Analysis)

静态时序分析采用穷尽分析方法，提取整个电路的所有时序路径，通过计算信号沿在路径上的延迟找出违背时序约束的错误，如，检查建立时间和保持时间是否满足要求，分析得到最大路径延迟和最小路径延迟，考察一个同步设计并确定它的最高工作频率。

这种方法运行速度很快，占用内存很少，克服了动态时序验证的缺陷，适于 VLSI SOC 的验证，可以节省多达 20%的设计时间。Quartus Ⅱ 的静态时序分析采用 TimeQuest。

6.3.1 LFSR 计数器与二进制计数器设计

在 FPGA 设计中，如果没有递增/递减顺序计数的要求，如定时，最好使用线性反馈移位寄存器 LFSR(Linear Feedback Shift Registers) 计数器，而不是二进制计数器。因为 LFSR 速度很快，占用资源少，性能好，而普通二进制计数器在翻转的时候，会产生毛刺。如果这个时候计数器的值被作为其他信号的输入，那么结果必将是错误的，同时，LFSR 的扇入比较小，而二进制计数器的扇入比较大，从规模上或者在性能上，LFSR 计数器都要优于二进制计数器。

示例 6 - 7 实现了计数 8 的 LFSR 计数器和二进制计数器，为了便于比较，我们采用一个实体两个结构体配置形式的 VHDL 语言描述。

示例 6 - 7　二进制计数器与 LFSR 计数器

```
Library Ieee;
Use Ieee.Std_Logic_1164.All;
Use Ieee.Std_Logic_Unsigned.All;

Entity clkcount Is
    Port( clk: In Std_Logic;
          dout: Out Std_Logic_Vector(4 Downto 0));          --二进制计数器
End Entity clkcount;

Architecture bincount Of clkcount Is
Signal count:Std_Logic_Vector(4 Downto 0):="00000";         --防止仿真死锁
Begin
    Process(clk) Is
    Begin
        If (clk'Event And clk='1')Then
            If(count="11111" ) Then  count<="00000";
            Else count <= count +1;
            End If;
        End If;
    End Process;
    dout <= count;
End Architecture bincount;

Architecture lfsr Of clkcount Is                            --lfsr,m 序列
Signal rq: Std_Logic_Vector(4 Downto 0):="00001";           --防止仿真死锁
Begin
    p1: Process(clk) Is
    Begin
        If clk'Event And clk='1' Then
            rq(4)<=rq(0) Xnor rq(3);                        --防止死锁,加了“非”
            rq(3)<= Not rq(4);                              --防止死锁,加了“非”
            rq(2 Downto 0)<=rq(3 Downto 1);
        End If;
    End Process p1;
```

```
    dout<=rq;
End Architecture lfsr;

Configuration clkcount_cfg Of clkcount Is          --配置语句
    For bincount                                   --为 clkcount 指定结构体 bincount
    End For;
End Configuration clkcount_cfg;
```

6.3.2 LFSR 计数器与二进制计数器的比较

对于时序电路设计主要从时序、速度和面积(占用资源)来评价。可以通过查看设计报告以及利用 TimeQuest Timing Analyzer 查看速度等来比较,有时可以通过 RTL Viewer 较直观地看出来。二进制计数器仿真如图 6-14 所示,LFSR 计数器仿真如图 6-15 所示。

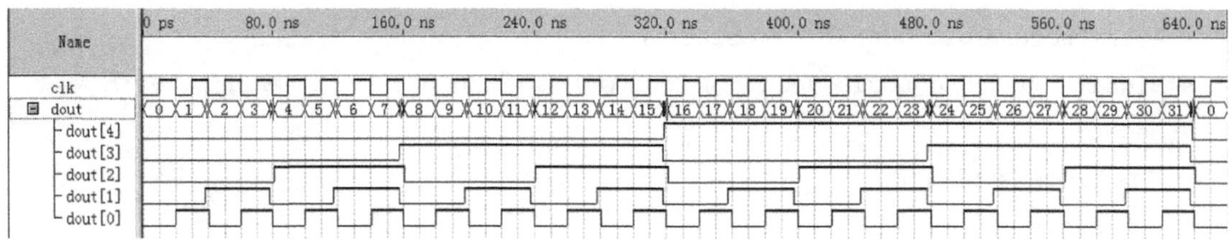

图 6-14　二进制计数器仿真

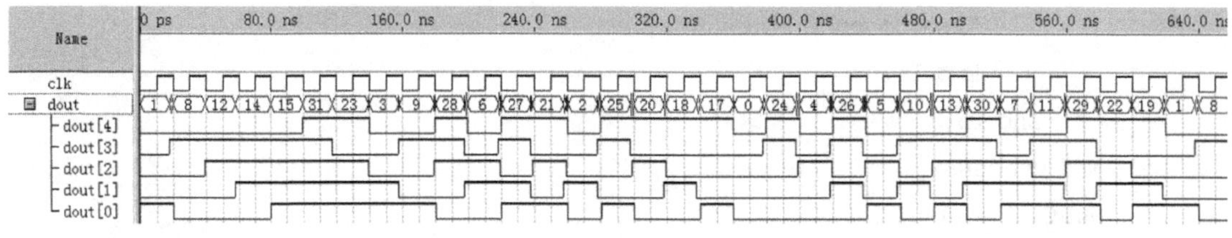

图 6-15　LFSR 计数器仿真

(1)仿真说明。

从仿真可以看出,二进制计数器是计数值是连续的,其低位到高位分别是时钟的 2 分频、4 分频、8 分频、16 分频、32 分频。即 2^N计数,本例中 LFSR 计数器可产生 2^N-1 的计数。LFSR 计数器计数值是不连续的,不可兼做分频器。

Modelsim 仿真与 Quartus Ⅱ仿真的一个较大的区别是,Modelsim 的信号与变量没有缺省的初值,而 Quartus Ⅱ的信号与变量是有缺省的初值的,所以我们在上述实体的 2 个结构体的信号说明部分都赋予了初值。

(2)RTL Viewer 与 Technology Map Viewer。

从图 6-16～图 6-19 可以看出 LFSR 计数器的综合结果明显简洁。

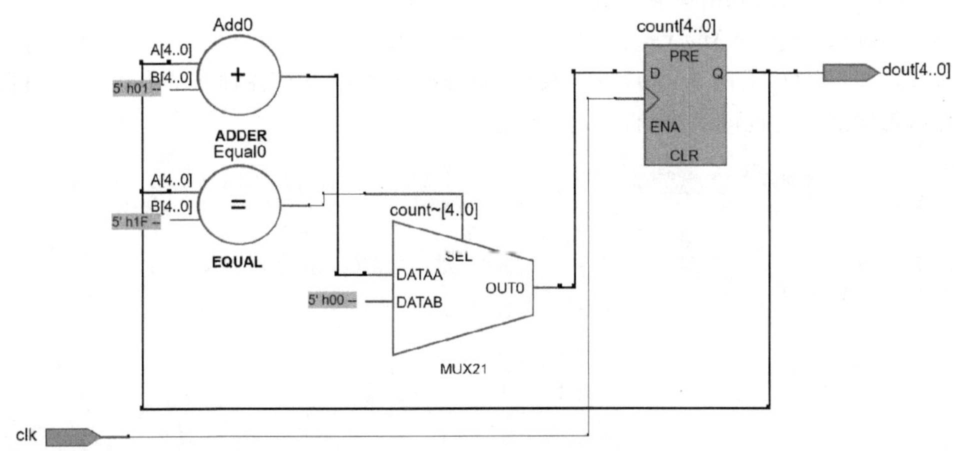

图 6-16 二进制计数器的综合结果

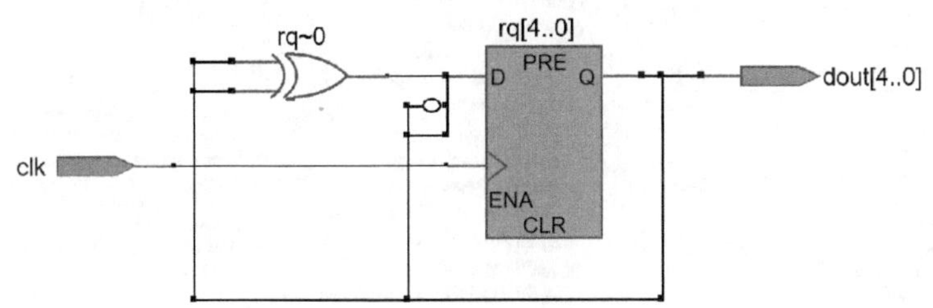

图 6-17 LFSR 计数器的综合结果 RTL Viewer

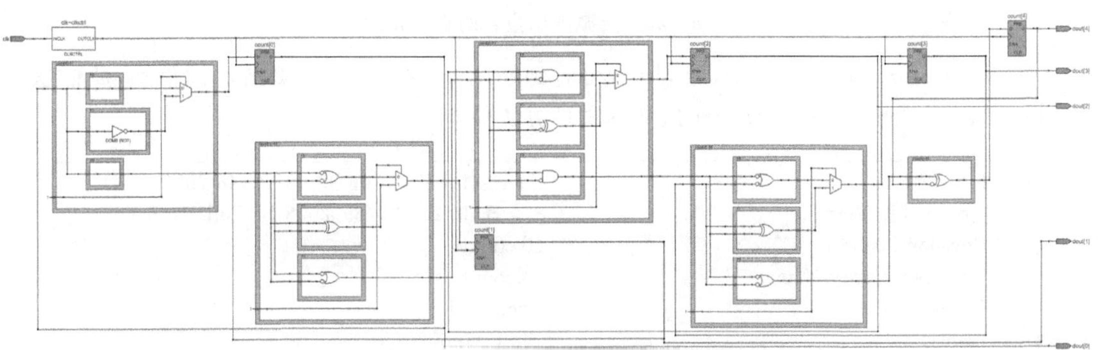

图 6-18 二进制计数器的综合结果 Technology Map Viewer

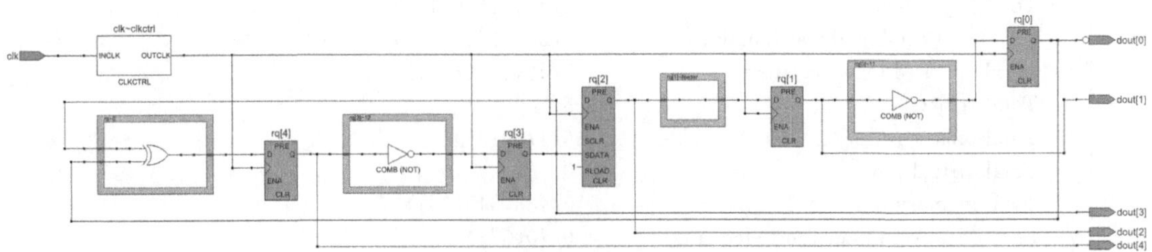

图 6-19 LFSR 计数器的综合结果 Technology Map Viewer

(3)Quartus Ⅱ芯片两种计数器报告信息的比较。

在 Compile Design→Analysis & Synthisis→Design assistant→View Reports→在 Compilaction Report 窗口(见图 6-20)→Fitter→Summary 可以查到各种布局布线后占用资源的简要信息,每次编译后也可以看到,一定要注意。

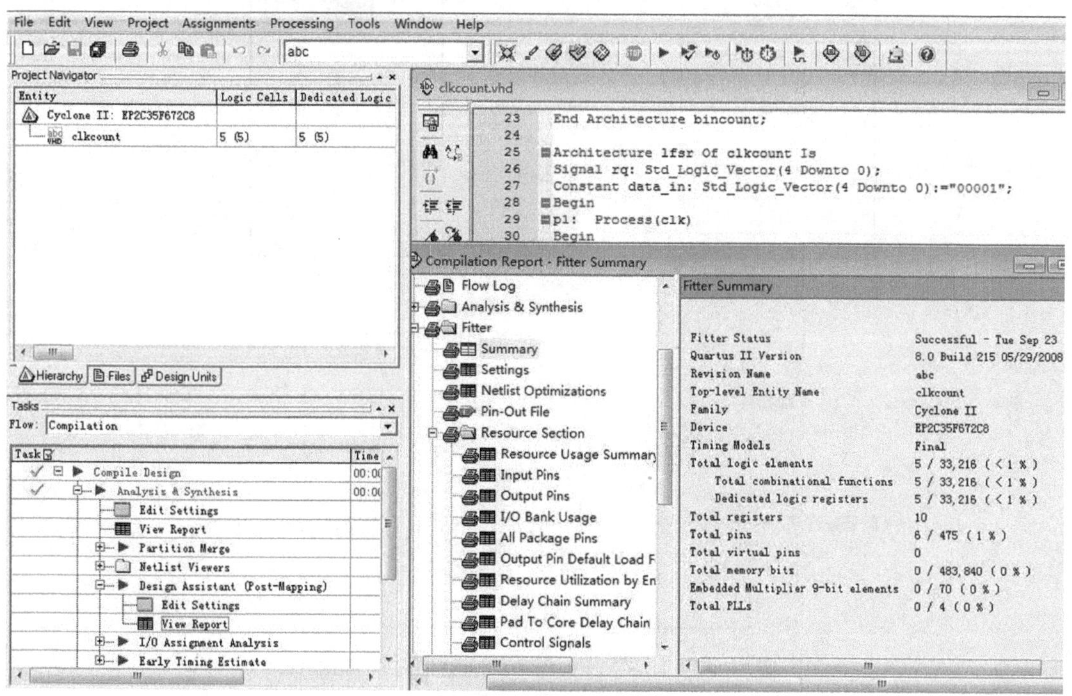

图 6-20 资源占用报告的查看界面

编译 bincount 和 Ifsr 结构体成功后,可看到如图 6-21、图 6-22 所示的信息,此信息也可以通过查看 Compilation Report 以文本文件方式保存。

Flow Status	Successful-Tue Aug 30 08:11:46 2015
Quartus Ⅱ Version	8.0 Build 215 05/29/2008 SJ Full Version
Revision Name	clkcount
Top-level Entity Name	clkcount
Family	Cyclone Ⅱ
Device	EP2C35F672C8
Timing Models	Final
Met timing requirements	Yes
Total logic elements	5/33,216(<1%)
Total combinational functions	5/33,216(<1%)
Dedicated logic registers	5/33,216(<1%)
Total registers	5
Total pins	6/475(1%)
Total virtual pins	0
Total memory bits	0/483,840(0%)
Embedded Multiplier 9-bit elements	0/70(0%)
Total PLLs	0/4(0%)

图 6-21 二进制 32 计数器资源占用报告

Flow Status	Successful-Tue Aug 30 08:11:46 2015
Quartus Ⅱ Version	8.0 Build 215 05/29/2008 SJ Full Version
Revision Name	clkcount
Top-level Entity Name	clkcount
Family	Cyclone Ⅱ
Device	EP2C35F672C8
Timing Models	Final
Met timing requirements	Yes
Total logic elements	5/33,216(<1%)
Total combinational functions	3/33,216(<1%)
Dedicated logic registers	5/33,216(<1%)
Total registers	5
Total pins	6/475(1%)
Total virtual pins	0
Total memory bits	0/483,840(0%)
Embedded Multiplier 9-bit elements	0/70(0%)
Total PLLs	0/4(0%)

图 6-22　LFSR(m 序列)32 计数器资源占用报告

在 Resource Section 栏目下还有更详细的信息报告项。如查看 Resource Usage Summary 中，如图 6-23 中所示占用 5 个逻辑单元，更详细的标注为占用普通模式的逻辑单元 1 个，算术单元的逻辑单元 4 个……。

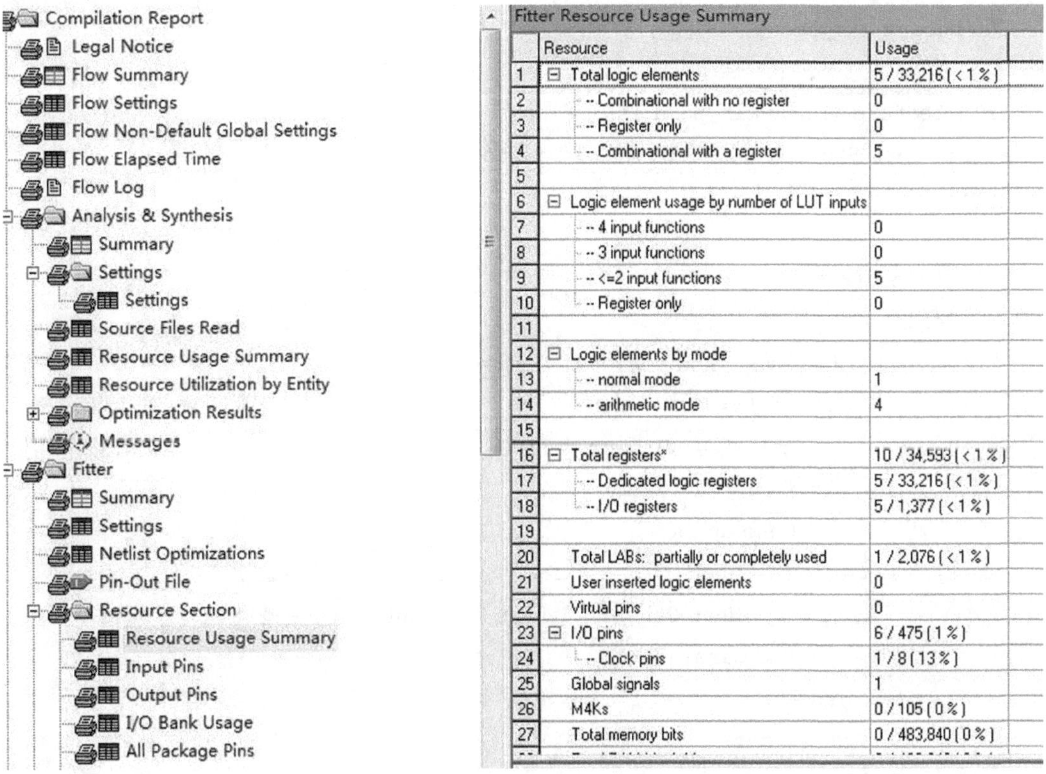

- Compilation Report
  - Legal Notice
  - Flow Summary
  - Flow Settings
  - Flow Non-Default Global Settings
  - Flow Elapsed Time
  - Flow Log
  - Analysis & Synthesis
    - Summary
    - Settings
      - Settings
    - Source Files Read
    - Resource Usage Summary
    - Resource Utilization by Entity
    - Optimization Results
    - Messages
  - Fitter
    - Summary
    - Settings
    - Netlist Optimizations
    - Pin-Out File
    - Resource Section
      - Resource Usage Summary
      - Input Pins
      - Output Pins
      - I/O Bank Usage
      - All Package Pins

Fitter Resource Usage Summary

	Resource	Usage
1	⊟ Total logic elements	5 / 33,216 (< 1 %)
2	-- Combinational with no register	0
3	-- Register only	0
4	-- Combinational with a register	5
5		
6	⊟ Logic element usage by number of LUT inputs	
7	-- 4 input functions	0
8	-- 3 input functions	0
9	-- <=2 input functions	5
10	-- Register only	0
11		
12	⊟ Logic elements by mode	
13	-- normal mode	1
14	-- arithmetic mode	4
15		
16	⊟ Total registers*	10 / 34,593 (< 1 %)
17	-- Dedicated logic registers	5 / 33,216 (< 1 %)
18	-- I/O registers	5 / 1,377 (< 1 %)
19		
20	Total LABs: partially or completely used	1 / 2,076 (< 1 %)
21	User inserted logic elements	0
22	Virtual pins	0
23	⊟ I/O pins	6 / 475 (1 %)
24	-- Clock pins	1 / 8 (13 %)
25	Global signals	1
26	M4Ks	0 / 105 (0 %)
27	Total memory bits	0 / 483,840 (0 %)

图 6-23　Resource Usage Summary(资源应用汇总)

在 Quartus Ⅱ 中使用 Tools→TimeQuest Timing Analyzer 可以看到各种有关时序的信息。

LFSR(m 序列)序列计数器速度 F_{max} 为 326.48 MHz，二进制计数器速度 F_{max} 为 281.21 MHz。

在做设计比较时，可以从如下方面来看：

①功能、时序仿真(可以借助 Modelsim 等)。

②综合结果的 RTL Viewer，Technology Map Viewer。

③综合布局布线后的报告(占用资源、扇入扇出)。

④TimeQuest Timing Analyzer 静态时序分析。

示例 6-8　参数化的二进制加减计数器

```
Library Ieee;
Use Ieee.Std_Logic_1164.All;
Use Ieee.Std_Logic_Unsigned.All;

Entity clk_para Is
    Generic( N: Positive:=12);                          --计数器位宽
    Port( clk,en,up,rst_b : In Std_Logic;
          q : Out Std_Logic_Vector(N-1 Downto 0));
End Entity clk_para;

Architecture rtl Of clk_para Is
Signal cnt: Std_Logic_Vector (N-1 Downto 0);            --计数值 N-1
Begin

    Process(clk) Is
    Begin
        If (clk'Event And clk='1') Then
            If(rst_b='0') Then                          --同步复位，低电平有效
                cnt <= (Others =>'0');
            Elsif en='1' Then
                Case up Is
                    When '1' => cnt <= cnt +1;          --递增
                    When Others=> cnt <= cnt-1;         --递减
                End Case;
            End If;
            q<= cnt;
        End If;
    End Process;
```

```
End Architecture rtl;
```

此参数二进制计数电路也可设定做幂次分频电路，如图 6－24 所示。

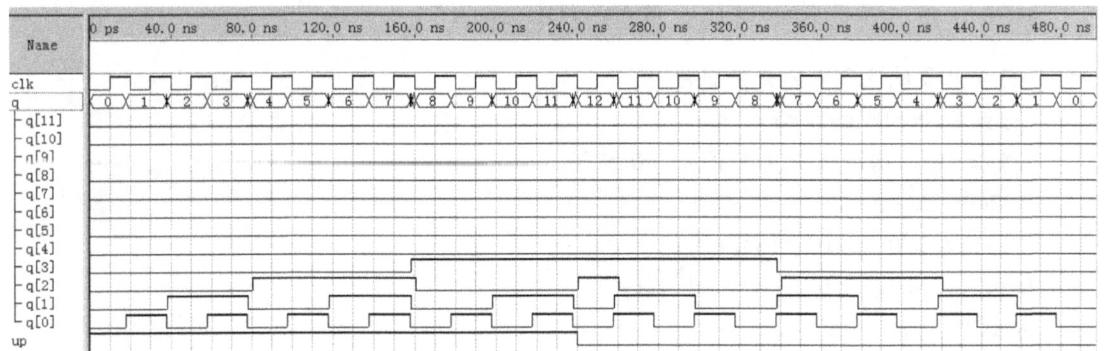

图 6－24 参数化的二进制加减计数器电路仿真

6.4 分频相关电路与设计规范

6.4.1 2 的幂次分频

时钟信号 clk 的 2 分频、4 分频、8 分频、16 分频，即 2 的幂次分频的情况是最简单的分频电路，只需要一个计数器，将计数端输出分别引出，分频是等占空比的，参见示例 6－8 二进制计数器仿真图 6－24。

在进行硬件设计的时候，需要的分频信号不一定都是一个占空比 50%，即 1∶1的信号，可以采用计数器的方法来产生不是 1∶1的分频信号。将输入的时钟信号进行 16 分频，分频信号为 1∶16，也就是说，其中高电位的脉冲宽度为输入时钟信号的一个周期。

如果将进位引出分频则非等占空比的，进位的产生参见示例 6－5，模 10 计数器。

6.4.2 偶数等占空比分频与设计规范

设计任务：

给出二的偶数等占空比分频原始分频程序 6－9 clkgen1. vhd，进行程序的改进与复用。

设计要求：

①自定义程序规范，要求将程序 6－9 改为更便于复用和更易于管理的形式；

②用更高 VHDL 版本书写程序，并具有自己的写作规范；

③分别采用常量 Constant 方式、类属 Generic 方式、自定义库实现。

④设 clk 为 50 MHz 分别产生 1 us(1 MHz)，1 ms，1 s 的分频应用。

设计提示：

文件头规范：文件头必须包括设计者，设计时间，模块功能，修改，版本记录等内容。

Quartus Ⅱ工程文档管理：

①工程文件，如，在 E:/QTEST/irlcd12864/pro 文件夹中。

②Pro 文件夹中放置工程文件；并列的 Src 文件夹中放置 VHDL 文件；Mylib 文件夹中放

置程序包文件等，参见章节 3.8。

注意：在 Assignments→Settings，Libraries 指定库文件和 Files 中添加子文件夹中的文件。

示例 6－9　偶数等占空比分频（VHDL 87 版）

```
--简单信号赋值方式的分频程序
--分频程序 clkgen1.vhd
--版本 V1.1，这个版本没有套用工程规范
LIBRARY IEEE;
USE IEEE.STD_LOGIC_1164.ALL;
Use Ieee.Std_Logic_Unsigned.All;
ENTITY clkgen1 IS
PORT(clkin:In Std_Logic ;
clkout:    OUT STD_LOGIC);
END;
ARCHITECTURE even OF clkgen1 IS                        --VHDL 程序对大小写不敏感
SIGNAL coun:integer range 0 to 5000000;                --区分大小写是为了可读性
SIGNAL clk1:STD_LOGIC;
BEGIN
PROCESS(clkin)
BEGIN
IF(clkin'EVENT AND clkin='1')THEN                      --VHDL 书写格式自由
IF(coun=5000000)THEN
coun<=0;
clk1<=Not clk1;
else
coun<=coun+1;
END IF;
END IF;
END PROCESS;
clkout<=clk1;
END even;
```

设计任务实现如下：

示例 6－10　采用常量 Constant 指定分频（VHDL 93 版）

```
--采用常量 Constant 指定分频(VHDL 93 版)clkgen2.vhd
--版本 V1.2
--修改方式是增加一个常量说明，通过改变常量可以改变程序的分频
```

```
--自定义规范：关键字首字母大写，常数说明大写，采用了嵌套缩进形式
--输入信号前缀 i_，输出信号前缀 o_，提高可读性
--以后都采用本程序规范
--修改时间：2015.11.1
--修改人：xxx
```

```
Library Ieee;
Use Ieee.Std_Logic_1164.All;
Use Ieee.Std_Logic_Unsigned.All;

Entity clkgen2 Is
    Port(i_clk_50m:In Std_Logic ;                    --50 MHz 系统时钟
         o_clk: Out Std_Logic);                      --分频输出
End Entity clkgen2;

Architecture clkgen2_arch Of clkgen2 Is
Constant DIV_PARAM:Integer:=50000000;                --分频=DIV_PARAM×2
Signal coun:Integer Range 0 To DIV_PARAM;
Signal clk1:Std_Logic;
Begin
    Process(i_clk_50m) Is
    Begin
        If(i_clk_50m'Event And i_clk_50m ='1') Then
            If(coun= DIV_PARAM)Then
                coun<=0;
                clk1<=Not clk1;
            Else
                coun<=coun+1;
            End If;
        End If;
    End Process;
    o_clk<=clk1;
End Architecture clkgen2_arch;
```

示例 6－11　采用类属 Generic 指定分频(VHDL 93 版)

```
--采用类属 Generic 指定分频(VHDL 93 版)clkgen3.vhd
--版本 V1.3
```

```
--修改方式是增加一个 Generic 说明,改变 Generic 的常量,改变程序的分频
--修改时间:2015.11.5
--修改人:xxx
----------------------------------------------------------------------------------------
Library Ieee;
Use Ieee.Std_Logic_1164.All;
Use Ieee.Std_Logic_Unsigned.All;

Entity clkgen3 Is
    Generic (DIV_PARAM:Integer:=50000000);                  --分频=DIV_PARAM×2
    Port(i_clk_50m:In Std_Logic ;                           --50 MHz 系统时钟
         o_clk: Out Std_Logic);                             --分频输出
End Entity clkgen3;

Architecture clkgen3_arch Of clkgen3 Is
Signal coun:Integer Range 0 To DIV_PARAM;
Signal clk1:Std_Logic;
Begin
    Process(i_clk_50m) Is
    Begin
        If(i_clk_50m'Event And i_clk_50m ='1')Then
            If(coun= DIV_PARAM)Then
                coun<=0;
                clk1<=Not clk1;
            Else
                coun<=coun+1;
            End If;
        End If;
    End Process;
    o_clk<=clk1;
End Architecture clkgen3_arch;
```

示例 6-12a　分频程序例化调用(VHDL 93 版)

```
----------------------------------------------------------------------------------------
--分频程序例化调用(VHDL 93 版)reuse_clk.vhd
--版本 V1.1
--调用 clkgen3,改变 Generic 常量,实现 1 MHz(us),1 ms,1 s 的分频应用
--修改时间:2015.11.5
```

```
--修改人:xxx
-----------------------------------------------------------------------------------------
Library Ieee;
Use Ieee. Std_Logic_1164. All;

Entity reuse_clk Is
    Port ( i_clk_50m: In Std_Logic;                          --50 MHz 系统时钟
           o_clk_us:Out Std_Logic;                           --us 分频数据输出端
           o_clk_ms:Out Std_Logic;                           --ms 分频数据输出端
           o_clk_s:Out Std_Logic);                           --s 分频数据输出端
End Entity reuse_clk;

Architecture reuse_clk_arch Of reuse_clk Is
    Component clkgen3 Is                                     --分频元件声明
        Generic(DIV_PARAM:Integer:=2);                       --默认是 4 分频
        Port(i_clk_50m:In Std_Logic;
             o_clk:Out Std_Logic);
    End Component clkgen3;
Begin
    gen_us: clkgen3 Generic Map(25)                          --分频元件例化
                                                             --50 分频,产生 us 脉冲
        Port Map (i_clk_50m =>i_clk_50m,o_clk=>o_clk_us);
    gen_ms: clkgen3 Generic Map(25000)
                                                             --50 000 分频,产生 ms 脉冲
        Port Map(i_clk_50m =>i_clk_50m,o_clk =>o_clk_ms);
    gen_s:clkgen3 Generic Map(25000000)
                                                             --50 000 000 分频,产生 s 脉冲
        Port Map(i_clk_50m =>i_clk_50m,o_clk =>o_clk_s);
End Architecture reuse_clk_arch;
```

示例 6-12b　分频用默认 Work 库调用(VHDL 2008 版)

```
/*-----------------------------------------------------------------------------------------
分频用默认 Work 库调用(VHDL 2008 版)reuse_clk1. vhd
版本 V1. 2
调用 clkgen3,改变 Generic 常量,实现 1MHz(us),1ms,1s 的分频应用
例化采用位置映射方式
采用 Work 库,自定义程度包 clkgen_lib
修改时间:2015. 11. 5
```

```
修改人:xxx
------------------------------------------------------------------------------------------------------*/
--自定义程序包说明 clkgen_lib.vhd
Library Ieee;
Use Ieee.Std_Logic_1164.All;

Package clkgen_lib Is                                   --自定义程序包,clkgen_lib
    Component clkgen3 Is                                --分频元件调用声明
        Generic(DIV_PARAM:Integer:=2);                  --默认是 4 分频
        Port(i_clk_50m:In Std_Logic;
             o_clk:Out Std_Logic);
    End Component clkgen3;
End Package clkgen_lib;                                 --包定义结束

--应用程序 reuse_clk1.vhd
Library Ieee;
Use Ieee.Std_Logic_1164.All;
Use Work.clkgen_lib.All;                                --应用自定义程序包

Entity reuse_clk1 Is
    Port(i_clk_50m: In Std_Logic;                       --50 MHz 系统时钟和复位信号
         o_clk_us:Out Std_Logic;                        --us 分频数据输出端
         o_clk_ms:Out Std_Logic;                        --ms 分频数据输出端
         o_clk_s:Out Std_Logic);                        --s 分频数据输出端
End Entity reuse_clk1;

Architecture reuse_clk1_arch Of reuse_clk1 Is
Begin
    gen_us:clkgen3 Generic Map(25)                      --50 分频,产生 us 脉冲
        Port Map(i_clk_50m,o_clk=>o_clk_us);
    gen_ms:clkgen3 Generic Map(25000)                   --50 000 分频,产生 ms 脉冲
        Port Map(i_clk_50m,o_clk_ms);
    gen_s:clkgen3 Generic Map(25000000)                 --50 000 000 分频,产生 s 脉冲
        Port Map(i_clk_50m,o_clk =>o_clk_s);
End Architecture reuse_clk1_arch;
```

示例 6-12c 分频用自定义库和自定义包(VHDL 2008 版)

```
/*------------------------------------------------------------------------------------------------------
```

```
分频用自定义库和自定义包(VHDL 2008 版)reuse_clk2.vhd
版本 V1.3
调用 clkgen3,改变 Generic 常量,实现 1 MHz(us),1 ms,1 s 的分频应用
例化采用位置映射方式
应用自己的库 mylib,自定义程序包 clkgen_lib
修改时间:2015.11.5
修改人:xxx
------------------------------------------------------------------------*/
Library Ieee,mylib;
Use Ieee.Std_Logic_1164.All;
Use mylib.clkgen_lib.All ;                             --应用自己定义库 mylib

Entity reuse_clk2 Is
    Port(i_clk_50m: In Std_Logic;                      --50 MHz 系统时钟
         o_clk_us:Out Std_Logic;                       --us 分频数据输出端
         o_clk_ms: Out Std_Logic;                      --ms 分频数据输出端
         o_clk_s: Out Std_Logic);                      --s 分频数据输出端
End Entity reuse_clk2;

Architecture reuse_clk2_arch Of reuse_clk2 Is
Begin
    gen_us:clkgen3 Generic Map(25)                     --50 分频,产生 us 脉冲
        Port Map(i_clk_50m, o_clk_us);
    gen_ms:clkgen3 Generic Map(25000)                  --50 000 分频,产生 ms 脉冲
        Port Map(i_clk_50m,o_clk_ms);
    gen_s:clkgen3 Generic Map(25000000)                --50 000 000 分频,产生 s 脉冲
        Port Map(i_clk_50m, o_clk_s);
End Architecture reuse_clk2_arch;
```

分频器是数字系统设计中的基本电路,对于系统中的主要时钟和时序要求高的场合,通常会采用 PLL 来提供。PLL 可以倍频、分频等,但是 PLL 的数量与池围毕竟有限,根据不同设计的需要,会进一步进行偶数分频或奇数分频或半整数分频等,有时要求等占空比,有时要求非等占空比。

通常开发系统都提供一个高频的晶振,比如系统时钟 f_0 为 12 MHz,采用计数进位引出分频,如需产生 f 为 240 kHz 信号,则根据分频系数计算:

$$N = f_0/f = 12\ \text{MHz}/240\ \text{kHz} = 50$$

分频的方法不同,具体的分频系数也有所不同。

6.4.3　等占空比奇数分频与半整数分频

实现奇数分频的办法不是唯一的。本章示例 6 - 13 将给出异步和同步 2 种奇数分频

方式。

(1)示例 6－13a 调用分频元件作为一个模 Mod 计数器/分频器，再与 2 分频和异或门进行异步连接实现奇数分频和半整数分频。

通用奇数/半整数分频器的原理如图 6－25 所示。

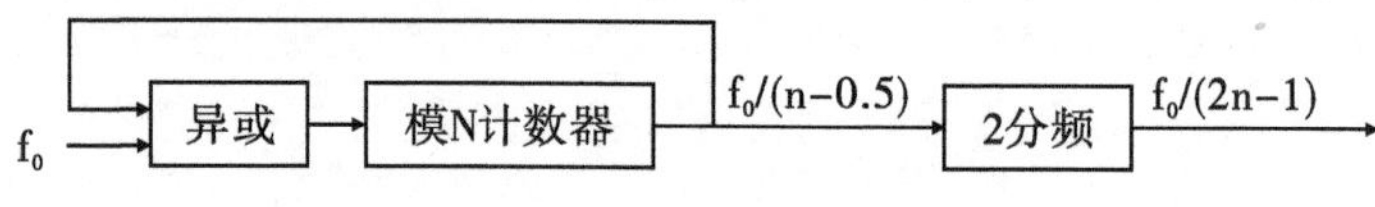

图 6－25　通用奇数/半整数分频器的原理

$$\text{Mod}=\frac{\text{分频数}}{2}+0.5 \tag{6-1}$$

对于 7 分频，根据公式(6－1)，Mod=4。即需要模 4 的计数器/分频器。

(2)示例 6－13b 采用错位异或实现同步奇数分频。奇数分频需要 2 个分频计数值。

第 1 个分频计数值，$n_1=(N-1)$；

第 2 个分频计数值，$n_2=(N-1)/2$；

示例 6－13　等占空比奇数分频

```
Library Ieee;
Use Ieee.Std_Logic_1164.All;
Use Ieee.Std_Logic_Unsigned.All;
Use Work.All;

Entity divideodd Is
    Generic(N: Positive:=7);                 --7 分频
    Port(clkin:In Std_Logic;                 --时钟输入信号
         clkout:Out Std_Logic;               --分频信号
         clkoutd:Out Std_Logic);             --7 分频信号输出
End Entity divideodd;
```

示例 6－13a　异步等占空比奇数分频

```
Architecture asynodd Of divideodd Is         --7 分频结构体 1
Signal count,clkc,clkout0: Std_Logic;
Begin

    Process(count) Is
    Begin
        If (count'Event And count='1') Then
            clkout0 <= Not clkout0 ;         --2 分频
        End If;
```

```
    End Process;

    clkc<= clkout0 Xor clkin;                 --异或
    gen_us: clkgen3 Generic Map(1)         --(N/2—1)分频元件 clkgen3 ,示例 6－11
        Port Map (clkc,count);
    clkout<=count;                            --3.5 分频信号输出
    clkoutd<= clkout0;                        --7 分频信号输出
End Architecture asynodd;
```

示例 6－13a 具体连接如图 6－25 和图 6－26 所示。电路的时钟没有连接到同一个时钟端,因而是异步电路。

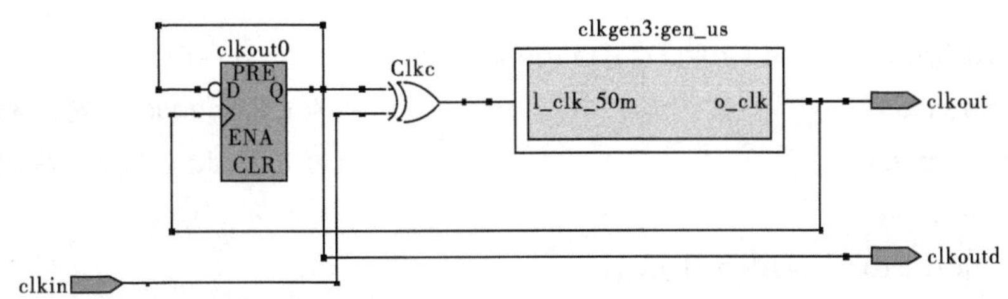

图 6－26　示例 6－13a 的综合结果

如图 6－27 所示,示例 6－13a 实现了奇数 7 分频和半整数 3.5 分频。

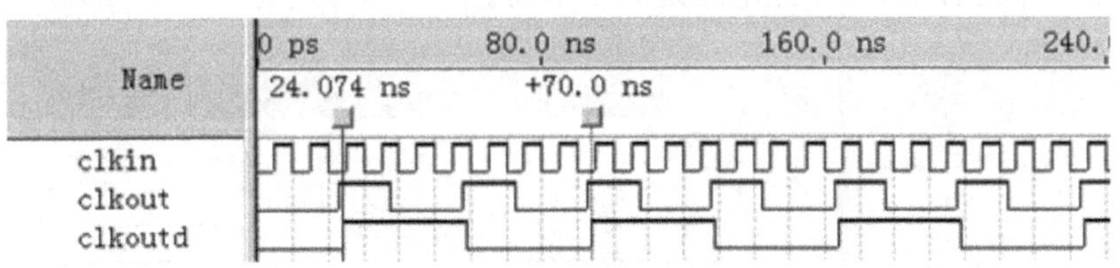

图 6－27　示例 6－13a 7 分频与 3.5 分频电路仿真

示例 6－13b　同步等占空比奇数分频

```
Architecture synodd Of divideodd Is              --7 分频结构体 2
Signal coun: Integer Range 0 To N－1;
Signal temp1,temp2: Std_Logic;
Begin
    Process(clkin) Is
    Begin
        If(clkin'Event And clkin='1') Then
            If(coun= N—1) Then coun<=0;              --coun= 6
                temp1<=Not temp1;
            Else coun<=coun+1;
```

```
            End If;
        End If;
        If(clkin'Event And clkin='0') Then
            If(coun=(N-1)/2) Then                    --3
                temp2<=Not temp2;
            End If;
        End If;
    End Process;
    clkoutd<= temp2;
    clkout<=temp1 Xor temp2;
End Architecture synodd;

Configuration divideodd_cfg Of divideodd Is              --配置语句
    For asynodd                              --为 divideodd 指定结构体 asynodd
    --For synodd                             --为 divideodd 指定结构体 synodd
    End For;
End Configuration divideodd_cfg;
```

如图 6-28 所示的是示例 6-13 第 2 个结构体 7 分频电路仿真波形，可以看到奇数分频是两个程序中 2 个不同分频值错位异或得到的，查看其程序及其综合结果，电路的各时序元件时钟连接到了一个时钟端，可知此电路是同步电路。

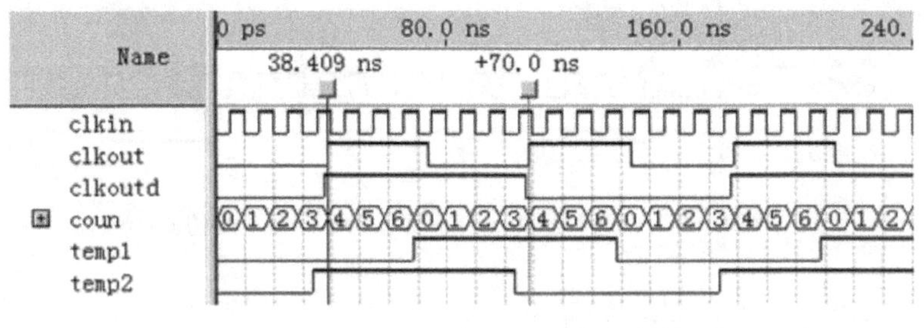

图 6-28　示例 6-13b　7 分频电路仿真

示例 6-14　脉冲扣除电路与倍频电路

通用奇数/半整数分频器的 2 分频和异或门的组合又称脉冲扣除电路。计数/分频模块带入脉冲扣除电路就会被扣除一个脉冲。如把示例 6-13 例化后再带入脉冲扣除电路结构，就会得到 13 分频和 6.5 分频。6.5 分频的结果可以看作 7 分频被扣除 1 个脉冲。

脉冲扣除电路结构的一个极端示例就是倍频，其电路连接如图 6-29 所示，倍频仿真如图 6-30 所示。

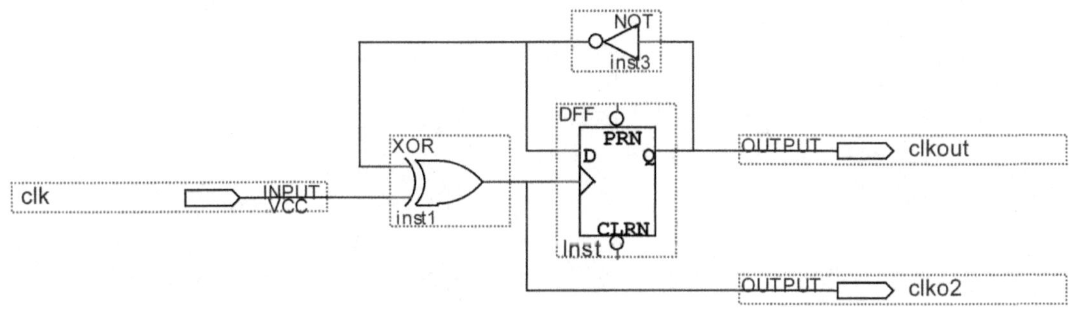

图 6-29　倍频电路

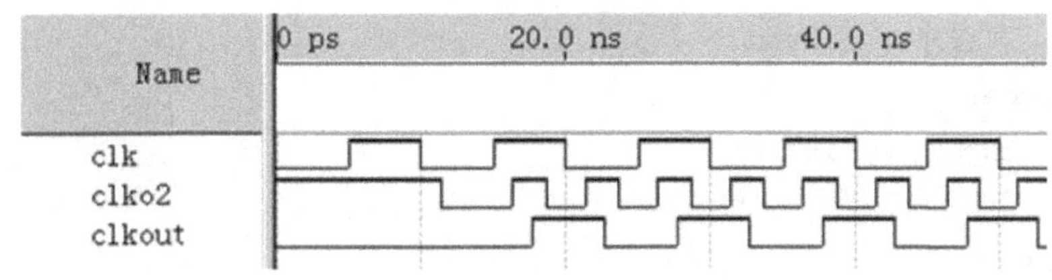

图 6-30　倍频电路仿真

脉冲扣除电路结构在数字锁相环设计以及通信传输需要码速调整等方面有应用。

示例 6-15　微分与同步电路设计

设计要求：

(1)键盘延时简单去抖。

(2)同步单周期脉冲输出。

(3)同步半周期窄脉冲输出。

(4)双边沿检测电路。

微分电路设计又叫同步脉冲电路或者边沿检测电路，可用于按键信号的去抖动和数字通信等。

如图 6-31 所示，out1a，out1b 对应的电路为双边沿检测电路，检测上升沿和下降沿，在数据上沿之后，out1b 输出同步单周期脉冲，在数据下降沿之后，out1a 输出同步单周期脉冲，两路信号再与时钟与，则在数据上升沿和下降沿输出同步半周期窄脉冲分别为 out2b 和 out2a。

在数字通信时，外部输入的同步信号常常需要进行同步化(与系统时钟同步)和整形(将输入信号整形为一个时钟周期长的信号脉冲)，这个输出窄脉冲，既包含了'0'，'1'数据信息，又附带了发送端的同步时钟信息，所以这个电路也常用于数据通信。

在键盘输入 din 以后，out0 给出简单的键盘延时去抖输出。

图 6-31 中第 1 个触发器是为了防止电路可能在不同时钟域中使用产生亚稳态，提高电路的可靠性。

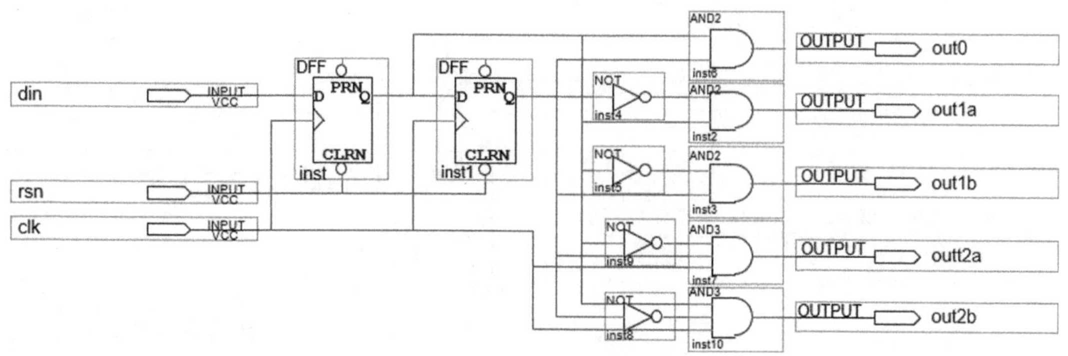

图 6-31 示例 6-15 总设计电路

```
--示例 6-15 程序,diff.vhd
Library Ieee;
Use Ieee.Std_Logic_1164.All;

Entity diff Is
    Port(din,clk:In Std_Logic;
         out0:Out Std_Logic;                    --延时简单去抖
         out1a,out1b:Out Std_Logic;             --同步单周期脉冲输出
         out2a,out2b:Out Std_Logic);            --同步半周期窄脉冲输出
End Entity diff;

--示例 6-15 程序,结构体 RTL 级描述
Architecture diff_a Of diff Is
Signal q0,q1:Std_Logic;
Begin
    Process(clk) Is
    Begin
        If(clk'Event And clk='1')Then
            q0 <= din;                          --第 1 个触发器延时
            q1 <= q0;                           --第 2 个触发器延时
        End If;
    End Process;
    out0<=q1 And q0;
    out1a<= q1 And Not q0;                      --下降沿检测单周期脉冲输出
    out1b<= Not q1 And q0;                      --上升沿检测单周期脉冲输出
    out2a<= clk And q1 And Not q0;              --下降沿检测窄脉冲输出
    out2b<=clk And Not q1 And q0;               --上升沿检测窄脉冲输出
End Architecture diff_a;
```

```
--示例 6-15 程序,同步电路结构体行为级描述
Architecture diff_b Of diff Is
Signal clear: Std_Logic:='1';
Signal s: Std_Logic:='1';
Begin
    Process (din) Is
    Begin
        If clear='1' Then
            s<='0';
        Elsif din'Event And din= '1' Then      --上升沿
            s<='1';
        End If;
    End Process;

    Process (clk) Is
    Begin
        If clk'Event And clk= '0' Then         --下降沿
            If s='1' Then
                out1b<='1';
                clear<='1';
            Else
                out1b<='0';
                clear<='0';
            End If;
        End If;
    End Process;
End Architecture diff_b;
```

程序说明:此程序中,握手协议式的进程间的通信方法非常值得借鉴。程序仿真结果见图6-32。

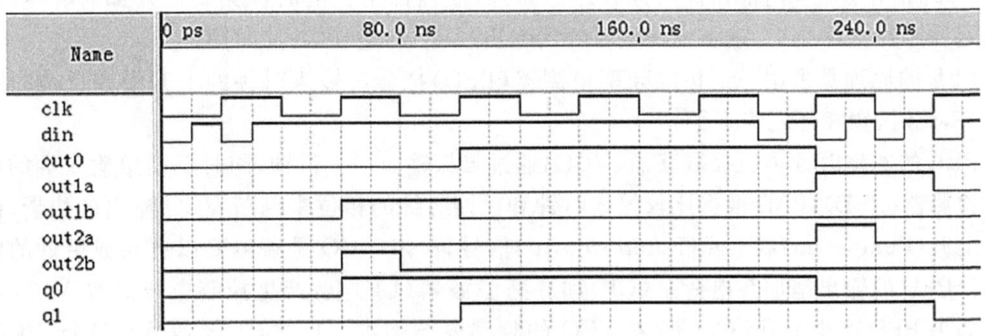

图 6-32 微分与同步电路设计仿真

6.5　系统的定时设计

定时设计与系统成败与性能密切相关，精确稳定的定时是必须密切关注的。

分频电路常用的用途就是作为时钟定时。前面介绍的分频电路有两个缺点：其一是只能给出频率与占空比，而没有对输出信号(时钟)的延时控制，因而不能进行精确的相位控制；其二是频率比较低，低于时钟源频率。后面章节我们给出了 DDS 设计，可以比较精确地控制频率与相位，但是如果不利用 Altera 的锁相环 ALTPLL 倍频，也与前面的分频电路一样只能实现比较低的频率。ATLPLL 具有对源时钟倍频和分频、相位偏移、可编程占空比和外部时钟输出功能，支持 LVDS 高速 I/O，可进行系统级的时钟管理和偏移控制，Cycolne Ⅱ 的系统结的时钟速度性能可达 402.5 MHz，它的问题是频率范围比较窄。

FPGA 高速并行的特点，使之很适合做 DSP。Altera 提供 FIR 滤波器，FFT 快速傅里叶变换、NCO 数控振荡器等 IP 核。NCO 数控振荡器核的性能通常要好于我们自行设计的 DDS，我们只有对器件进行更加极致的综合研究才能接近或达到系统提供的 IP 的性能。

我们希望享有更多的 Open IP 核，这也在于我们自行设计 IP 的能力，前提是我们要用好可达到极致的资源。

ATLPLL 对于芯片而言是珍稀的极致资源，Cycolne Ⅱ 可以具有 68416 逻辑单元(LEs)，622 个可用引脚，1.1 Mbits 嵌入存储器，150 个 18×18 乘法器等，有 4 个 PLL，Cycolne Ⅳ 的逻辑单元(LEs)可以到达 150 K，嵌入存储器达 6.3 Mbits，同时还具有嵌入硬 IP 核，360 个 18×18 DSP 乘法器单元等，最多有 8 个 PLL，其中普通应用目的 PLL 有 4 个。Cycolne Ⅱ 有 16 个高速时钟线，Cycolne Ⅳ 有 30 个高速时钟线。这些时钟线具有高扇出的特性。

通常将 ATLPLL 用于对时序要求非常高的主要时钟，就整个系统设计而言，分频或者是定时将会几种方法混用。

6.5.1　FPGA 锁相环 PLL

Altera FPGA 不同代的 Cyclone 芯片 LEs 基本相同，其锁相环(PLL，Phase Lock Loop)和全局时钟网络，对于不同代的 FPGA 其内部有些区别，尤其是 PLL，PLL 的发展比较快，而发展比较快的往往都是重视程度高的热点问题。如图 6-33 所示，就其最基本的原理不同代 PLL 之间是相通的。Cyclone Ⅲ 之前所用的 PLL 没有可配置功能。

PLL 常用于同步内部器件时钟和外部时钟，使内部工作的时钟频率比外部时钟更高，时钟延迟和时钟偏移最小，减小或调整时钟到输出(T_{CO})和建立(T_{SU})时间。

PLL 的原理是采用一个相位频率检测器(PFD)把参考输入时钟的上升沿和反馈时钟边沿对齐，此时，锁相环就锁定了。

PLL 的本地时钟由压控振荡器(VCO)通过自振输出一个时钟，同时反馈给输入端的频率相位检测器(PFD)，PFD 根据比较输入时钟和反馈时钟的相位来判断 VCO 输出的快慢，同时输出上升(Pump-up)或下降(Pump-down)信号，决定 VCO 是否需要以更高或更低的频率工作。PFD 的输出施加在电荷泵(CP)和环路滤波器(LF)上，产生控制电压设置 VCO 的频率。如果 PFD 产生上升信号，那么 VCO 的频率就会增加。反之，下降信号会降低 VCO 的频率。

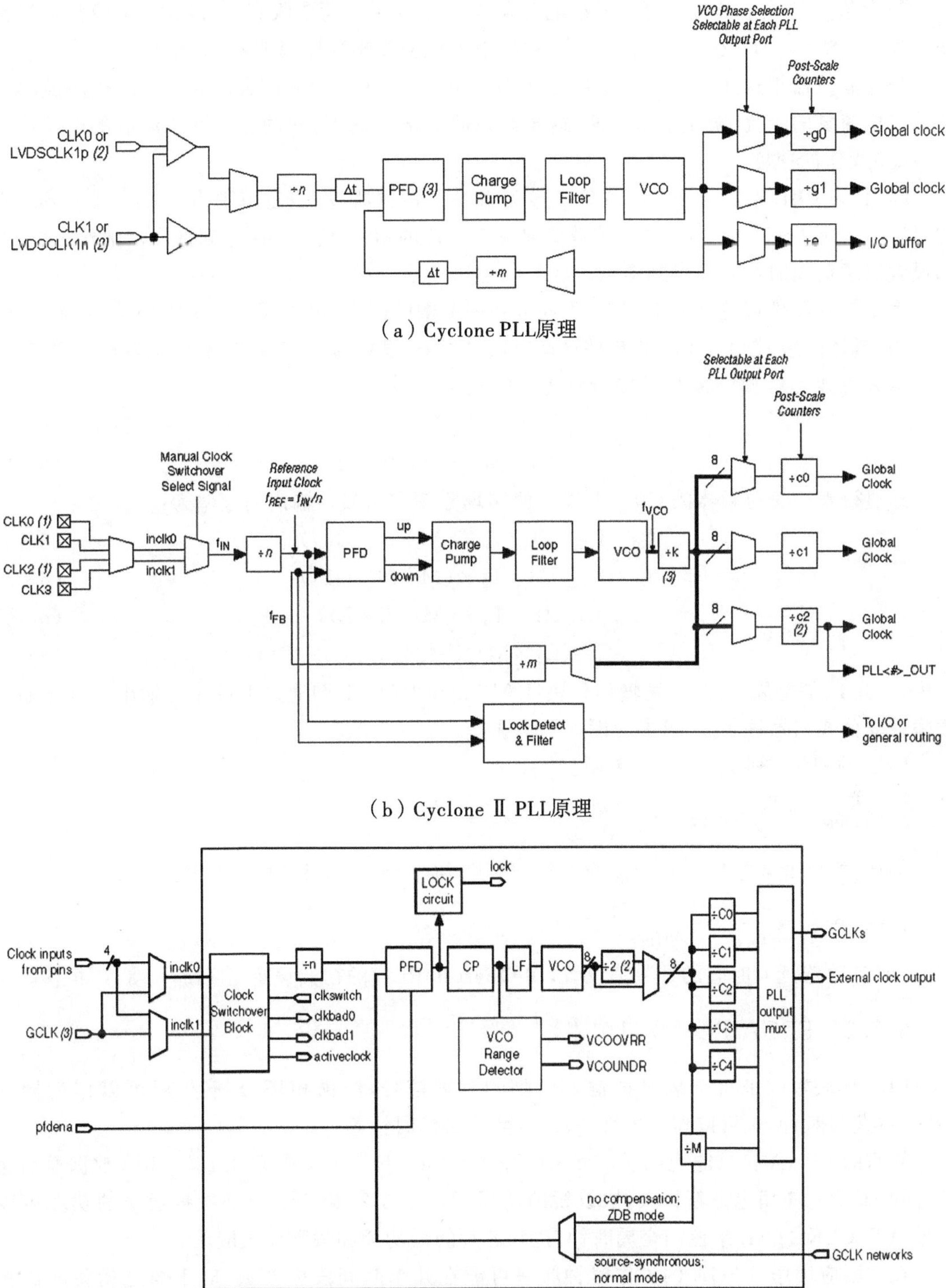

（a）Cyclone PLL原理

（b）Cyclone Ⅱ PLL原理

（c）Cyclone Ⅳ GX PLL原理

图 6-33　Cyclone PLL 原理

PFD 输出这些上升和下降信号给电荷泵(CP)。如果电荷泵收到上升信号,电流注入环路滤波器(I_{CP}增大)。反之,如果收到下降信号,电流就会流出环路滤波器(I_{CP}减小)。

环路滤波器把这些上升和下降信号转换为电压,作为 VCO 的偏置电压。环路滤波器还消除了电荷泵的干扰,防止电压过冲,这样就会最小化 VCO 的抖动。环路滤波器的电压决定了 VCO 操作的速度。

Cyclone PLL 中包含一个后置分频器 N 和一个倍频器 M,设定范围为 1~32, Cyclone Ⅳ 的设定范围是 1~512。具体设置的数值取决于芯片的型号,Quartus Ⅱ 软件自动地给出控制,如设置出了范围时会立即得到提示。

为了超越器件可达到的 1~512 可调节频率范围的限定,可以对图 6-31 中的 c0 到 c4 进行级联,具体的级联 Quartus Ⅱ 软件自动地做。如 c0 设置为 2,c1 设置为 4,级联的值是 8。

输入时钟 f_{in}经预分频 N 后得到参考时钟 f_{ref}:

$$f_{ref}=f_{in}/N \tag{6-2}$$

$$f_{vco}=f_{ref}\times M=f_{in}\times(M/N) \tag{6-3}$$

通过设置后置分频器的 G0、G1 和 E 值实现分频和倍频。输出的频率为:

$$f_{c0}=f_{VCO}/G0=f_{in}\times(M/(N\times G0)) \tag{6-4}$$

$$f_{c1}=f_{VCO}/G1=f_{in}\times(M/(N\times G1)) \tag{6-5}$$

$$f_{E}=f_{VCO}/E=f_{in}\times(M/(N\times E)) \tag{6-6}$$

$$f_{out}=f_{VCO}/C=(f_{ref}\times M)/C=(f_{in}\times M)/(N\times C) \tag{6-7}$$

式中,f_{c0}和 f_{c1}是全局时钟,为逻辑阵列块(LAB)提供时钟;f_E则通过 I/O 单元输出。具体输出频率计算公式与器件有关,可参考图 6-31 选择。

PLL 的相位调控,分细调与粗调。

细调:$\Phi_{fine}=\dfrac{T_{VCO}}{8}=\dfrac{1}{8f_{VCO}}=\dfrac{N}{8Mf_{REF}}$

例如,如果 f_{REF}是 100 MHz,N=1,M=8,则 f_{VCO}=800 MHz,Φ_{fine}=156.25 ps

粗调:$\Phi_{coarse}=\dfrac{c-1}{f_{VCO}}=\dfrac{(c-1)N}{Mf_{REF}}$

c 是计数器延时时间的计数值设置,在 Quartus PLL 的初始设置是 1,c−1 是 0°相移。

6.5.2　FPGA PLL 应用需求

PLL 模块接收来自全局时钟输入引脚的时钟信号,经锁相环分/倍频后可以作为异步 FIFO 的读写时钟,也可以作为外部 A/D 转换器采样时钟等。

新型的 FPGA 接口能力和通信能力很强,Cyclone Ⅳ GX 可以有 3.125 Gbps 数据通信速率,同时低功耗应用也是新特点,协议桥的功耗可在 1.5 W 以下。普通数据通道的功耗可以通过 ATLCLKCTRL 来控制全局时钟,达到芯片的低功耗和睡眠模式应用。

在实际应用中,FPGA 的工作时钟频率可能在几个时间段内变动,对于与之相关的锁相环,同步时,若 PLL 的输入时钟在初始设定的时钟频率的基础上变化不太大时,PLL 一般可以自己调整过来,并重新锁定时钟,获得正确的时钟输出;但是,若 PLL 的输入时钟频率较之原来设定的时钟频率变化较大时(比如,PLL 输入时钟频率由 50 MHz 变为 200 MHz),PLL 可能无法重新锁定时钟,其输出时钟频率将变为不确定的值。

对于后面这种情况，一般有两种处理方法：

方法一：针对不同的输入时钟使用不同的 PLL 分别进行配置，当输入时钟变化时，内部逻辑根据不同 PLL 的锁定情况，选择合适的时钟作为工作时钟。

方法二：利用 FPGA 开发厂商提供的 PLL 可重新配置宏（比如 Altera 的 ALTPLL_RECONFIG 宏模块），通过对其参数进行重新设定，然后，实时地重新配置 PLL，使其在新的输入时钟下可以正常锁定和工作。

对比两种方法，方法一的实现较为直观，但需要更多的 PLL 资源；方法二则通过对原来的 PLL 资源进行参数的重新配置，使其适应新的工作时钟，其实现较为复杂，但不需要额外的 PLL 资源。

FPGA 内的 PLL 能否实时地实现重新配置，与该 FPGA 是否提供相关的可重新配置机制有关，具体请参考相应厂商的 FPGA 的使用手册。

在 Quartus Ⅱ 软件中，利用 Altera 公司所提供的 MegeWizard Plug - In Manager 产生 PLL，并对参数进行配置。

在 Altera 公司提供的 FPGA Cyclone Ⅱ 中，锁相环分为增强型锁相环（EPLL）和快速锁相环（FPLL）两种。

EPLL 可为整个设计提供丰富的时钟资源，它有 6 个内部输出时钟和 4 个（或者 4 对差分信号）专用的片外输出时钟。

FPLL 同样可以提供内部使用的时钟，它的另一个主要功能是作为高速差分信号的随路时钟输入，同样输出高速采样时钟和控制信号给内部的源同步接口 SERDES 电路。

6.5.3　可重配置锁相环的使用

可重配置锁相环 PLL_RECONFIG 是 LPM 模块，在 Quartus Ⅱ 的 Tools 选项中“MegaWizard Plug - In Manager”→IO→ALTPLL_RECONFIG，生成可重配置的 PLL 时，会自动生成一个后缀为 .mif 文件（或 .hex 文件），此文件是一个扫描链的位图文件，包含了当前 PLL 配置的扫描链信息。可以根据这些信息，调整 PLL 的扫描链中相应的计数器，来达到调整 PLL 的目的。

图 6 - 34 给出了 EPLL 重配置的时序波形。由图可看出，PLL 重配置的过程主要是将扫描链信息从最高位（bit173）到最低位（bit0）依次连续地读入可重配置 PLL。

6.5.4　PLL 的重配置模块

从图 6 - 34 可以看出，PLL 的配置过程是按位按序连续进行的，因此，直接操作起来不方便。为此，Altera 提供了 PLL 的重配置模块，让配置过程变得简单。它让用户可以按照自己的需要，对需要修改的计数器单独进行修改，并且修改顺序是随机的，不用遵照扫描链的固有排列顺序进行。在所有修改完成后，仅需启动一个时钟周期的重配置信号（图 6 - 35 中的 reconfig），此 PLL 的重配置模块就会按照图 6 - 34 所示的时序自动地对 PLL 完成重配置。

PLL 的可重配置模块可以通过 Altera 的 Quartus Ⅱ 开发工具的 Megawizard Plug - In Manager 工具生成，与 PLL 的生成过程类似。

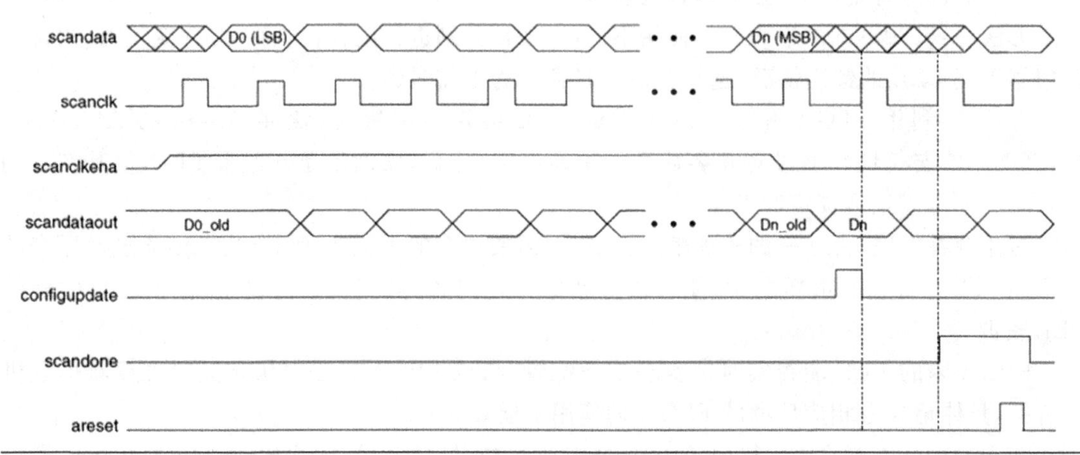

图 6-34 PLL 的重配置时序波形

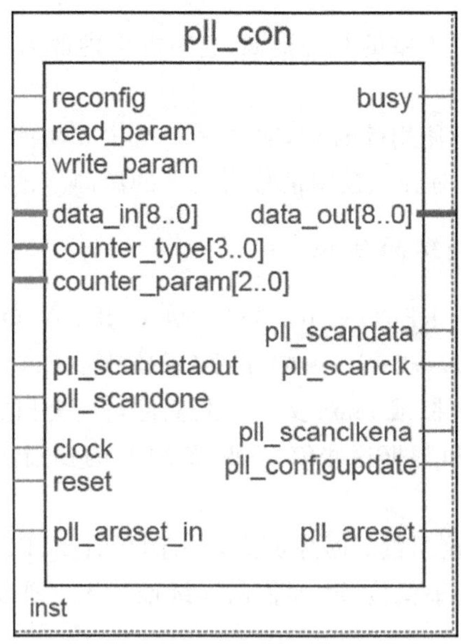

图 6-35 Cycone Ⅳ可重配置 PLL

6.5.5 PLL 重配置模块的端口说明

我们可以通过对 altpll_reconfig 模块的输入端口进行适当的驱动，来完成 PLL 的重新配置。表 6-1，表 6-2 介绍了该模块的输入/输出端口。

表 6-1　altpll_reconfig 模块的输入端口

端口名	是否必需	描　述
clock	是	用于加载单独的参数，在 PLL 重配置时，用于驱动 PLL。该端口必须被连接到一个有效时钟，且最高频率为 100 MHz
data_in[8:0]	否	9 位总线，在写参数时，其上的数据作为输入。有的参数并没有用到全部的 9 位，此时，只有从 bit0 开始的若干有效数据被使用。当该端口未连接时，其默认值为 0
counter_type[3:0]	否	4 位总线，用于选择哪一类的计数器将被更新
counter_param[2:0]	否	3 位总线，对 counter_type 给定的计数器，用于选择具体哪个参数将被更新
read_param	否	此参数有效时，由 counter_type[3:0]和 counter_param[2:0]所指定位置的参数，将从扫描链(Scanchain)中读出并放置到 data_out[]总线上。此信号在上升沿被采样。需要注意的是，此信号应该只保持一个时钟周期，以避免参数在相邻时钟周期被重复读取 busy 信号将随着 read_param 的有效而变为有效，只有当 busy 信号变为无效时，data_out[]总线上的值才是有效的，而且，只有当 busy 信号变为无效时，才可以加载下一个参数
write_param	否	该参数有效时，data_in[]总线上的数据将被写入由 counter_type[3:0]和 counter_param[2:0] 所指定的在扫描链中的位置。与 read_param 相似，该信号在上升沿被采样，且只保持一个时钟周期 busy 信号将随着 write _param 的有效而变为有效，当 busy 信号有效时，data_in[]总线上的数据将被忽略。只有当 busy 信号变为无效时，才可以开始写入下一个参数
reconfig	是	此信号表明 PLL 将开始按照扫描链中的设置进行重新配置。设备在时钟 clock 的上升沿采样此信号。同样地，该信号应该只保持一个时钟周期，以避免在重配置完成后 PLL 被重复加载 reconfig 被检测到有效时，busy 信号会紧跟着变为有效。pll_scanwrite 信号也会随着变为有效，以开始从扫描链中加载新的设置到 PLL 中。在 PLL 的重配置期间，busy 信号保持有效。一旦 busy 信号无效，参数就可能被重新修改 Altera 推荐将计数器和相位偏移设置好后，再启动 reconfig 信号，这样仅将 scandata 写入扫描链一次
reset	是	异步复位输入信号，用于初始化状态机使之处于合法状态。在第一次使用前，状态机必须被复位，否则将无法保证状态的合法性
pll_scandataout	否	此输入信号是由 altpll 示例 的 scandataout 端口所驱动的。scandataout 信号是来自扫描链移位寄存器的直接输出，可以使用此信号来读出扫描链中已有的内容
pll_scandone	否	此输入信号是由 altpll 示例 的 scandone 端口所驱动的。此信号变高时，表示重配置已经完成

表 6 - 2 altpll_reconfig 模块的输出端口

端口名	是否必需	描 述
data_out[8:0]	否	9 位总线,用于用户读回参数数据。在将 read_param 值设为高有效,并指定 counter_type[]和 counter_param[]值时,参数将从扫描链中读出到此总线上,然后,当 busy 信号变为无效时,此总线上的数据为合法值。
busy	否	此信号有效表明状态机处于忙状态,此时,状态机可能在从扫描链中读一个参数,或向扫描链中写入一个参数,或是在重配置 PLL。此信号有效时,状态机将忽略它的输入,并且直到此信号变为无效,状态机才能改变。
pll_scanclk	是	此信号用于驱动要重配置的 PLL 上的 scanclk 端口
pll_scanread	是	此信号用于驱动要重配置的 PLL 上的 scanread 端口
pll_scanwrite	是	此信号用于驱动要重配置的 PLL 上的 scanwrite 端口
pll_scandata	是	此信号用于驱动要重配置的 PLL 上的 scandata 端口

习 题

6 - 1 将程序 6 - 1b 改为 Case 语句来实现,编译综合后,与程序 6 - 1b 对比最高频率、占用资源等情况。

6 - 2 设计一个数字钟,编译综合后,查看相应的报告,了解占用资源等情况,下载测试。

6 - 3 设计一个数字秒表,编译综合后,查看相应的报告,了解占用资源等情况。

6 - 4 设计一个频率计,采用 6 位动态扫描 LED 显示,编译综合后,查看相应的报告,了解占用资源等情况。简易频率计的设计提示:主要模块是,分频 1,测频闸门用 1s;分频 2,LED 显示、键盘用;模 10 的同步级联;LED 译码;COM 动态扫描。

```
clk_div1:Process(clk)                              --1s
Begin
    If  Rising_edge(clk) Then countdiv<=countdiv+1; End If;
End Process clk_div1;
clk_div2:Process(clk) Is
Begin                                              --1kHz
End Process clk_div2;                              --分频的闸门控制
ctrl:Process(countdiv) Is
Begin
    If  countdiv=N Then clr<='1'; Else clr<='0'; End If;
    countdiv=N-1 Then en<='1'; Else en<='0';  End If;
End Process ctrl;
```

6－5　设计一个抢答器,编译综合后,查看相应的报告,了解占用资源等情况。

6－6　用 Modelsim 仿真,尝试更高效率的程序调试与仿真实现。

6－7　采用基于平台的方法设计,说明如何在测试过程中进行设计与测试资源的共享。

扩展学习与总结

1. VHDL 中所使用的名字或名称应该遵循哪些规则?

2. VHDL 有哪些对象? 他们有何特点? 是如何定义的?

3. VHDL 有哪些数据类型? 范围是多少? 如何定义?

4. 查阅资料学习流水线在系统级的应用,对自己设计的 Pipeline 给出一个应用,计算吞吐率、加速比、效率、流水线的最佳段数。

7 存储器的设计与应用

本章的主要内容与方法：

(1)LPM 模块的应用和 ROM 的应用。包括查表法乘法器，数字合成正弦波，压缩查表。

(2)多种波形发生器与 Modelsim 仿真。包括基于 DDS 的 Ask，Fsk，Psk 设计与 Modelsim 仿真。

(3)Sram 的设计。

(4)3 种 FPGA 引脚的 3 种设定方式。

(5)3 种在线测试工具嵌入式逻辑分析仪等。

(6)针对综合的属性 Attribute 应用举例。

7.1 应用 ROM 设计实现乘法器

设计要求：

采用 2 种 ROM 的设计方法实现二进制 4 位×4 位，即十进制 0×0 到 15×15 的乘法器。用一个实体两个结构体的方式给出 VHDL 程序代码，仿真 2 种乘法器，并进行设计讨论。2 种存储器采用查表法实现，具体方式：

(1)采用厂家 IP 核 LPM_ROM 宏单元设计实现；

(2)自定义 2 维数组来实现 ROM。

设计提示：

查表法乘法的基本设计思路就是将乘积直接存放在存储器 ROM 中，将操作数作为地址访问存储器，得到的输出数据就是乘法运算的结果。

在乘法表设计中，被乘数为高地址位，乘数为低地址位，对应内容为乘积。

示例 7－1　乘法器的存储器实现

```
--乘法器顶层文件 mulm.vhd
--通过 QuartusⅡ宏单元向导 Magawizard Plug-In Manager
--调用 LPM_ROM 模块，名为 mulrom.vhd
Library Ieee;
--Library altera;
Use Ieee.Std_Logic_1164.All;
Use Ieee.Std_Logic_Unsigned.All;
--Use Altera.Altera_Syn_Attributes.All;
Use Work All;
```

```
--结构体 rtl1 采用专门的 ROM 存储乘积
Entity mulm Is
  Port(a,b:In Std_Logic_Vector(3 Downto 0);              --被乘数、乘数
       clk: In Std Logic;                                --结构体 rtl1 的时钟信号
       q:Out Integer  Range 0 To 255);                   --乘积
End Entity mulm;
--结构体 rtl1 采用 LPM_ROM 来存储数据
Architecture rtl1 Of mulm Is
Begin
  U1:mulrom Port Map(address=>a&b, clock=>clk,q=>q);
                                    -- mulrom 的地址、时钟、输出为 address,clock,q
                                    -- 被乘数为高地址位,乘数为低地址位
End Architecture rtl1;
--结构体 rtl2 采用 2 维数组来存储数据--
Architecture rtl2 Of mulm Is
Type mrom Is Array (0 To 255) Of Integer Range 225 Downto 0;
--Signal rom : mrom;
--Attribute Romstyle Of rom : Signal Is "M4K";
Constant pro_rom:mrom:=
(0,0,0,0,0,0,0,0,0,0,0,0,0,0,0,0,
0,1,2,3,4,5,6,7,8,9,10,11,12,13,14,15,
0,2,4,6,8,10,12,14,16,18,20,22,24,26,28,30,
0,3,6,9,12,15,18,21,24,27,30,33,36,39,42,45,
0,4,8,12,16,20,24,28,32,36,40,44,48,52,56,60,
0,5,10,15,20,25,30,35,40,45,50,55,60,65,70,75,
0,6,12,18,24,30,36,42,48,54,60,66,72,78,84,90,
0,7,14,21,28,35,42,49,56,63,70,77,84,91,98,105,
0,8,16,24,32,40,48,56,64,72,80,88,96,104,112,120,
0,9,18,27,36,45,54,63,72,81,90,99,108,117,126,135,
0,10,20,30,40,50,60,70,80,90,100,110,120,130,140,150,
0,11,22,33,44,55,66,77,88,99,110,121,132,143,154,165,
0,12,24,36,48,60,72,84,96,108,120,132,144,156,168,180,
0,13,26,39,52,65,78,91,104,117,130,143,156,169,182,195,
0,14,28,42,56,70,84,98,112,126,140,154,168,182,196,210,
0,15,30,45,60,75,90,105,120,135,150,165,180,195,210,225);
Begin
   q<=pro_rom(Conv_Integer(a&b ));
End Architecture rtl2;
```

结构体 1ROM 中的数据,15 * 15 乘法的 Mif 文件如下:

```
DEPTH = 256;           % Memory depth And width are required %
Width = 8;             % Enter a decimal number %
ADDRESS_RADIX =DEC;    % address And value radixes are optional %
DATA_RADIX = DEC;      % Enter Bin,DEC,HEX,or  OCT; unless %
                       % Otherwise specified, radixes = HEX %
% Specify values for addresses, which can be single address or range %
CONTENT
  Begin
0: 0 0 0 0 0 0 0 0 0 0 0 0 0 0 0 0;
16: 0 1 2 3 4 5 6 7 8 9 10 11 12 13 14 15;
32: 0 2 4 6 8 10 12 14 16 18 20 22 24 26 28 30;
48: 0 3 6 9 12 15 18 21 24 27 30 33 36 39 42 45;
64: 0 4 8 12 16 20 24 28 32 36 40 44 48 52 56 60;
80: 0 5 10 15 20 25 30 35 40 45 50 55 60 65 70 75;
96: 0 6 12 18 24 30 36 42 48 54 60 66 72 78 84 90;
112: 0 7 14 21 28 35 42 49 56 63 70 77 84 91 98 105;
128: 0 8 16 24 32 40 48 56 64 72 80 88 96 104 112 120;
144: 0 9 18 27 36 45 54 63 72 81 90 99 108 117 126 135;
160: 0 10 20 30 40 50 60 70 80 90 100 110 120 130 140 150;
176: 0 11 22 33 44 55 66 77 88 99 110 121 132 143 154 165;
192: 0 12 24 36 48 60 72 84 96 108 120 132 144 156 168 180;
208: 0 13 26 39 52 65 78 91 104 117 130 143 156 169 182 195;
224: 0 14 28 42 56 70 84 98 112 126 140 154 168 182 196 210;
240: 0 15 30 45 60 75 90 105 120 135 150 165 180 195 210 225;
End;
```

在实际设计中，如果对性能要求高，我们推荐采用厂家提供的专用乘法器，这些专用乘法器速度比较快，面积比较小。比如厂家提供 18×18 的乘法器，如果我们需要设计两个多于 18 位的乘法器，可以把多个内嵌的乘法器连在一起用。设 A，B 为 32 位，C，D，E，F 为 16 位，并有

$$A=C\times 2^{16}+D$$

$$B=E\times 2^{16}+F$$

则

$$A\times B= C\times E\times 2^{32}+(D\times E+C\times F)\times 2^{16}+D\times F$$

这样 32 位乘法器可以由 4 个内嵌的乘法器与加法器来实现。

7.2 LPM_ROM 初始化文件 MIF 格式

MIF(Memory Initialization File)，用来配置 ROM 的数据，格式见表 7-1 所示。

表 7-1　MIF 格式

第一部分	文件内容	说明
第二部分	Width＝6； DEPTH＝64；	数据线宽度以及存储单元数目
第三部分	Address_Radix＝HEX； DATA_ Radix ＝HEX；	地址和存储数据采用的数制，如： "HEX"，16 进制，"Bin"，2 进制； "OCT"，8 进制，"DEC"，10 进制
第四部分	CONTENT Begin	存储内容的起点标志
第五部分	0：00 1：00 …… 3f：00	存储内容，格式为："地址：数据"
第六部分	End；	结束标志

7.3　ROM 应用与波形发生器

设计要求：

(1)正弦波形发生器用直接数字合成(DDS)的方法实现，即生成正弦数据存入 ROM，由地址计数器查表法实现正弦波形；通过计算实现上下锯齿波、三角波、正弦波、方波。ROM 采用两种方法实现：一种是片上 LPM_ROM，另一种是 VHDL Case 语句查表生成分布式 ROM。

(2)采用 Modelsim 连接 Quartus Ⅱ器件库进行仿真。

(3)学习多种方式定义引脚，下载，在示波器上观察 DA(AD7302)输出波形。

(4)采用 Quartus Ⅱ在线工具嵌入逻辑分析仪 Signal Tap Ⅱ观察各种波形：

①采用在线工具存储器编辑器(In System Memery Content Editor)修改 ROM，导入，导出，利用嵌入逻辑分析仪观察波形；

②采用在线工具源和探针(In-System Sources and Probes)观察波形。

(5)利用正弦波形的对称性，用 1/4 正弦数据表查表法实现正弦波形，并用 Modelsim 连接 Quartus Ⅱ器件库进行时序仿真。

7.3.1　设计信号波形的选取

常见波有方波、锯齿波、三角波、正弦波。传输采用正弦波；三角波和正弦波经过比较可以获得 SPWM 波；锯齿波可以作为示波器的激励波。

设计和评价一个通信系统，最重要的是它的有效性和可靠性(抗干扰性)。在无线传输过程中信噪比除与信号功率噪声功率的大小有关，还取决于调制方式，改变调制方式可以改善系统性能。因此在选择传输波形上面就显得极其重要。

比较三种波的传输方式。由傅里叶变换可知，任何连续周期信号可以由一组适当的正弦曲线组合而成。三种波形进行傅氏变换的理论推导如下。

1. 方波的频谱

周期为 T、高为 1 的连续方波，时域表达式为

$$f(t) = \sum_{k=-\infty}^{\infty} P_{\frac{T}{2}}(t-kT) \tag{7-1}$$

对应的傅氏级数为

$$F_n = \frac{1}{2}S_a(\frac{n\pi}{2}) = \frac{\sin(\frac{n\pi}{2})}{n\pi} \tag{7-2}$$

其中 S_a 是通信领域常用的抽样函数，$S_a(x) = \sin x/x$。

当 $n = 0$ 时，$F_0 = 1/2$；$n = 1$ 时，$F_1 = 1/\pi$；$n = 2$ 时，$F_2 = 0$；$n = 3$ 时，$F_3 = -1/3\pi$；$n = 4$ 时，$F_4 = 0$；$n = 5$ 时，$F_5 = 1/5\pi$……。由傅氏级数可知，周期方波仅有奇次谐波分量为 1，偶数谐波均为 0。基波幅度最大，其余奇次谐波的幅度随谐波次数的增加而递减。

2. 三角波的频谱

周期为 T 的连续三角波，时域表达式为

$$f(t) = \frac{4t}{T}P_{\frac{T}{2}}(t)\delta_T(t) \tag{7-3}$$

进行傅氏变换，可得对应的傅氏级数为

$$F = \frac{-8j}{(n\pi)^2}\sin(\frac{n\pi}{2}) \tag{7-4}$$

由傅氏级数可见，周期三角波的频谱与方波类似，仅有奇次谐波分量，谐波幅度也随谐波次数的增加而递减。但周期三角波的谐波幅度收敛更快，因为幅度值与谐波次数的平方成反比。

3. 正弦波的频谱

周期为 T 的连续正弦波，时域表达式为

$$f(t) = \sin\omega_0 t \tag{7-5}$$

傅氏变换式为

$$F_n = j\pi[\delta(\omega+\omega_0) - (\omega-\omega_0)] \tag{7-6}$$

其中 $\omega_0 = \frac{2\pi}{T}$。由傅氏变换式可见，正弦波只有单一谱线。

经研究论证及事实证明，谐波会使设备老化得更快，严重时可导致设备损坏，并且其对邻近通信系统会产生干扰，轻者产生噪声，降低通信质量；重者导致信号丢失，使通信系统无法正常工作。因此信号产生的谐波越多，其通信质量越不好，所以我们要关心采用 0 谐波分量的周期正弦波作为传输信号的波形产生。

示例 7－2　多种波形发生器顶层设计

分别将 Switch 设置为 001，010，000，011，101 可以得到下锯齿波、三角波、正弦波、方波、上锯齿波。多种波形发生原理如图 7－1 所示。

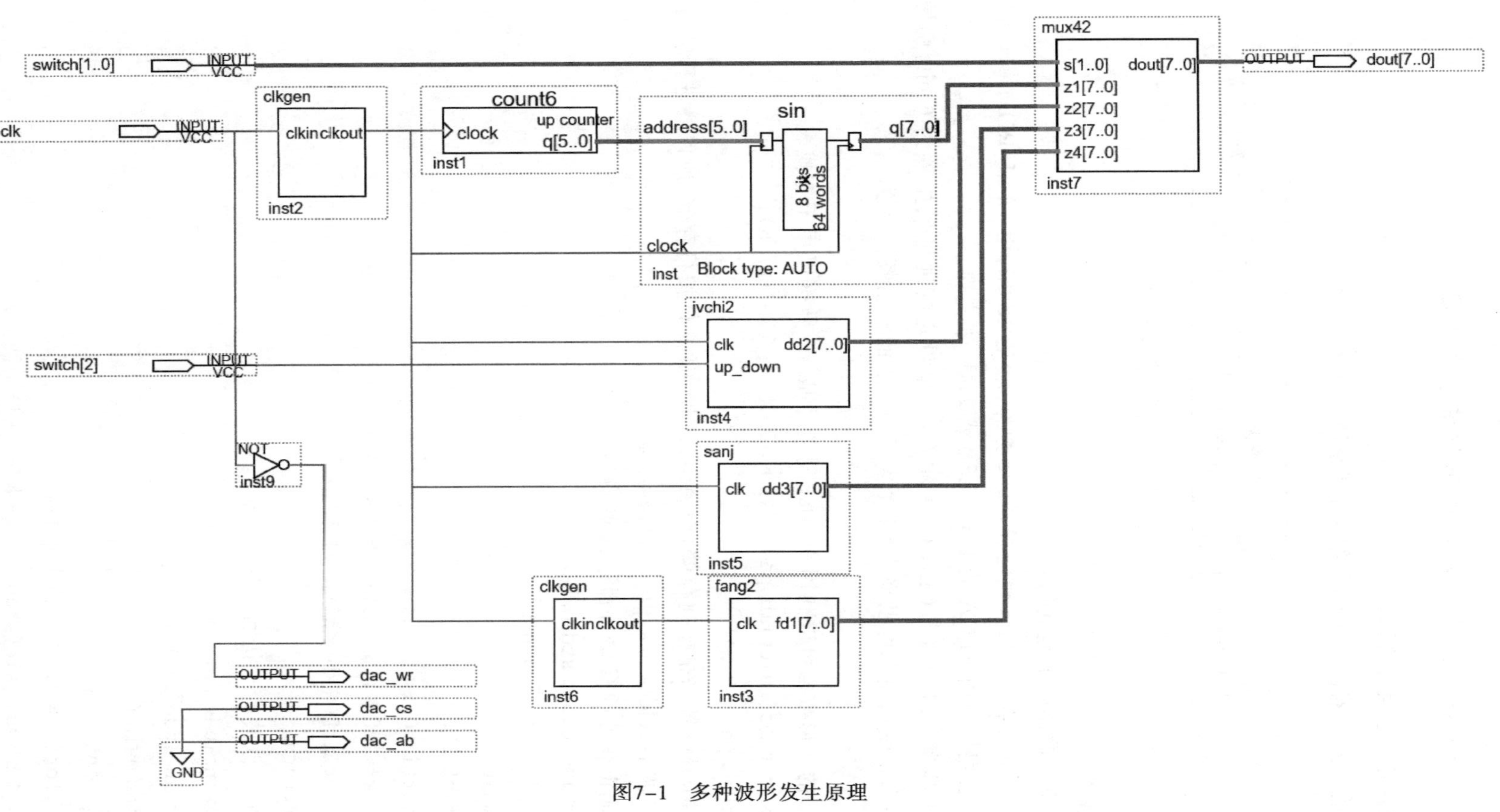

图7-1 多种波形发生原理

7.3.2　LPM 片上 ROM 实现正弦信号发生器

1. 正弦信号发生器原理

正弦信号发生器由 3 个部分组成，其原理如图 7－2 所示。

图 7－2　正弦信号发生器原理

顶层文件 singt.vhd 的 3 个部分：

(1)6 位计数器产生地址信号。

(2)存储正弦信号(6bits 地址线，8bits 数据线)的 ROM，由 LPM_ROM 模块实现，LPM_ROM 模块底层可以由 FPGA 的 EAB、ESB 或 M4K 来实现。

地址发生器的时钟频率 CLK 假设为 f_0，这里设定的地址发生器为 6 bit，则周期为 $2^6=64$，所以一个正弦周期内可以采样 64 个点，DAC 后的输出频率 f 为

$$f=f_0/64$$

sin(x)的波形数据可以利用 Matlab 计算生成 sin 数据以用于查找表：

```
x=round((sin(linspace(0,2*pi,64))+1)*127.5);
reshape(x,8,8)'
```

因为 Matlab 数据是按列存储，而 Altera 的 Mif 文件按行存储，所以要进行转置。

sin(x)波形数据也可以由 C 语言程序来生成。

①生成正弦波的 C 语言程序(256 点)：

```
#include <stdio.h>
#include "math.h"
main()
{int i,k;
for(i=0;i<256;i++){
  k=128+128*sin(360.0*i/256.0*3.1415926/180);
  printf(%d:%d;\n",i,k);}
return;}
```

②生成正弦波的 C 语言程序(64 点)：

```
#include<stdio.h>
#include"math.h"
int main ()
{  int i,s;
  for( i=0;i<64;i++);
  { s=abs(sin(atan(1.0)*8/64*i)*255);
  printf("%d,",s); }
```

```
    return 0; }
```

(3)利用 AD7302 的接口输出波形。

信号发生器所输出的信号经并行 DAC 器件输出到示波器上。并行 8 位 DAC 器件采用 Analog Devices 公司的 AD7302(或 AD7801)芯片,工作电压为 2.7～5.5 V,功耗小,适合于电池驱动应用,还具有高速寄存器和双缓冲接口逻辑以及可与并行微处理器和 DSP 兼容的接口。

2. DA 芯片的原理与 FPGA 器件的设置

AD7302 功能框图如图 7-3 所示。

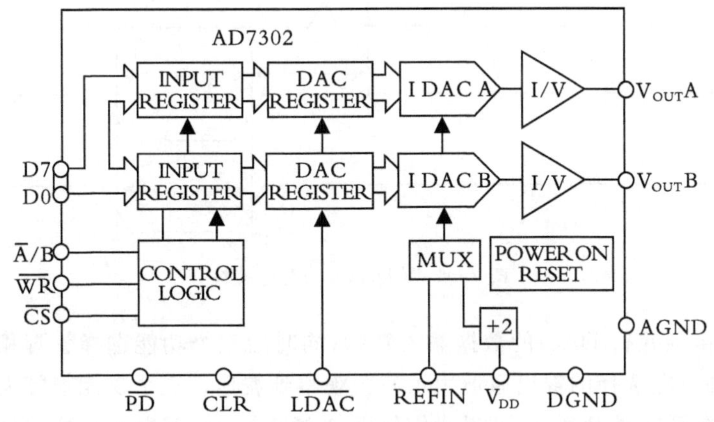

图 7-3 AD7302 功能框图

AD7302 引脚与功能如表 7-2 所示。

表 7-2 AD7302 引脚与功能

序号	名称	引脚功能说明
1-8	D7-D0	并行数据输入,由/CS/WR 控制 8 位数据加载到输入寄存器
9	/CS	片选,低有效
10	/WR	写输入,低有效。与/CS 和 $\overline{A}$/B 联合写所选的 DAC 寄存器
11	$\overline{A}$/B	DAC 选择,用于选择写 DAC A 还是 DAC B
12	/PD	低有效。减少电流消耗的低电源模式
14	/CLR	异步清零,低有效。这个输入为低时,DAC 寄存器加载为 0,DAC 输出清为 0 V
15	VDD	电源输入,可工作在 2.7～5.5 V
16	REFIN	外部参考输入,可用于两个 DAC 的参考。参考输入的范围是 1 V 到 $V_{DD}/2$。如果 REFIN 直接联到 V_{DD},内部则选择为 $V_{DD}/2$
18	$V_{OUT}B$	DAC B 的模拟输出
19	$V_{OUT}A$	DAC A 的模拟输出
20	DGND	数字地

AD7302 的时序特性如图 7-4 所示。

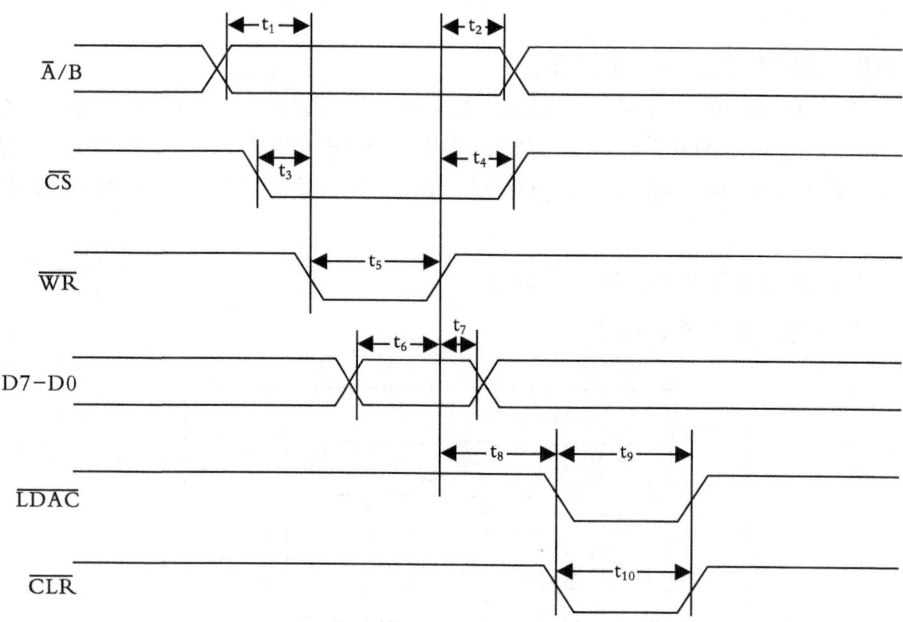

图 7-4　AD7302 的时序特性

8 位数据连接到并行 DAC 的数据输入管脚，同时给三个功能选择管脚相应的输入信号，DAC 正常工作即可在选择的输出通道用示波器观察到波形。三个功能管脚为 A_n/B，低电平选择 A 通道，高电平选择 B 通道(这里我们使用 A 通道)；wr，低电平脉冲有效，脉冲频率需要比数据读取频率快一倍以上，可以用 clk 信号取反以后产生，即产生一或两个写信号，采集 8 位数据；cs，片选信号，低电平有效。AD7302 典型的转换时间为 1.2 us。

在原来设计的 singt1.bdf 中添加三个输出引脚和一个非门，用来实现上述对 D/A 控制，其中 dac_ab 为 D/A 模拟输出通道选择信号输出管脚；dac_cs 为 D/A 片选信号输出管脚；dac_wr为 D/A 数据写入使能信号输出管脚，如图 7-5 所示。

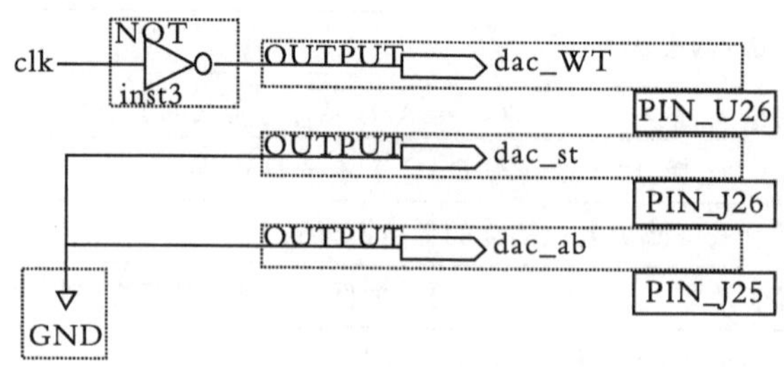

图 7-5　AD7302 驱动连接

点击分析编译通过后，然后再点击，综合编译。

选菜单 Assignments→Settings，进入 Settings 对话框，如图 7-6 所示。

Settings 对话框是 Quartus Ⅱ 软件比较重要的一个设置工具，几乎和项目所有相关的设置工作都可以在这里找到，并做出相应的修改。主要包括工程的配置，分析综合工具的配置，编译过程的设置，目标器件相关的配置。当然，其中的大部分设置如果不是工程有什么特殊要

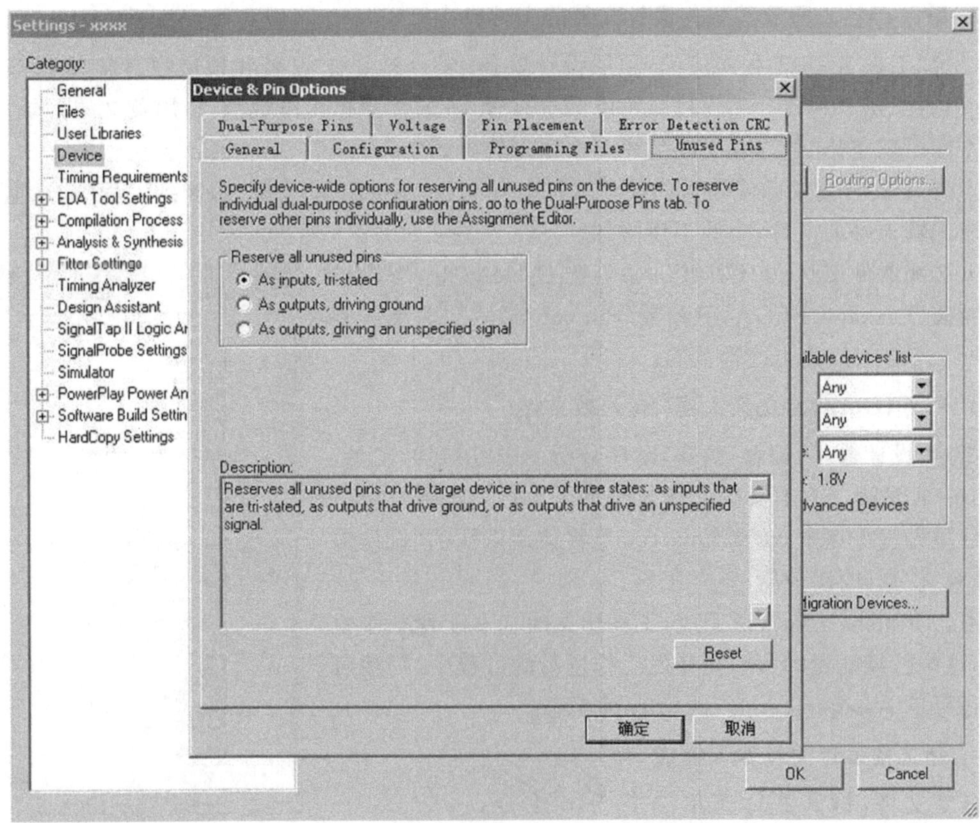

图 7－6　Settings→Device→Unused Pin

求，使用默认的配置就可以了。

Settings→ Device，其中包含和目标器件相关的设置，在任何工程中都需要进行配置。我们在新建工程向导中所配置的器件在这里可以修改。

Device & Pin Options 主要包括：

产生何种编程烧写文件，如何处理未使用的器件管脚，如何处理多用途管脚，产生何种非JTAG 下载方式的下载文件等。

注意 Unused Pins 的设置。系统为了防止外部电路与 FPGA 的相互干扰，系统 Unused Pins 的缺省设置为接地，不设置设计结果可能总是 0 输出。

要将未使用的管脚配置为输入高阻态，As Input Tri-Stated。

将设计 *.sof 文件（调试时文件）或 *.pof（调试结束后文件）下载到实验平台上，在 D/A 的输出通道，可以用示波器观察波形，也可以利用内嵌式逻辑分析工具，验证正弦波的设计效果。

7.3.3　正弦信号发生器的具体实现

1. 工程创建

进入 Quartus Ⅱ 开发软件，选择 File，点击 New Project Wizard。弹出工程向导对话框，选择 Next。

2. 输入存放工程及其相关设计文件的文件夹

指定“工程名”和工程对应的“顶层设计实体名”。这里工程名和顶层设计实体名都取作“singt”，再点击 Next。（工程目录可以随意设置，一般按照标识符规范，采用英文的目录，文件名中不能出现空格，否则在工程编译时会出现错误。）

点击 Add 将先期已经输入的设计文件（*. bdf; *. vhd; *. v 等）添加到工程中，这里我们没有事先输入好的文件，因此不用添加。

如选择器件 EP2C35F672C8，系列选 Cyclone Ⅱ，可以先给出封装(Package)FBGA，引脚数(Pin count)672，速度级别(Speed grade)C8，缩小搜索范围。

3. 打开(Open)建立新工程(New)或文件

指定“设计输入，综合，仿真，时序分析……”用到的工具，Quartus Ⅱ对第三方工具的支持比较完善，可以选择 Modelsim，这里先直接点击 Next，Quartus Ⅱ将使用默认的“设计输入，综合，仿真，时序分析……”工具。

点击 Finish，工程新建完成，工程相关的基本配置工作也完成，这些已经配置的参数在开发工作进行的过程中，仍然可以通过菜单 Assignments→Settings 来修改。

4. 新工程设计文件输入模式

新建文件，打开 File 菜单点击 New 命令，选择 Device Design Files 子类中的 Block Diagram/Schematic File 点击 OK，创建一个图形文件，如图 7－7 所示。

图 7－7　创建一个图形文件

文件名为 wave. bdf。

5. LPM_ROM 波实现形的方法

(1)调用 LPM_ROM 模块，Tools→Mega Wizard Plug-In Manager 或在图形输入界面双击屏幕空白处→Mega Wizard Plug-In Manager，选择 Create a new custom megafunction variation，点击 Next，如图 7－8 所示。

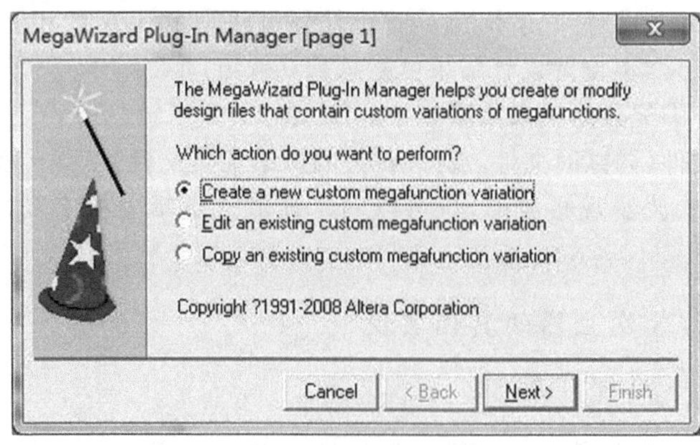

图 7－8　进入 Mega Wizard Plug-In Manager

(2)设置 LPM_ROM 模块。

设置输出文件的格式为 VHDL,输出文件名为 data_rom. vhd;选择 Memory Compiler 中的 ROM:1-PORT。点击 Next,如图 7-9 所示。

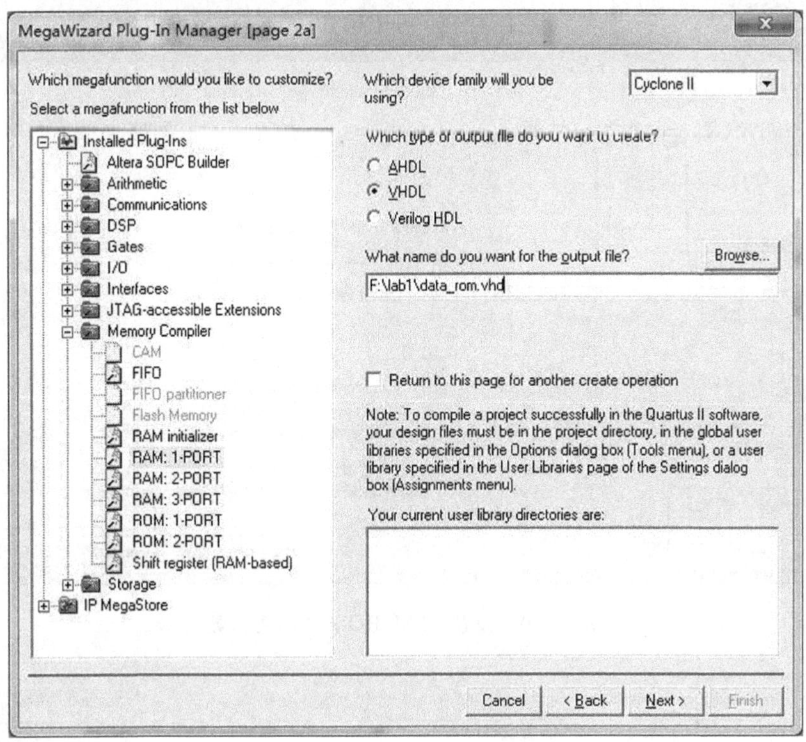

图 7-9　选用 LPM_ROM

(3)进一步设置 LPM_ROM 的相关参数,如图 7-10 所示:

ROM 输出的总线宽度为 8bits,该查找表共有 64 个数据;选择 Single clock 方式,RAM 类型 M4K,点击 Next。

(4)取消'q 'output port 选项,点击 Next,如图 7-11 所示。

(5)构成 ROM 中初始化数据文件的方式有两种:Memory Initialization File(. mif)格式和 Hexadecimal(Intel-Format)File(. hex)格式,选择 . mif 格式。

内存初始化的数据文件指定为 sin. mif。

选择 Allow In-System Memory Content Editor to…,表示允许 Quartus Ⅱ将通过 JTAG 口对下载于 FPGA 中的此 ROM 进行在系统的测试和读写。这个 ROM 的 ID 名为 ram1,如图 7-12 所示。

其他页面采用默认设置,点击 Finish。

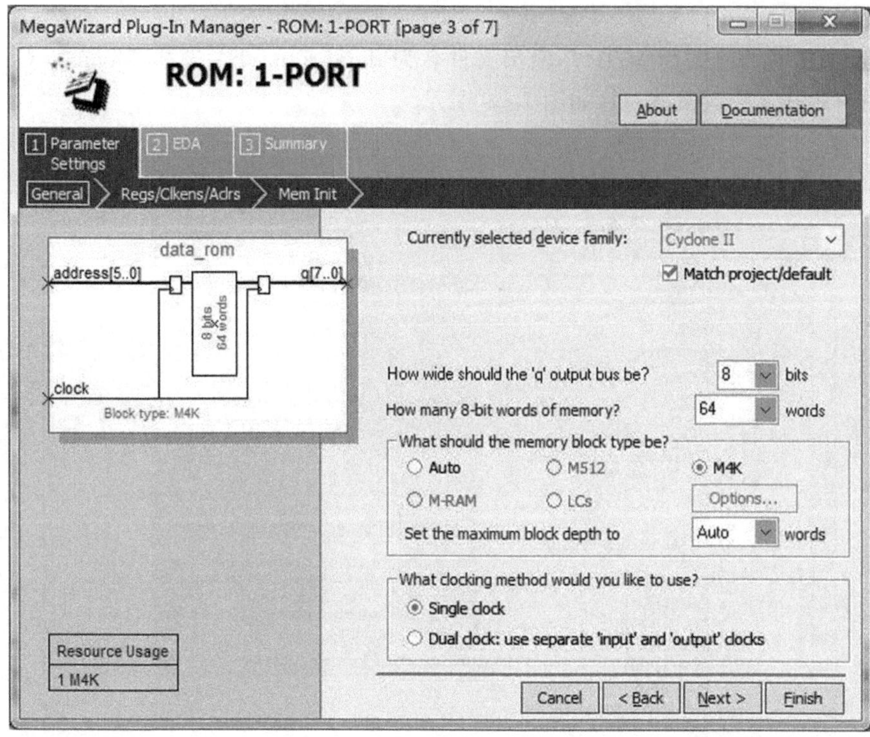

图 7-10 设置 LPM_ROM 的相关参数

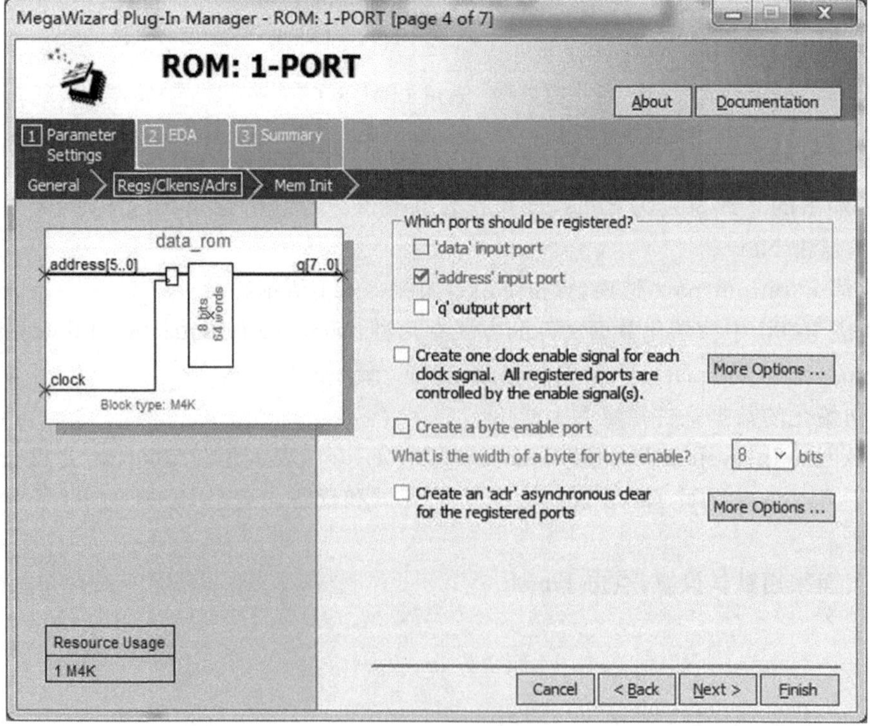

图 7-11 设置 LPM_ROM

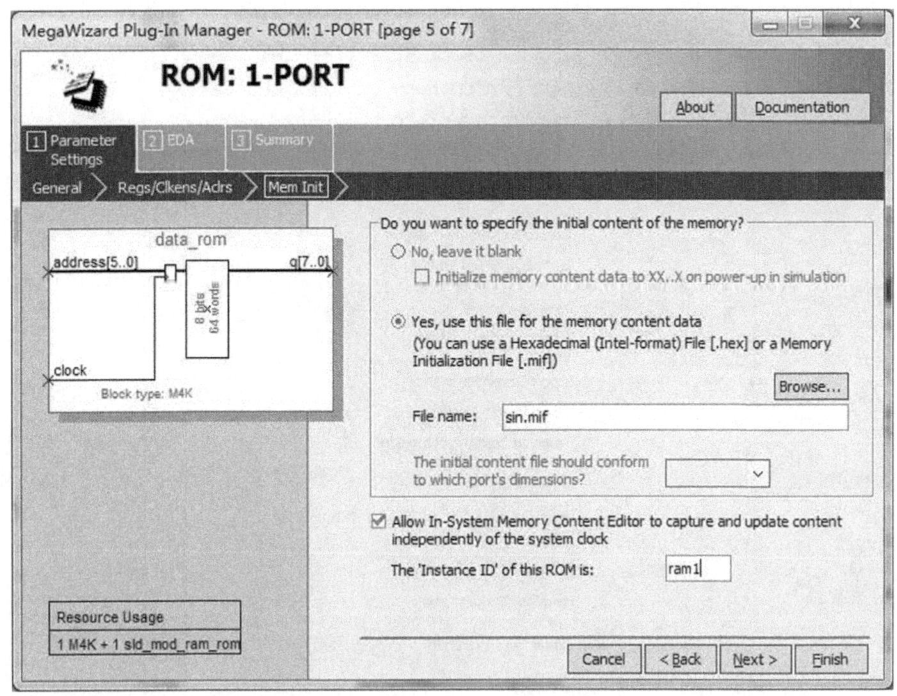

图 7 - 12　指定 LPM_ROM 初始化文件

6. 定制 ROM 存储 sin 波形数据 . mif 文件

点击 Quartus Ⅱ 的 File→New 项，选择 Memory Initialization File。

图 7 - 13 中 Number of words 对应查找表中查找数据的个数，为 64；Word size 对应 sin 输出波形的数据宽度，为 8。设置完成后点击 OK。

复制计算出的正弦信号数据或直接输入到 mif 表格中，如图 7 - 14 所示。

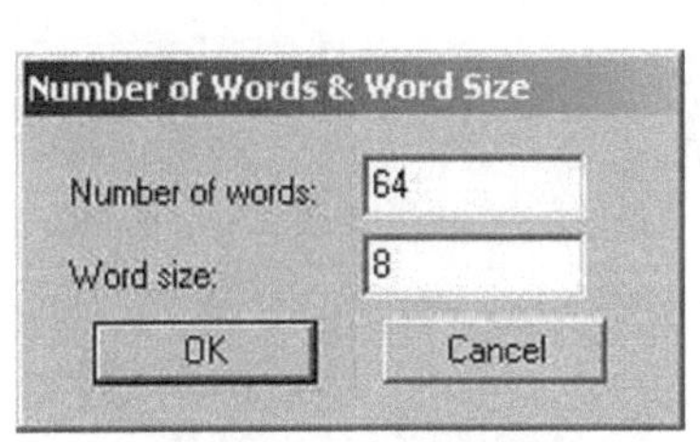

图 7 - 13　ROM 设置

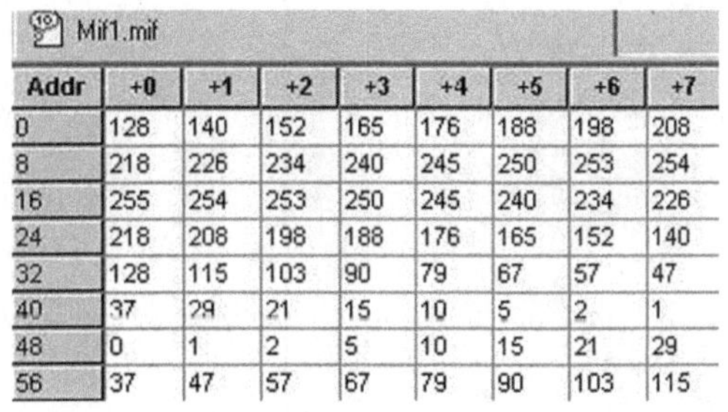

Mif1.mif

Addr	+0	+1	+2	+3	+4	+5	+6	+7
0	128	140	152	165	176	188	198	208
8	218	226	234	240	245	250	253	254
16	255	254	253	250	245	240	234	226
24	218	208	198	188	176	165	152	140
32	128	115	103	90	79	67	57	47
40	37	29	21	15	10	5	2	1
48	0	1	2	5	10	15	21	29
56	37	47	57	67	79	90	103	115

图 7 - 14　ROM mif 数据

有多种方法得到数据，可以通过 Excel 得到这些数据，也可以编程得到这些数据，或者在 Matlab/Simulink 的 DSP Builder 下完成 ROM 波形数据文件的编写。然后将文件保存为 sin. mif(与 ROM 定制中指定的一致)。

7. 顶层设计文件保存

点击 File→Save，文件名为 wave. bdf。

进行如图 7 - 15 所示的操作将 wave. bdf 设置成顶层文件。

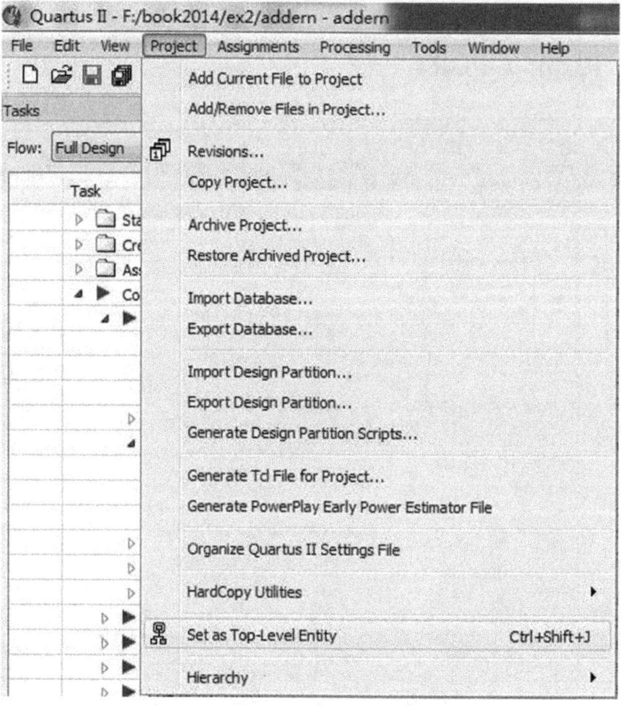

图 7-15　设置成顶层文件的操作

地址计数器也可以采用调用 LPM 模块的方式。调用方法与调用 ROM 类似，选择 Arithmatic 中的 LPM_Counter。64 个数需要建立 6 位计数器。

在 Altera 的 Primitive 库 pin 中调用 input 和 output 管脚，如图 7-16 所示。

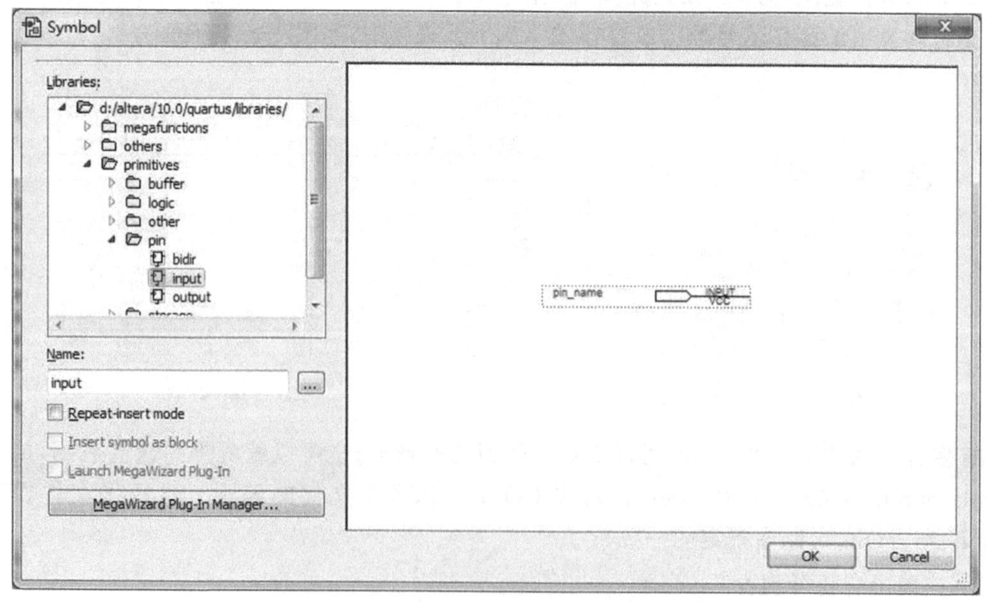

图 7-16　输入输出管脚调用

输入管脚命名为 clk，输出管脚命名为 dout[7..0]，如图 7-17 所示。

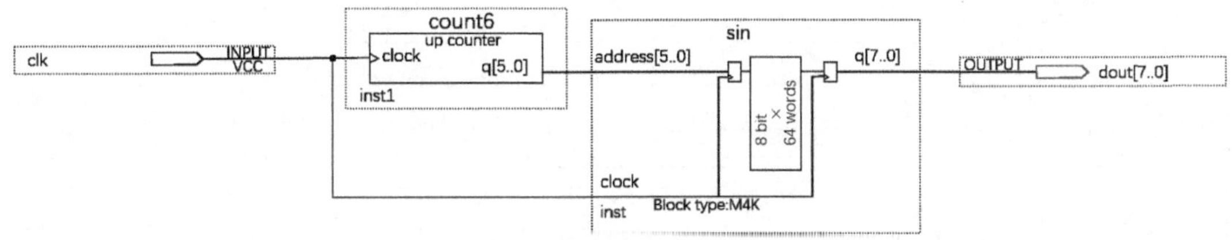

图 7-17　正弦信号发生器原理图

8. 编译、综合等

Quartus Ⅱ编译器由一系列处理模块组成，如设计工程的查错、逻辑的综合、结构的综合、输出结果的编辑配置、时序分析等。在编译前，可以设置一些参数使得编译器采取一些特别的综合和适配技术（如时序驱动技术等），提高工程编译的速度，优化器件的资源利用率等。点击快捷图标 Start Compilation 进行全程编译，如图 7-18 所示。

编译的时候，Quartus Ⅱ会给出编译的一些相关信息，如果出错，则根据这些提示进行排错，直至无误。编译完成后出现框图如图 7-19 所示，并给出编译报告，查看报告的方法见 6.3 节。

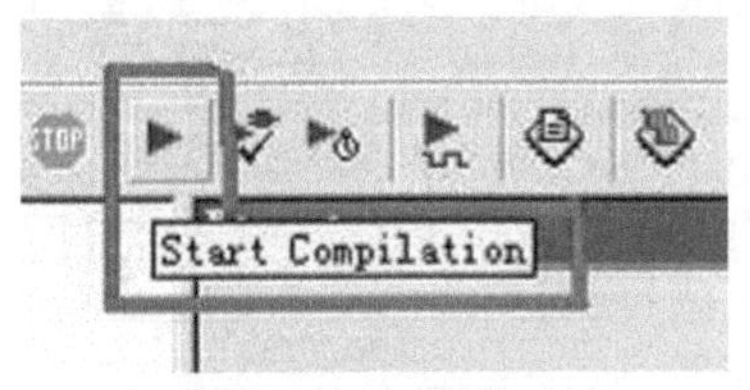

图 7-18　全程编译

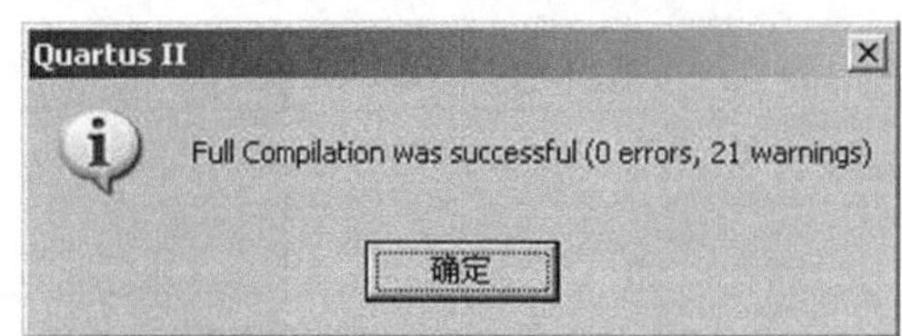

图 7-19　编译完成

7.4　FPGA 引脚分配

引脚分配方法只要有自动分配法、图形界面人工指定法和文件人工指定法。

自动分配引脚可以在 FPGA 器件没有外接其他元件时，或者仅对接有元件的 FPGA 器件进行仿真测试时采用。实际下载测试，对接有元件的 FPGA 器件的情况，须通过文件人工指定，人工指定后还要再做仿真参看设计细节，确认设计。基于平台的设计，一般都要采用人工指定。

7.4.1　在图形界面人工指定

(1)编译(Compilation)后，点击菜单 Assignments→Pins，进入管脚分配编辑器。在 Location 填引脚名称，如图 7-20 所示。

为了观察设计的工程是否正常工作，将输出引脚分配到实验箱的 8 个 LED 上。

如确定引脚分别为：主频时钟 clk 接 B13；8 位输出数据总线 dout[7..0]对应的引脚编号分别为 AC10、W11、W12、AE8、AF8、AE7、AF7、AA11。

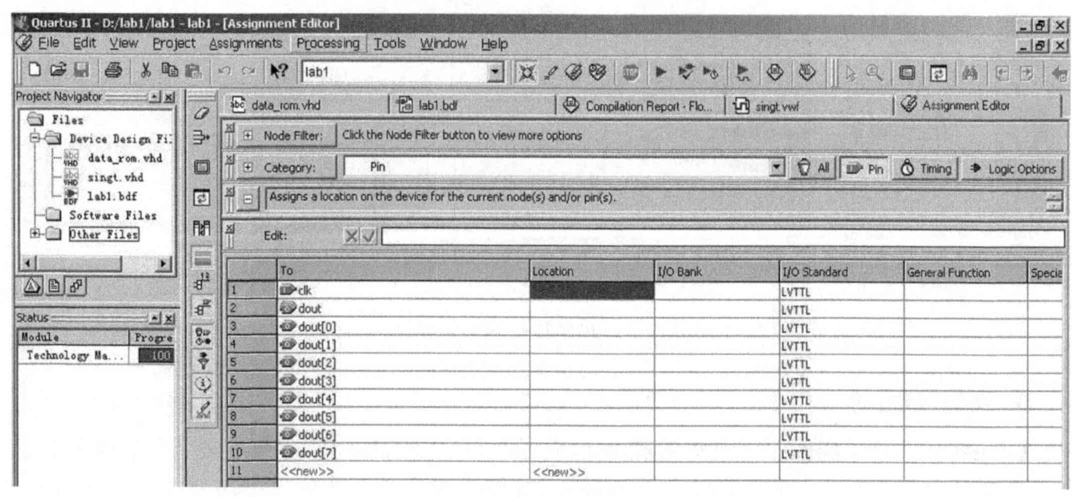

图 7 - 20　Assignments→Pins 管脚分配

引脚名称是在编号前加 PIN_前缀，例如 PIN_AC10。

(2)还可以选择 Assignments 菜单中的 Assignment Editor 项，即进入 Assignment Editor 编辑器窗，如图 7 - 21 所示。在 Category 栏中选择 Pin，或直接单击右上侧的 Pin 按钮。

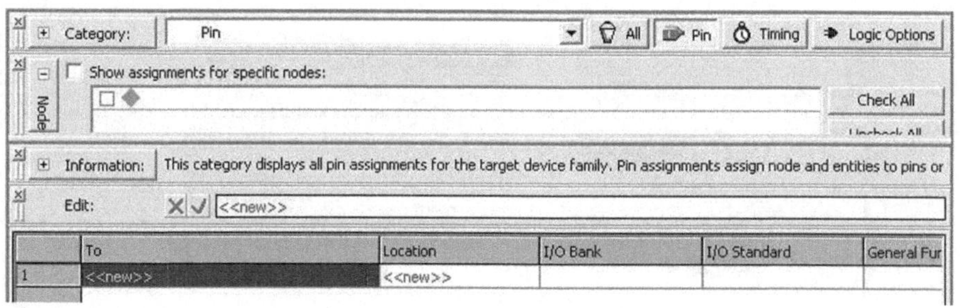

图 7 - 21　Assignment Editor 管脚分配(1)

双击 To 栏的＜＜new＞＞，在出现的下拉栏中选择并单击本工程要锁定的端口信号名(如 clk)，如图 7 - 22 所示。

再双击 Location 栏下的＜＜new＞＞，输入引脚号或在下拉菜单寻找相应的引脚，如图 7 - 23 所示；如对应 clk，选择 B13 脚，由此重复输入，至所有管脚都分配完成；也可以在＜＜new＞＞处，右键选择，Node Finder，把所有 pin 列出，Export 到 .csv 文件补充修改。

在 Assignment Editor 中还可以指定引脚接上拉电阻等，这种电特性的设置非常重要。

(4)引脚分配锁定后，必须再编译(启动 Start Compilation)一次，才能将引脚锁定信息编译进编程下载文件中。此后就可以准备将生成好的 .sof 文件下载到实验系统的 FPGA 中去了。如果希望设计在断电后能够保存，则要改换下载端口，如 AS 口，下载 .pof 文件。

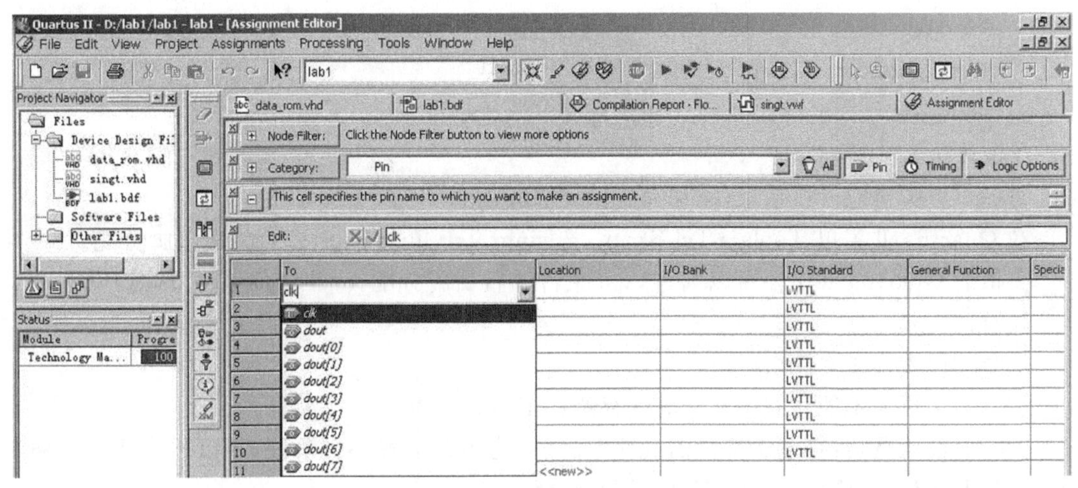

图 7 - 22　Assignment Editor 管脚分配(2)

	To	Location	I/O Bank	I/O Standard	General Funct
1	clk	PIN_B13	4	LVTTL	Dedicated Clo
2	dout[0]	PIN_AC10	8	LVTTL	Column I/O
3	dout[1]	PIN_W11	8	LVTTL	Column I/O
4	dout[2]	PIN_W12	8	LVTTL	Column I/O
5	dout[3]	PIN_AE8	8	LVTTL	Column I/O
6	dout[4]	PIN_AF8	8	LVTTL	Column I/O
7	dout[5]	PIN_AE7	8	LVTTL	Column I/O
8	dout[6]	PIN_AF7	8	LVTTL	Column I/O
9	dout[7]	PIN_AA11	8	LVTTL	Column I/O
10	<<new>>	<<new>>			

图 7 - 23　Assignment Editor 管脚分配(3)

7.4.2　反标注法引脚自动分配

编译(Compilation)后，点击菜单 Assignments→Pins，进入引脚分配编辑器。

第一次设计 Locations 栏下面应该是空的。关闭引脚分配编辑器。点击菜单 Assignments→Back Annotate Assignments 进行反标注，再进入引脚分配编辑器，就会看到 Quartus II 自动分配的引脚。引脚信息在工程名 . qsf 文件。

如果引脚分配编辑器 Location 下面不是空的，而我们希望清空，可以单击进入 Assignments→Remove→Assignments 选择 Pin 选项，Location & Routing Assignments，删除现有引脚再做反标注。

7.4.3　引脚分配等信息的文件处理

1. 在 . tcl 和 . qsf 文件中指定引脚分配

引脚分配还可以采用建立 . tcl 文件的方法。选择 Quartus Ⅱ→Projec→Generate Tcl File For Project 选项，生成 . tcl 文件。

如 singt. tcl，在 singt. tcl 文件中根据开发板或实验箱给定的信息添加引脚说明。对于前面所说的引脚 clk 指定到引脚 PIN_B13，dout[0] 指定到引脚 PIN_AC10，添加引脚说明如下：

```
#clk
set_location_assignment PIN_B13 -to clk
set_location_assignment PIN_AC10 -to dout[0]
……
```

每一个引脚对应一行信息，添加完后保存 singt. tcl。

在 Quartus Ⅱ菜单 Tools→Tcl Script，选 singt. tcl，单击 Run，即可完成引脚分配。

类似的 Quartus Ⅱ中还可以编写 . qsf 文件进行引脚分配，. qsf 文件不需要运行。

注意：

①"#"代表 Tcl 文件注释；

② Tcl 文件名不能含有中文且其保存路径不能含有中文。

"#"的设计技巧：在基于平台的设计方法中，尤其在了解平台阶段，可以把常用的引脚都列出来加上注释，需要用哪个引脚只需打开注释即可。

2. 在 . tcl 中执行引脚处理命令

尽量防止运行修改过 Tcl 文件后，出现"?"的管脚。"?"可能是在引脚分配时有原来的引脚分配存在，解决的办法只要在 Tcl 文件中添加如下语句命令：

```
remove_all_instance_assignments -name *
```

该语句的含义就是删除当前存在的管脚分配。

通常 FPGA 开发对于不使用的管脚要处理为输入三态，在 . tcl 的语句为

```
set_global_assignment -name RESERVE_ALL_UNUSED_PINS "As Input tri-stated"
```

3. 在 . tcl 指定文件的链接位置

Quartus Ⅱ的编译系列工程文件默认在 Work 库，默认是指当前工作目录，但是在工程文档中，通常会将工程文件放在 pro 子目录，源文件放在 src 子目录，仿真文件放在 sim 子目录，我们只需在 *. tcl 中加入如下语句，就可以实现源文件与工程文件分别存放的目的。

```
set_global_assignment -name VERILOG_FILE ../scr/led_block.v
set_global_assignment -name VHDL_FILE ../scr/lcd_block.vhd
```

在 Assignments→Setting→Files 中，找到 led_block. v，lcd_block. vhd 文件的路径，选中，添加，之后生成的 *. tcl 文件中也会有上面两句。

4. 用 . txt，. csv，. qsf 文件导入导出引脚

可以通过 Assignments→Export Assignments 导出引脚文件，也可以通过 Assignments→Import Assignments 导入引脚文件。

举例：. txt 文件导入引脚

①使用记事本或类似软件新建一个引脚文件 . txt 文件（或 . csv 文件）。按如图 7－24 所示格式编写管脚分配内容。注意：To 和 Location 两个关键字中间有一个半角逗号。

②在 Quartus Ⅱ软件中，选择"Assignments→Import Assignments"，导入 pin. txt 或者 pin. csv 文件即可。pin. csv 是 Excel 格式的表格文件。

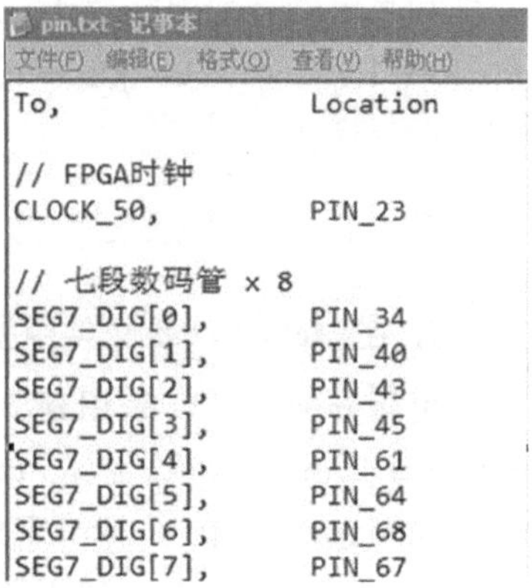

pin.txt - 记事本
文件(F) 编辑(E) 格式(O) 查看(V) 帮助(H)

```
To,              Location

// FPGA时钟
CLOCK_50,        PIN_23

// 七段数码管 x 8
SEG7_DIG[0],     PIN_34
SEG7_DIG[1],     PIN_40
SEG7_DIG[2],     PIN_43
SEG7_DIG[3],     PIN_45
SEG7_DIG[4],     PIN_61
SEG7_DIG[5],     PIN_64
SEG7_DIG[6],     PIN_68
SEG7_DIG[7],     PIN_67
```

图 7－24　pin. txt 文件

③在 Quartus Ⅱ软件中，选择“Assignments→Pins”标签(或者点击按钮)，打开 Pin Planner，可以验证管脚是否分配正确。

7.5 多种波形设计与嵌入逻辑分析仪测试

1. 打开工程

在 Quartus Ⅱ中，点击 File→Open Project…，选择顶层工程文件 singt. qpf。点击 Start Analysis & Synthesis 编译：Tools→Assignments Editor 进行引脚定义。然后进行综合，综合编译结束后，就可以进行 Signal Tap Ⅱ的相关配置了。也可以使用快捷键打开文件，如图 7-25所示。

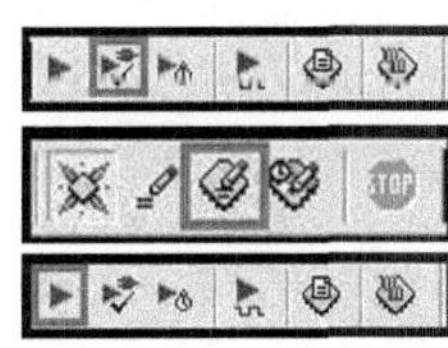

图 7-25 快捷键

2. Signal Tap Ⅱ配置

选择 File 菜单→New→Signal Tap Ⅱ File，进行 Signal Tap Ⅱ的相关配置。

(1)在 Instance Manager 窗口，Instance 栏为实例名称，点击鼠标右键对 Instance 进行 Create、Delete、Rename 操作，将名称(Rename)改为“singt”，如图 7-26 所示。

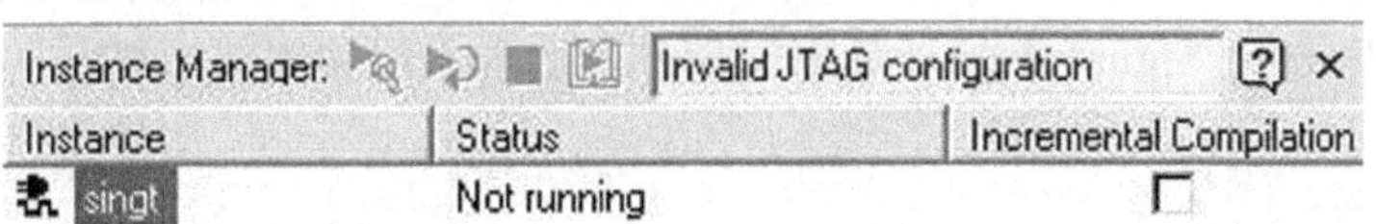

图 7-26 重命名(Rename)

(2)在 JTAG Chain Configuration 窗口，点击 Hardware 栏的 Setup…按钮，弹出 Hardware Setup 对话框，如图 7-27 所示。

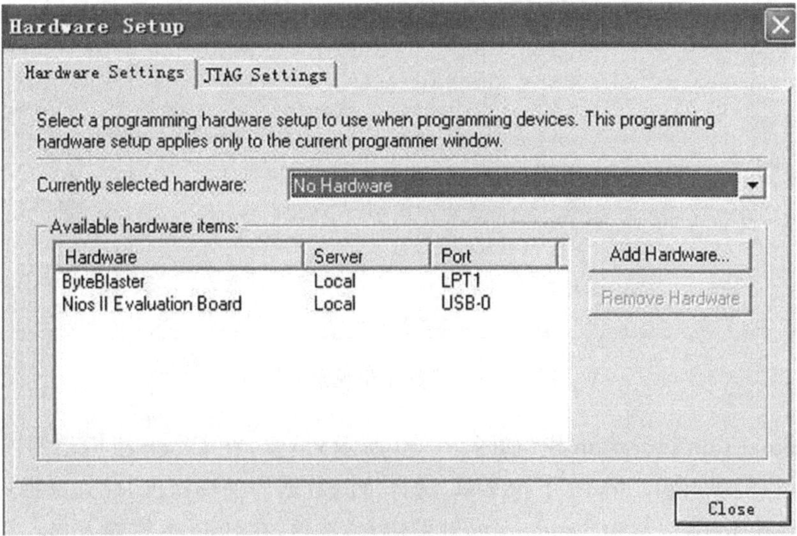

图 7-27 连接硬件

双击 Available hardware items 栏中的 Nios Ⅱ Evaluation Board 选项，然后选 Close 关闭。此时在 JTAG Chain Configuration 窗口的 Hardware 栏将显示 Nios Ⅱ Evaluation Board 字样，有的开发板是 USB blaster，同时 Device 栏将显示@1:EP2C35(0x020B40DD)字样，表示自动识别到的 FPGA 芯片，如图 7 - 28 所示。

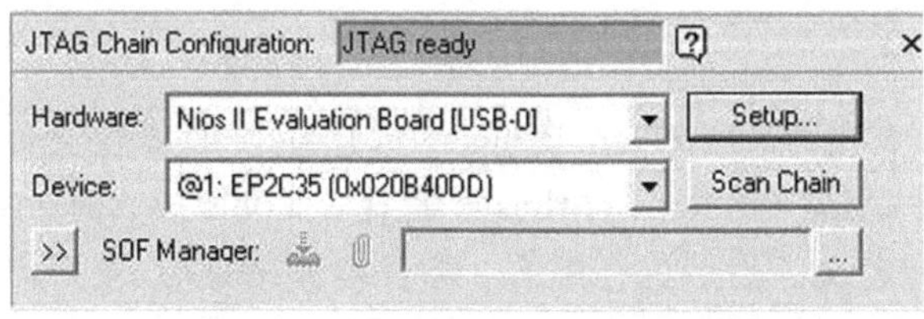

图 7 - 28 自动识别到的 FPGA 芯片

(3)在 SOF Manager 栏，点击 Browse 按钮，在弹出的 Select Programming File 对话框中选中 singt1. sof 文件。

(4)在 singt 栏，分为 Data 和 Setup 窗口，Data 为数据显示窗口，Setup 为配置窗口。在 Setup 窗口，双击空白处，弹出 Node Filter 对话框，如图 7 - 29 所示，点击 List 列出信号列表，双击 dout 信号，此时在 Selected Nodes 栏列出了已选信号 dout(添加希望观察的信号，一般不选 clk，而是将 clk 作为 Signal Tap Ⅱ 的采样信号)，点击 OK 关闭对话框。

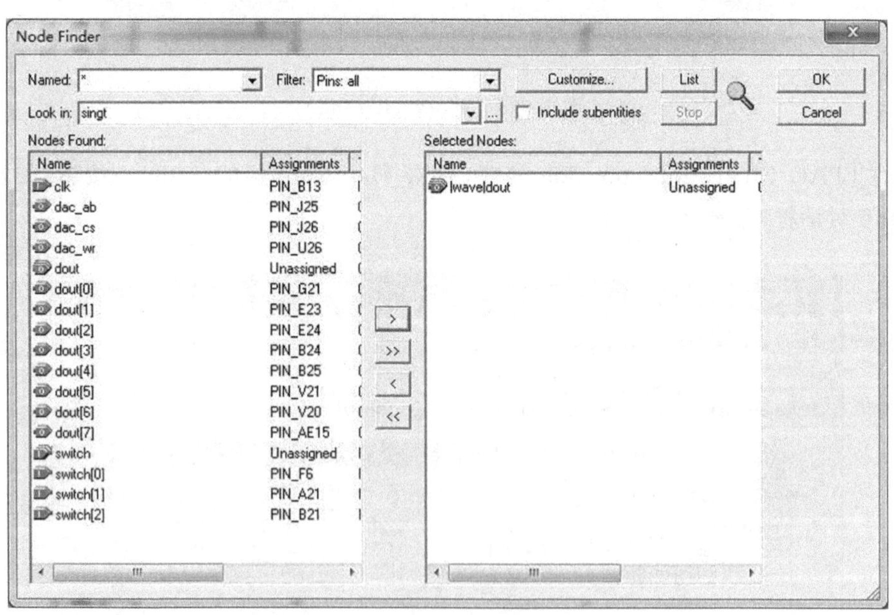

图 7 - 29 添加希望观察的信号

(5)在 Signal Configuration 栏(如图 7 - 30 所示)设置，在 Clock 栏中填写工程的时钟名 clk。在 Data 栏的 Sample depth 下拉菜单，选择待测数据的抽样深度，此处选择 8 K，RAM 的类型，选择 M4K/M9K。选中 Trigger in(触发输入)选项，在 Source 栏输入 clk，在 Pattern 栏，选择 Rising Edge。其他设置如图 7 - 31 所示。

Signal Configuration:
Clock: clk
Data
Sample depth: 8 K　RAM type: M4K/M9K
Segmented: 2 4 K sample segments
Trigger
Trigger flow control: Sequential
Trigger position: Pre trigger position
Trigger conditions: 1
Trigger in
Source: clk
Pattern: Rising Edge
Trigger out
Target:
Level: Active High
Latency delay:

图 7－30　Signal Tap 的设置

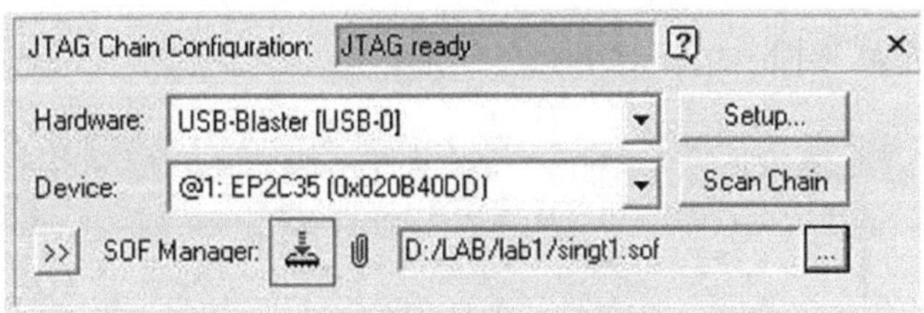

图 7－31　工程文件下载

Signal Tap Ⅱ的设置完成后，保存设置，名字任取，此处保留默认名称 stp1. stp。

综合编译完成以后，点击 Program Device 按钮（如图 7－31 所示）进行芯片烧写。选择“singt. sof”工程文件下载到 FPGA 芯片。

烧写完成后，点击工具栏按钮的 Auto Analysis 按钮运行逻辑分析仪。此时在芯片内部运行的信号数据，已经通过 JTAG 链读取到计算机，并在 Data 窗口显示，如图 7－32所示。

log: 2006/07/08 20:22:22 #0　　click to insert time bar

Type	Alias	Name	-1024	-992	-960	-928	-896	-864	-832	-800	-768
		dout	8Ch	98h	A5h	B0h	BCh	C6h	D0h	DAh	

图 7－32　正弦波发生器的 Signal Tap 测试结果

为了便于观察，在 dout 栏点击鼠标右键，在弹出的对话框中选择 Bus Display Format 栏的 Unsigned Line Chart 选项，如图 7-33 所示。

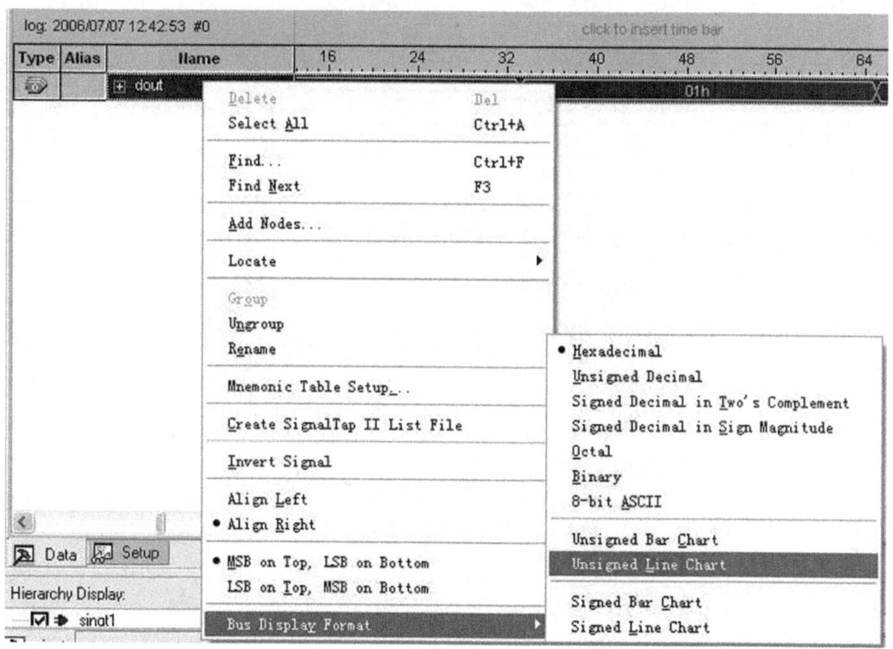

图 7-33 显示格式转换为 Unsigned Line Chart

此时 dout 显示格式如图 7-34 所示。

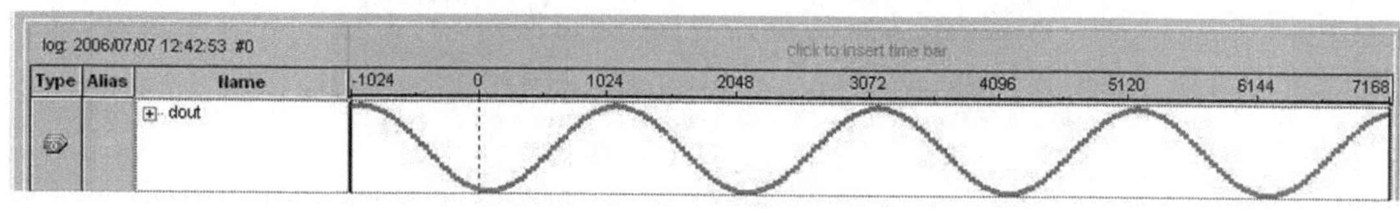

图 7-34 正弦波发生器的 Signal Tap 测试结果

在波形处点击鼠标左键、右键可以对波形放大、缩小。三角波和锯齿波等的测试改变开关重新采样即可完成。

在设计数字合成的正弦波时，有时会将正弦波的 4 个象限的数据压缩成 1 个象限的数据数据来存储。设计从正弦波的最大值开始存数，用 2 位数来确定象限，配合取反操作可以得到 4 个象限的数据，这样就可以节省存储器，相应的，设计中其他波形也从高位开始生成。

示例 7-3 分布式 ROM 法实现正弦波

```
-- 正弦波 sincom.vhd
-- 分布式 ROM 法比较适合 CPLD 实现，其特点是与或逻辑比较多
Library Ieee;
Use Ieee.Std_Logic_1164.All;
```

```
Use Ieee.Std_Logic_Unsigned.All;
Entity sincom Is
  Port ( clk: In Std_Logic;
        dout: Out Integer Range 255 Downto 0 );
End Entity sincom;

Architecture dacc Of sincom Is
Signal q  : Integer Range 63 Downto 0 ;
Signal d  : Integer Range 255 Downto 0 ;
Begin
  Process(clk) Is
  Begin                          --ROM 地址计数器
    If clk'Event And clk = '1' Then
      If q < 63 Then   q <= q + 1;
      Else q <= 0 ;
      End If;
    End If;
  End Process;

  Process(q) Is                  --ROM 数据
  Begin
    Case q Is
      When 00=>d<=255; When 01=>d<=254; When 02=>d<=252;
      When 03=>d<=249; When 04=>d<=245; When 05=>d<=239;
      When 06=>d<=233; When 07=>d<=225; When 08=>d<=217;
      When 09=>d<=207; When 10=>d<=197; When 11=>d<=186;
      When 12=>d<=174; When 13=>d<=162; When 14=>d<=150;
      When 15=>d<=137; When 16=>d<=124; When 17=>d<=112;
      When 18=>d<= 99; When 19=>d<= 87; When 20=>d<= 75;
      When 21=>d<= 64; When 22=>d<= 53; When 23=>d<= 43;
      When 24=>d<= 34; When 25=>d<= 26; When 26=>d<= 19;
      When 27=>d<= 13; When 28=>d<= 8; When 29=>d<= 4;
      When 30=>d<= 1; When 31=>d<=0; When 32=>d<= 0;
      When 33=>d<= 1; When 34=>d<= 4; When 35=>d<=8;
      When 36=>d<= 13; When 37=>d<= 19; When 38=>d<= 26;
      When 39=>d<= 34; When 40=>d<= 43; When 41=>d<= 53;
      When 42=>d<= 64; When 43=>d<= 75; When 44=>d<=87;
      When 45=>d<= 99; When 46=>d<=112; When 47=>d<=124;
      When 48=>d<=137; When 49=>d<=150; When 50=>d<=162;
```

```
      When 51=>d<=174; When 52=>d<=186; When 53=>d<=197;
      When 54=>d<=207; When 55=>d<=217; When 56=>d<=225;
      When 57=>d<=233; When 58=>d<=239; When 59=>d<=245;
      When 60=>d<=249; When 61=>d<=252; When 62=>d<=254;
      When 63=>d<=255;
      When Others => Null;
    End Case;
    dout <= d ;
  End Process;
End Architecture dacc;
```

设计说明:编译结果报告，这种方式设计的正弦波采用了56个逻辑单元,9个引脚,即所谓分布式ROM,由逻辑单元实现,没有占用器件本身的嵌入存储器资源,这种ROM无法采用在线存储器编辑工具来修改。

还可以通过声明2维数组来说明一个正弦波。

示例7-4 三角波设计

```
Library Ieee;
Use Ieee.Std_Logic_1164.All;
Use Ieee.Std_Logic_Arith.All;
Use Ieee.Std_Logic_Unsigned.All;

Entity sanj Is
  Port(clk: In  Std_Logic;
       dd3 : Out Std_Logic_Vector(7 Downto 0));
End Entity sanj;

Architecture sanj_arch Of sanj Is
Begin

  Process(clk) Is
  Variable num : Std_Logic_Vector(7 Downto 0);
  Variable ff : Std_Logic;
  Begin
    If (clk'Event And clk='1') Then
      If ff = '0' Then
        If num="11111000" Then
          num:="11111111";
          ff:='1';
        Else
```

```
          num:=num+8;
        End If;
      Else
        If num="00000111" Then
          num:="00000000";
          ff:='0';
        Else
          Num:=Num-8;
        End If;
      End If;
    End If;
    dd3<=num;
  End Process;
End Architecture sanj_arch;
```

三角波发生器的 Signal Tap 测试结果如图 7-35 所示。

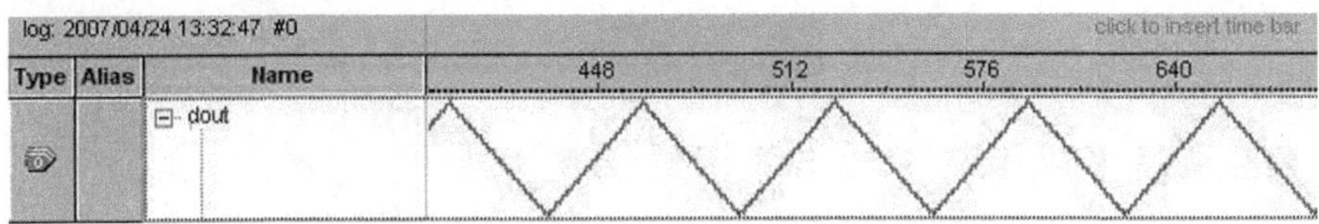

图 7-35　三角波发生器的 Signal Tap 测试结果

示例 7-5　锯齿波设计

```
Library Ieee;
Use Ieee.Std_Logic_1164.All;

Entity Jvchi Is
  Port(clk,up_down: In Std_Logic;
       dd2:Out Integer Range 255 Downto 0);
End Entity jvchi;

Architecturejvchi_arch Of jvchi Is
Signal d,temp:Integer Range 255 Downto 0;
Begin

  Process(clk) Is
  Begin
    If(clk'Event And clk='1') Then
      If temp<240 Then Temp<=temp+2;
```

```
      Else temp<=0;
      End If;
    End If;
  End Process;

  Process(temp,up_down) Is
  Begin
    If up_down='0' Then d<=temp;
    Else d<=240-temp;
    End If;
  End Process;
  dd2<=d;
End Architecture jvchi_arch;
```

锯齿波发生器的 Signal Tap 测试结果如图 7 - 36 所示。Modelsim 波形仿真默认显示方式如图 7 - 37 所示。

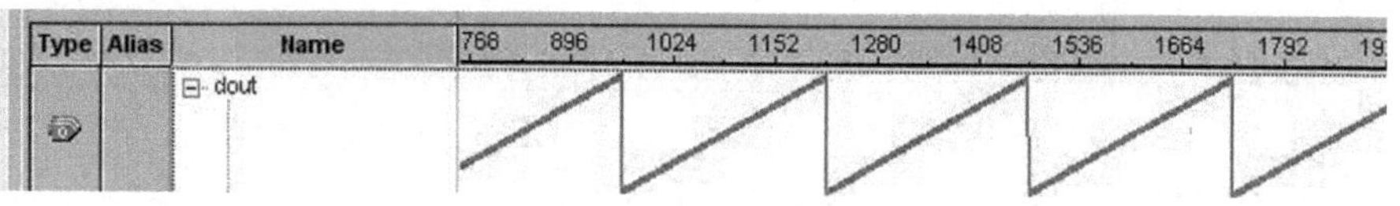

图 7 - 36　锯齿波发生器的 Signal Tap 测试结果

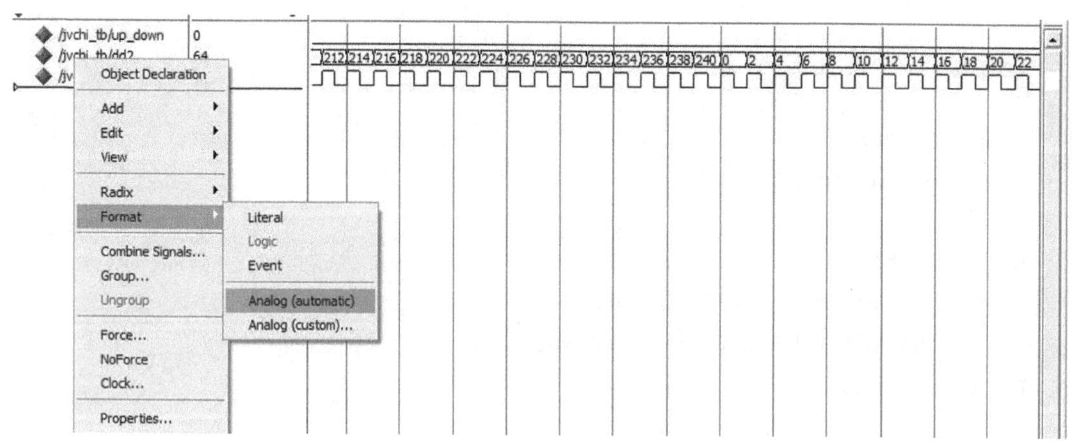

图 7 - 37　Modelsim 波形仿真默认显示方式

选择 Format→Analog(automatic)或者 Analog (custom)选项，在出现的对话框中，进行如图 7 - 38 所示的设置。设置好后，Modelsim 锯齿波波形仿真模拟显示如图 7 - 39 所示。

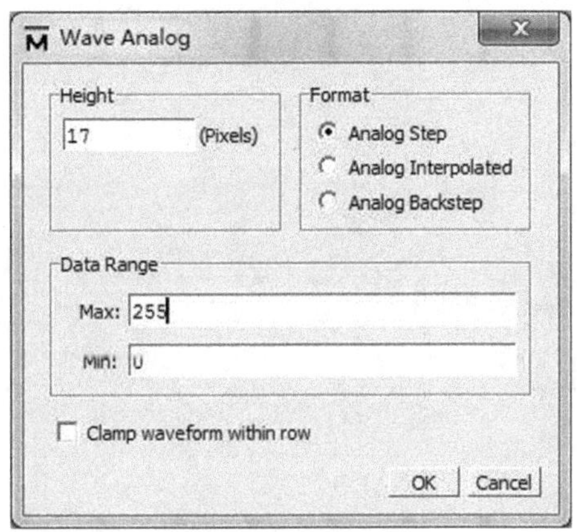

图 7 - 38 选择 Format→Analog(custom)

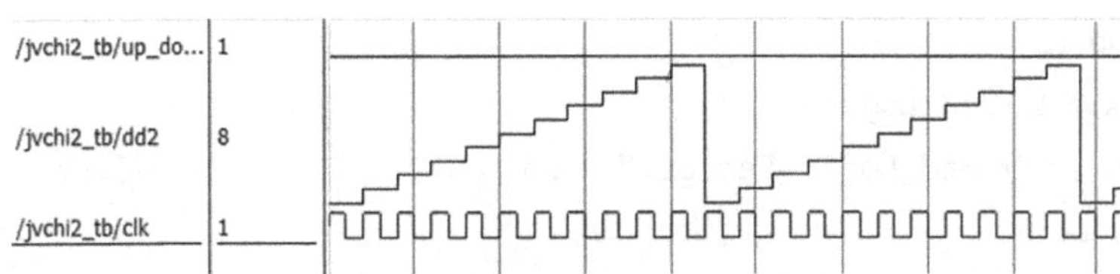

图 7 - 39 Modelsim 锯齿波波形仿真模拟显示

```
Library Ieee;                              --锯齿波仿真 Testbench
Use Ieee.Std_Logic_1164.All;
Entity jvchi_tb  Is
End Entity jvchi tb;

Architecture jvchi_tb_arch Of jvchi_tb Is
Signal up_down :Std_Logic :='0';
Signal dd2:Integer;
Signal clk:Std_Logic;
  Component Jvchi Is
    Port (up_down: In Std_Logic ;
          dd2: Out Integer ;
          clk: In Std_Logic );
  End Component jvchi;
Begin
  dut: jvchi
  Port Map (up_down => up_down,
            dd2 => dd2, clk => clk ) ;

  Process Is                               --这是一个用 Wait For 语句写时钟典型方法
  Begin
```

```
    clk<='1';
    Wait For 50ns;                          --时钟高电平 50 ns
    clk<='0';
    Wait For 50ns;                          --时钟低电平 50 ns
  End Process;                              --进程进行无限循环
End Architecture jvchi_tb_arch ;
```

示例 7－6　方波设计

```
Library Ieee;
Use Ieee.Std_Logic_1164.All;
Use Ieee.Std_Logic_Arith.All;
Use Ieee.Std_Logic_Unsigned.All;
Entity fang Is
  Port(clk : In Std_Logic;
       fdl : Out Std_Logic_Vector(7 Downto 0));    --8 位输出
End Entity fang;

Architecture fang_arch Of fang Is
Signal q:Integer Range 0 To 1;
Begin

  Process(clk) Is
  Begin
    If(clk'Event And clk='1') Then
      q<=q+1;                               --2 计数
    End If;
  End Process;

  Process(q) Is
  Begin
    Case q Is
      When 0=>fdl<="00000000";              --方波 0 电平
      When 1=>fdl<="11111111";              --方波 1 电平
      When Others=>Null;
    End Case;
  End Process;
End Architecture fang_arch;
```

示例 7－7　简单多种波形调频调幅(调相,正弦波)

设计要求:

(1)在此前多种波形产生的基础上增加相位控制与幅度控制,并用逻辑分析仪测试,或者通过 da 在示波器上测试验证。

(2)调频通过分频时钟来实现对各种信号调频。

(3)调幅通过对输出波形进行移位的方式进行调幅。

(4)调相对于正弦波,在地址计数器与 ROM 间加加法器来实现。

设计提示:

通过在程序中加程序包说明

```
Use Ieee.Std_Logic_Unsigned.All;
Use Ieee.Std_Logic_Arith.All;
```

可以应用一些数据类型转换函数。

示例 7－7a　简单调相

方法:在地址计数器与 ROM 中间增加加法器,给加法器两个常数,与原来地址计数器的值相加,实现 0 相位与 180 相位的变换,改变增加常数数目来增加可调性。

```
Library Ieee;
Use Ieee.Std_Logic_1164.All;


Entity mux21 Is
  Port( s: In Std_Logic;                                    --相位选择
        dout: Out Std_Logic_Vector(5 Downto 0));             --调相输出
End Entity mux21;

Architecture mux21_arch Of mux21 Is
Begin
  Process(s) Is
  Begin
    Case s Is
      When '0' =>dout<="000000";
      When '1' =>dout<="010000";
      When Others=>dout<=Null;
    End Case;
  End Process;
End Architecture mux21_arch;
```

示例 7－7b　简单调频

方法:通过 3 个拨码给出 8 种分频(由 8 位计数器产生 2 的幂次分频,接波形产生器的时钟)。

```
Library Ieee;
Use Ieee.Std_Logic_1164.All;

Entity mux81 Is
  Port( s: In Std_Logic_Vector(2 Downto 0);              --拨码,控制分频
        z: In Std_Logic_Vector(7 Downto 0);              --基波
        dout: Out Std_Logic);                            --调频波输出
End Entity mux81;

Architecture mux81_arch Of mux81 Is
Begin

  Process(s) Is
  Begin
    Case s Is
      When "000"=>dout<=z(0);
      When "001"=>dout<=z(1);
      When "010"=>dout<=z(2);
      When "011"=>dout<=z(3);
      When "100"=>dout<=z(4);
      When "101"=>dout<=z(5);
      When "110"=>dout<=z(6);
      When "111"=>dout<=z(7);
      When Others=>dout<=Null;
    End Case;
  End Process;
End Architecture mux81_arch;
```

示例 7-7c 简单调幅

```
To_BitVector(a)
To_StdLogicvector(a)

Library Ieee;
Use Ieee.Std_Logic_1164.All;
Use Ieee.Std_Logic_Unsigned.All;
Use Ieee.Std_Logic_Arith.All;

Entity shiftr Is
```

```
  Port(clk: In Std_Logic;
       sn,data: In Std_Logic_Vector( 7 Downto 0);              --幅度控制,基波
       c: Out Std_Logic_Vector( 7 Downto 0));                  --调幅波形输出
End Entity shiftr;

Architecture shiftr_arch Of shiftr Is
Bcgin
  Process(clk,sn) Is
  Begin
    If(clk'Event And clk='1') Then
      c<=To_Stdlogicvector(To_Bitvector(data) Srl Conv_Integer(sn));
    End If;
  End Process;
End Architecture shiftr_arch;
```

程序说明:用移位的方式进行调幅。Srl 只能应用于位矢量,因而要用到 Bit 位矢量与 Std_Logic_Vector 转换函数。函数中 Srl 的移位数要定义为整数,因而声明 Std_Logic_Unsigned 程序包,调用了 Conv_Integer(To_Std_logic_Vector),此函数可将 Std_Logic_Vector 转换为整数。

7.6　正弦信号发生器提高

设计要求:

(1)采用原来正弦信号发生器中 mif 的 1/4 来构成正弦信号发生器。

(2)采用在系统存储数据读写编辑器修改数据,并用 Signal TapⅡ来观察。

(3)要求采用三种在线硬件调试工具。

设计提示:

在真正设计数字合成的正弦波时,有时会将正弦波四个象限的四组数压缩成 1 个象限的组数来存储。这里从正弦波的最大值开始存数,用 2 位数来确定象限,通过对地址的加减计数,配合取反操作可以得到 4 个象限的数据,这样就可以节省存储器。相应的其他波形也从高位开始生成。

示例 7-8　由 1/4 象限正弦数据生成完整正弦波

1/4 象限数据实现设计框图如图 7-40 所示。

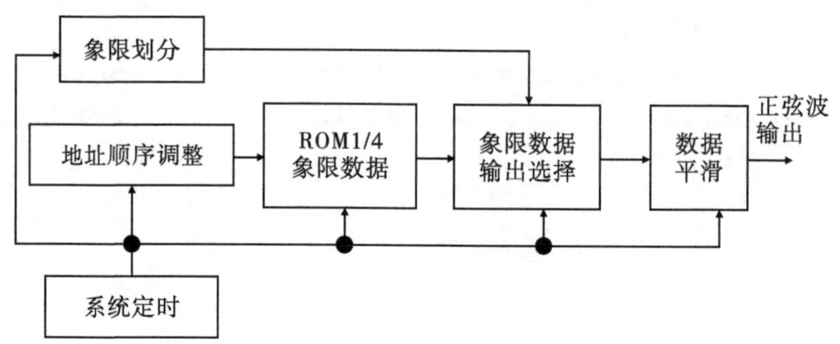

图 7-40　1/4 象限数据实现设计框图

示例 7-8a　count. vhd

/* 此程序提供加减计数控制信号，以及 4 个象限的输入控制信号，这个程序是在 LMP 模块自动生成的程序上增加与 LPM 模块参数设置无关的内容实现的，注意，LPM 模块参数生成程序一定要通过向导 Mage wizard Plug-in Manager 生成。在最初设计时为了快速实现设计，采用这个方法，即程序 7-8a，随即要再写此功能，所以有程序 7-8b。

LPM 参数模块生成的电路一般都与器件有关，每种器件的 LPM 生成的程序，通常是不能够移植到其他器件上的，甚至 Quartus Ⅱ 不同版本也会是不同的 LPM 参数模块，尽管这样，由于它的性能特点，一定要重视和深入了解。
*/

```
Library Ieee,Lpm;
Use Ieee.Std_Logic_1164.All;
Use Ieee.Std_Logic_Unsigned.All;
Use Lpm.All;

Entity count Is
  Port(clock: In Std_Logic ;
        s: Out Std_Logic_Vector (1 Downto 0);
        ud: Out Std_Logic:='0';
        q: Out Std_Logic_Vector (5 Downto 0));
End Entity count;

Architecture syn Of count Is
Signal sub_wire0: Std_Logic_Vector (5 Downto 0);
Signal c:Integer Range 0 To 63:=0;
  Component Lpm_Counter Is
    Generic (Lpm_Direction: String;
             Lpm_Port_Updown: String;
             Lpm_Type: String;
```

```
            Lpm_Width: Natural);
    Port (clock: In Std_Logic ;
          q: Out Std_Logic_Vector (5 Downto 0));
  End Component Lpm Counter;
Begin
  q <= sub_wire0(5 Downto 0);

  Process(clock,c)Is
  Begin
    If (clock'Event And clock='1') Then
      c<=c+1;
    End If;
    Case c Is
      When 0 To 15 => s<="00"; ud <= '0';                --象限划分
      When 16 To 31 => s<="01"; ud <= '1';
      When 32 To 47 => s<="10"; ud <= '0';
      When 48 To 63 => s<="11"; ud <= '1';
    End Case;
  End Process;

  Lpm_Counter_Component : Lpm_Counter
  Generic Map (Lpm_Direction => "Up",
               Lpm_Port_Updown => "Port_Unused",
               Lpm_Type => "Lpm_Counter",
               Lpm_Width => 4)
  Port Map (clock => clock,
            q => sub_wire0);
End Architecture syn;
```

示例 7-8b count416.vhd

```
--程序 7-8a 的替代程序,程序 7-8a 和程序 7-8b 功能一样
--程序提供加减计数控制信号,以及 4 个象限的输入控制信号
Library Ieee;
Use Ieee.Std_Logic_1164.All;
Use Ieee.Std_Logic_Unsigned.All;

Entity count416 Is
  Port(clk: In Std_Logic;
       counth: Out Std_Logic_Vector(1Downto 0);
```

```
        dout: Out Std_Logic);
End Entity count416 ;

Architecture bincount Of count416  Is
Signal count: Std_Logic_Vector(4 Downto 0);
Signal c: Std_Logic_Vector(3 Downto 0);
Signal count4: Std_Logic_Vector(1 Downto 0);
Signal cout,cout1: Std_Logic;
Begin

  Process(clk) Is
  Begin
    If (clk'Event And clk='1')Then
      If(count="11111" ) Then  count <= "00000";cout <= '1';
      Else  count <= count +1; cout <= '0';
      End If;
    End If;
  End Process;

  Process(clk) Is
  Begin
    If (clk'Event And clk='1') Then
      If c="1111" Then c<="0000"; cout1 <= '1';
      Else  c <= c +1; cout1 <= '0';
      End If;
    End If;
  End Process;

  Process(clk,cout1) Is
  Begin
    If (clk'Event And clk='1')Then
      If cout1='1' Then
        If(count4="11" ) Then  count4 <= "00";
        Else  count4 <= count4 +1;
        End If;
      End If;
    End If;
  End Process;
  counth <= count4;
```

```
  dout <= count(4);
End Architecture bincount;
```

示例 7-8c　LPM 模块加减计数

```
--count6. vhd 这是完全由 LPM 模块自动生成的程序,完成加减计数的功能
Library Ieee;
Use Ieee. Std_Logic_1164. All;

Library Lpm;
Use Lpm. All;

Entity count6 Is
  Port(clock: In Std_Logic ;
       updown: In Std_Logic ;
       q: Out Std_Logic_Vector (3 Downto 0));
End Entity count6;

Architecture syn Of count6 Is
Signal sub_wire0: Std_Logic_Vector (3 Downto 0);

  Component Lpm_Counter Is
    Generic (Lpm_Direction: String;
             Lpm_Port_Updown: String;
             Lpm_Type: String;
             Lpm_Width: Natural);
    Port (clock: In Std_Logic ;
          q: Out Std_Logic_Vector (3 Downto 0);
          updown: In Std_Logic );
  End Component  Lpm_Counter;

Begin
  q  <=sub_wire0(3 Downto 0);

  Lpm_Counter_Component : Lpm_Counter
  Generic Map (Lpm_Direction => "Unused",
               Lpm_Port_Updown => "Port_Used",
               Lpm_Type => "Lpm_Counter",
               Lpm_Width => 4)
  Port Map (clock => clock,
```

```
          updown => updown,
          q => sub_wire0);
End Architecture syn;
```

示例 7-8d　mux42. vhd

```
--此程序读正弦波的查表信号,然后按象限将数据依次送出
Library Ieee;
Use Ieee.Std_Logic_1164.All;
Use Ieee.Std_Logic_Unsigned.All;

Entity mux42 Is
  Port( clk:In Std_Logic;
        s: In Std_Logic_Vector(1 Downto 0);
        z: In Std_Logic_Vector(6 Downto 0);
        dout: Out Std_Logic_Vector(7 Downto 0));
End Entity mux42;

Architectureart Of mux42 Is
Signal dout1: Std_Logic_Vector(7 Downto 0):="11111111";
Begin
  Process(s) Is
  Begin
    If clk'Event And clk='0' Then
      Case s Is
        When "00" => dout <= dout1;
             dout1 <= ('0'&z);
        When "01" => dout <= dout1;
             dout1 <= '0'&z;
        When "10" =>dout <= dout1;
             dout1 <= Not('0'&z);
        When "11" => dout <= dout1;
             dout1 <= Not ('0'& z);
        When Others=>dout<="11111111";
      End Case;
    End If;
  End Process;
End Architecture art;
```

联调的 Modelsim 仿真结果如图 7-41 所示。表 7-3 为 QuartusⅡ正弦波 ROM 存储的 mif 文件。

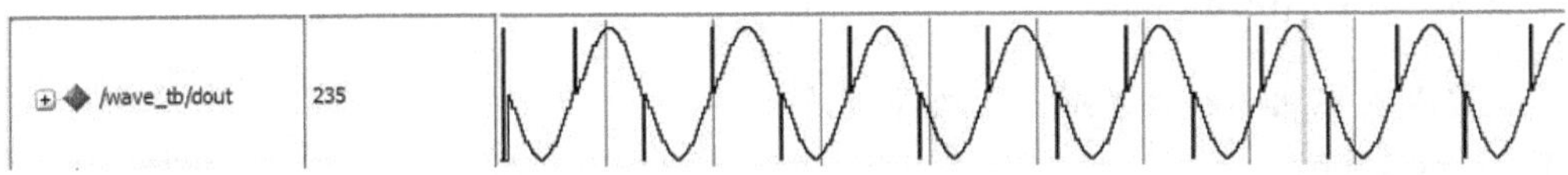

图 7-41 联调的 Modelsim 仿真结果

表 7-3 Quartus Ⅱ 正弦波 ROM 存储的 mif 文件

Addr	+000	+001	+010	+011	+100	+101	+110	111
0000	0	2	5	9	14	20	28	36
1000	46	56	67	78	90	102	115	127

如表 7-3 所示的数据是正弦波设计实验中四个象限 .mif 文件修改为一个象限的 .mif 文件，原来 64 个数，现在 16 个数。

```
% Quartus Ⅱ generated Memory Initialization File (.mif)
WIDTH=7;
DEPTH=16;

ADDRESS_RADIX=BIN;
DATA_RADIX=UNS;

CONTENT BEGIN
  00000  :  0;
  00001  :  2;
  00010  :  5;
  00011  :  9;
  00100  :  14;
  00101  :  20;
  00110  :  28;
  00111  :  36;
  01000  :  46;
  01001  :  56;
  01010  :  67;
  01011  :  78;
  01100  :  90;
  01101  :  102;
  01110  :  115;
  01111  :  127;
END;
```

由于上面设计在翻转处有错误出现，所以增加下面子程序 7-8e 纠正。

示例 7－8e　max1. vhd

```
--功能是对在相邻数据差别太大给予平滑
Library Ieee;
Use Ieee.Std_Logic_1164.All;
Use Ieee.Std_Logic_Arith.All;
Use Ieee.Std_Logic_Unsigned.All;

Entity max1 Is
  Port(clk : In Std_Logic;
       a: In  Std_Logic_Vector(7 Downto 0);
       dout: Out  Std_Logic_Vector(7 Downto 0));
End Entity max1;

Architecture max1_arch Of max1 Is
Type ta Is Array (0 To 3) Of Std_Logic_Vector(7 Downto 0);
Signal tt: ta;
Begin

  Process(clk) Is
  Begin
    If(clk'Event And clk='1')Then
      tt(0) <= a;
      tt(1) <= tt(0);
      tt(2) <= tt(1);
      tt(3) <= tt(2);
      If tt(3)>tt(2) Then
        If (tt(3)-tt(2))>"01111111" Then dout <= tt(1);
        Else dout <= tt(3); End If;
        If (tt(3)-tt(2))="01111111" Then dout <= tt(1);
        Else dout<=tt(3); End If;
      Else
        If (tt(2)-tt(3))>="01111111"  Then dout <= tt(1);
        Else dout <= tt(3); End If;
      End If;
    End If;
  End Process;
End Architecture max1_arch;
```

正弦波压缩格式修改后的顶层设计如图 7－42 所示。修改后的正弦波 Modelsim 仿真波形结果如图 7－43 所示。其 Signal TapⅡ测试结果如图 7－44 所示。

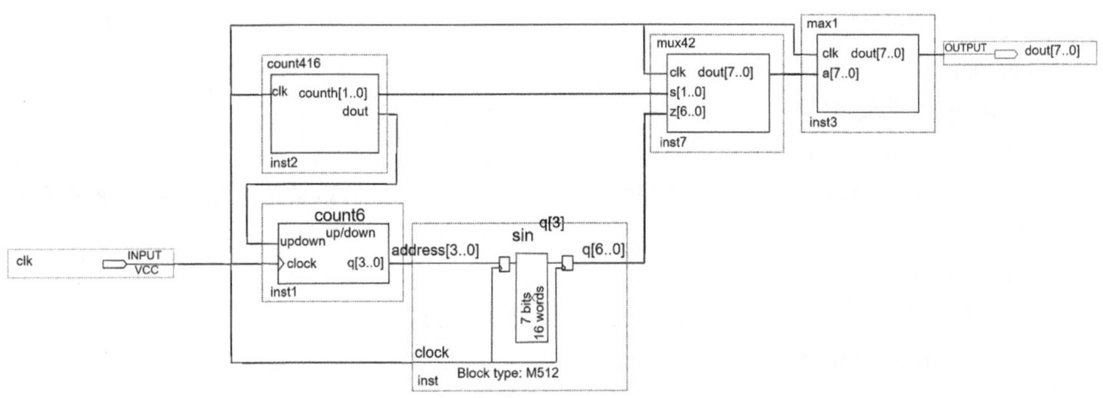

图 7－42　正弦波压缩格式修改后的顶层设计

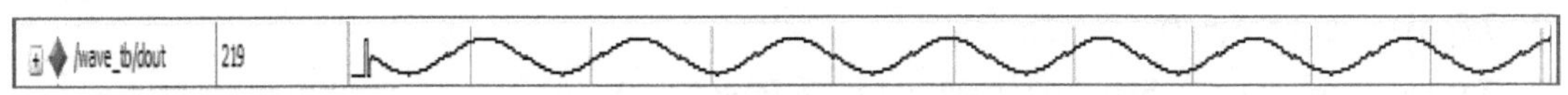

图 7－43　修改后的正弦波 Modelsim 仿真波形结果

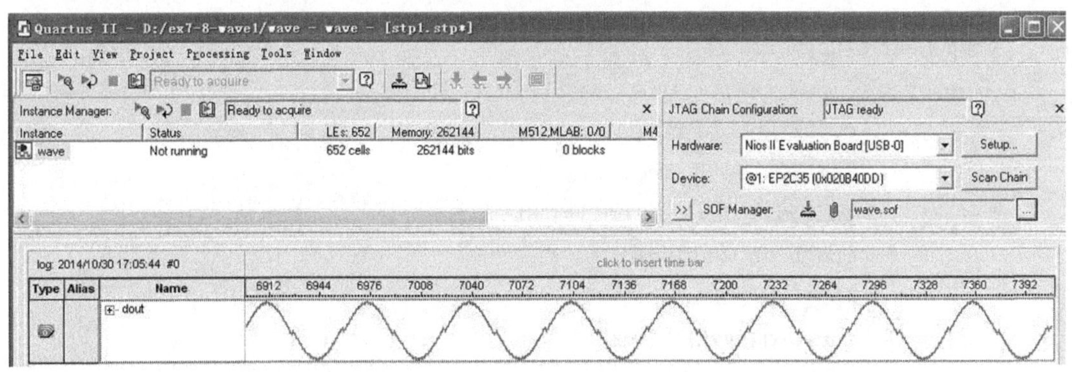

图 7－44　Signal Tap Ⅱ测试结果

7.7　利于属性 Attribute 指定综合

为了在线修改 ROM 数据，前面的正弦波 ROM 指定了采用 M4K 存储器，在生成 ROM 时会产生相应的 VHDL 程序，稍加修改，在这个程序里添加属性 Attribute，就可以指定 ROM 在芯片具体的块，如，是 M512 或是 M4K，具体依据芯片资源而定。下面对原来正弦波程序的修改，指定了“M4K”。有关综合实现的属性还有很多，比如，扇出、状态机编码等。查看 Quartus Ⅱ 的 Template 有一些参考。

示例 7－8f　sin. vhd

```
Library Ieee,Altera;                                          -- 加上 Altera 库
Use Ieee. Std_Logic_1164. All;
Library Altera_mf;
Use Altera_mf. all;
Use Altera. Altera_Syn_Attributes. All;              -- Altera_Syn_Attributes 程序包

Entity sin Is
  Port(address: In Std_Logic_Vector (5 Downto 0);
       clock: In Std_Logic := '1' ; ;
       q: Out Std_Logic_Vector (7 Downto 0));
End Entity sin;

Architecture syn Of sin Is
Signal sub_wire0: Std_Logic_Vector (7 Downto 0);
Attribute romstyle Of SYN:Architecture Is "M4K";   --用属性指定采用内嵌 ROM"M4K"

  Component altsyncram Is
    Generic (clock_enable_input_a: String;
             clock_enable_output_a: String;
             init_file: String;
             intended_device_family: String;
             lpm_hint: String;
             lpm_type: String;
             numwords_a: Natural;
             operation_mode: String;
             outdata_aclr_a: String;
             outdata_reg_a: String;
             widthad_a: Natural;
             width_a     : Natural;
             width_byteena_a: Natural);
    Port (clock0: In Std_Logic ;
          address_a: In Std_Logic_Vector (5 Downto 0);
          q_a: Out Std_Logic_Vector (7 Downto 0));
  End Component altsyncram;
Begin
  q <= sub_wire0(7 Downto 0);
  altsyncram_component : altsyncram
```

```
  Generic Map (clock_enable_input_a => "BYPASS",
               clock_enable_output_a => "BYPASS",
               init_file => "sint.mif",
               intended_device_family => "Cyclone II",
               lpm_hint => "Enable_Runtime_Mod=Yes, Instance_Name=sin",
               lpm_type => "altsyncram",
               numwords_a => 64,
               operation_mode => "ROM",
               outdata_aclr_a => "NONE",
               outdata_reg_a => "UNREGISTERED",
               ram_block_type => "M4K",
               widthad_a => 6,
               width_a => 8,
               width_byteena_a => 1)
  Port Map (clock0 => clock,
            address_a => address, q_a => sub_wire0);
End Architecture syn;
```

程序说明:有些属性说明综合器不支持,但采用它描述却可以给设计作更好的说明。

7.8 在线硬件调试的工具

Altera 公司推出了三种在线硬件调试的工具,分别是嵌入式逻辑分析仪 Signaltap II、在系统存储内容编辑器 In-System Memory Content Editor 和在系统信号源与探针 In-System Sources and Probes 。

7.8.1 在系统存储内容编辑器

In-System Memory Content Editor 是一种可以通过 JTAG 口在线读取或改写 FPGA 存储器数据的工具。前提是当设计使用的是 LPM 宏模块中的 RAM 或是 ROM,并且在模块调用时允许其对在线编辑,如图 7-45 所示,相应复选框要勾上,这里 ROM 的 ID 为 sin,原则上这些也可以用综合属性指定。

In-System Memery Content Editor 用法如下:

(1)连接 JTAG 线,在菜单 Tools 中选择 In-System Memory Content Editor 项,在右上角 Setup 中建立硬件。

(2)下载数据。参见图 7-46 右侧建立硬件和下载数据。

(3)读取存储器数据。右键单击上方数据文件 sin,在弹出菜单中选择 Read Data form In-System Memory 选项,即可以读出 RAM 中数据。

(4)写入数据到 RAM 中。方法如同读数据,先在数据框中直接修改数据,然后右击窗口左上角数据文件的 ram1,选择下拉菜单中 Write Data to In-System Memory 选项,这时就可以在 Signal Tap Ⅱ上看到修改过的波形了。

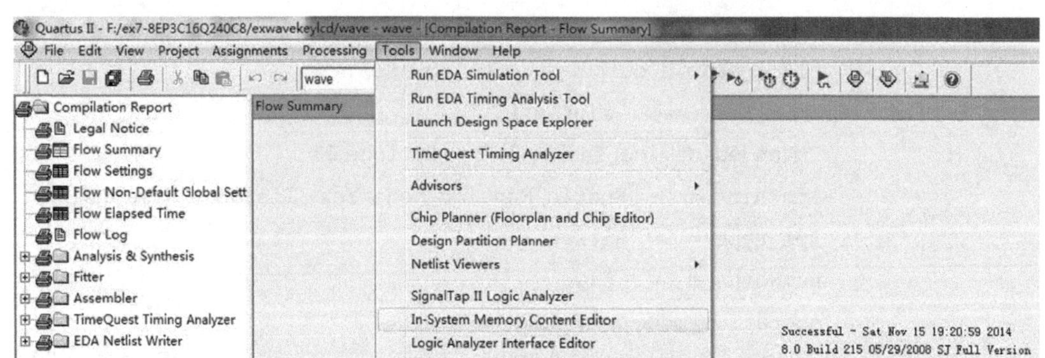

图 7-45　选择 In-System Memory Content Editor

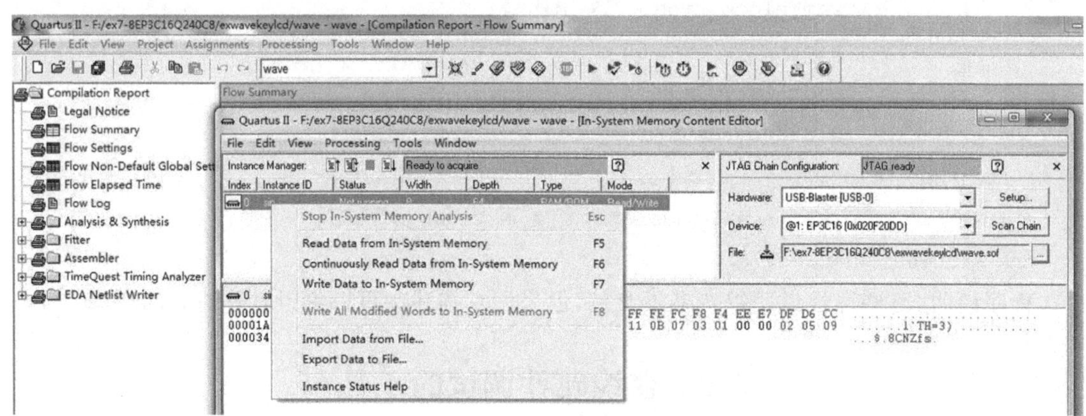

图 7-46　读取存储器数据

(5)修改数据。选中显示的 ROM 数据，随机修改两个数为 00。

(6)写入数据到存储器中。与(4)相同。

(7)打开嵌入逻辑分析仪重新采集，如图 7-47 所示。

(8)在线内容编辑器可重新导入 Import 原来的 mif 文件，在逻辑分析仪中采集可以看到修改以前的数据波形。

在线内容编辑器相当于一个任意波形发生器，既可以通过修改一个已有的波形给出新的波形，还可以在线输入不同的 . mif 数据来变换波形。

比如将图 7-47 逻辑分析仪采集的数据波形导出 Export 存储起来，比如作为低通滤波器验证时的输入波形，显然使验证更为便捷。

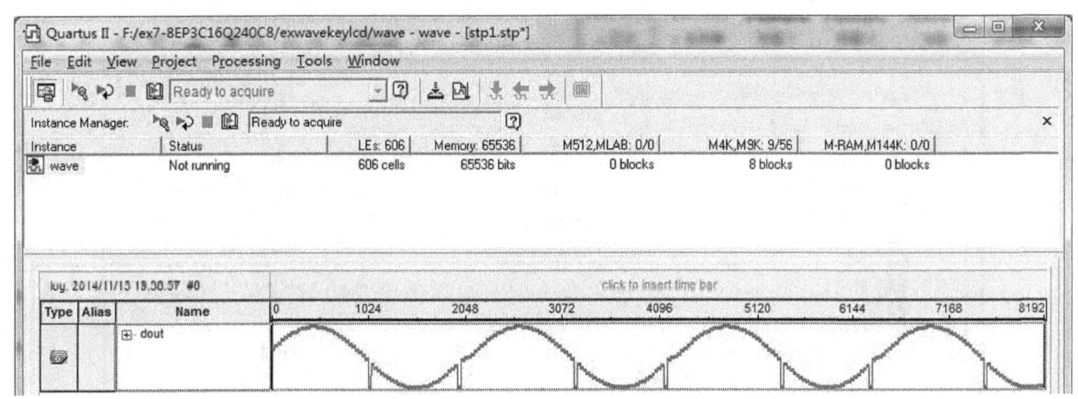

图 7 - 47　逻辑分析仪采集的数据

7.8.2　在系统信号源与探针测试

In - System Sources and Probes 可将测试置于设计之中，特点是实时采样测试。

在总图中添加测试模块 tprobe，接入后的顶层图如图 7 - 48 所示。

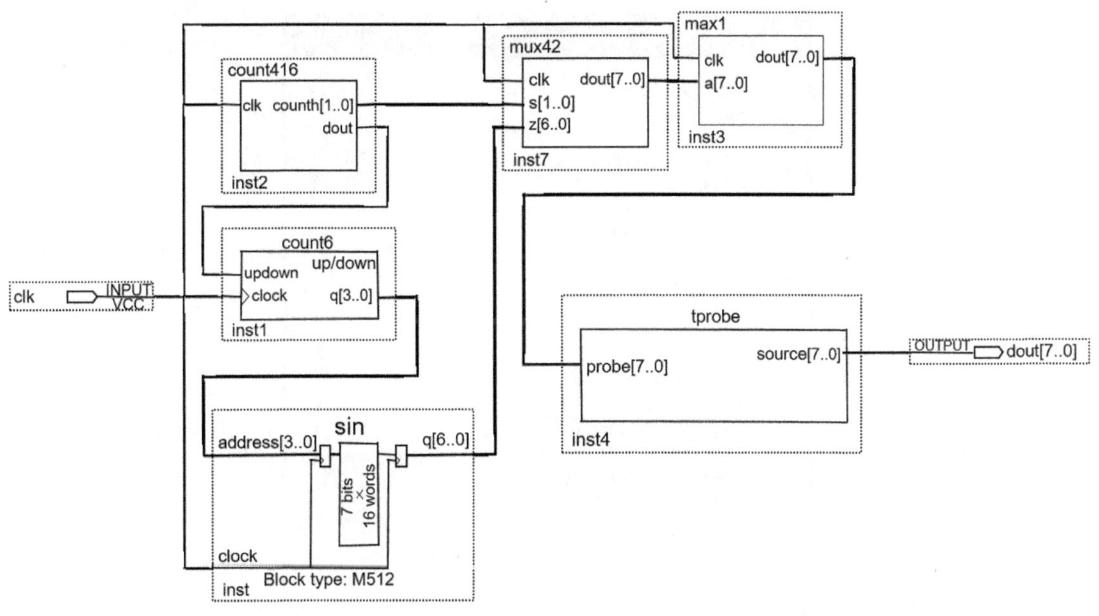

图 7 - 48　In-System Sources and Probes 接入后的顶层图

插入 LPM 模块的方法：在 JTAG - accessible Extensions 中选择 In-System Sources and Probes，加入后的结果如图 7 - 49 所示。

将生成的模块加入到设计中。在 File→New→选择 In-System Sources and Probes 即可，如图 7 - 50 所示。正弦数据测试结果如图 7 - 51 所示。

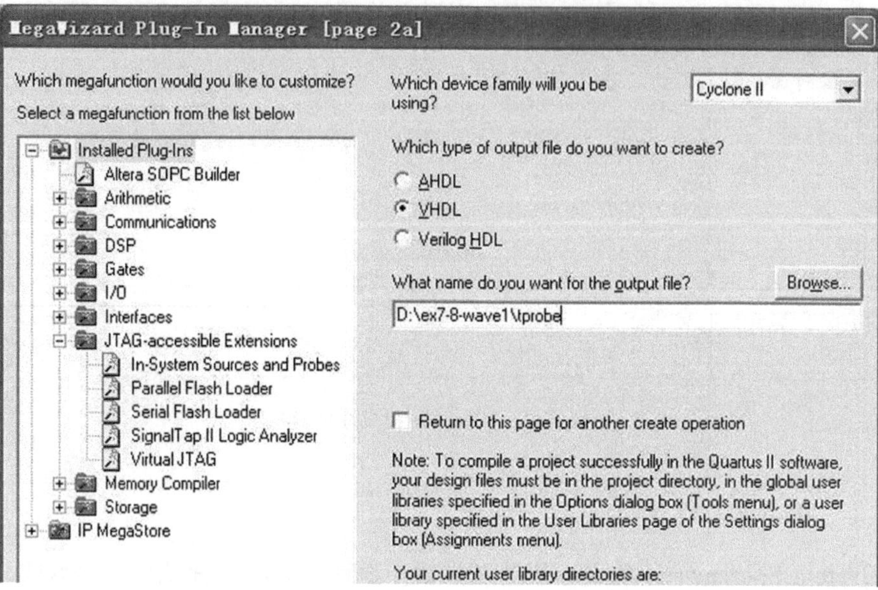

图 7-49　加入 In-System Sources and Probes 模块

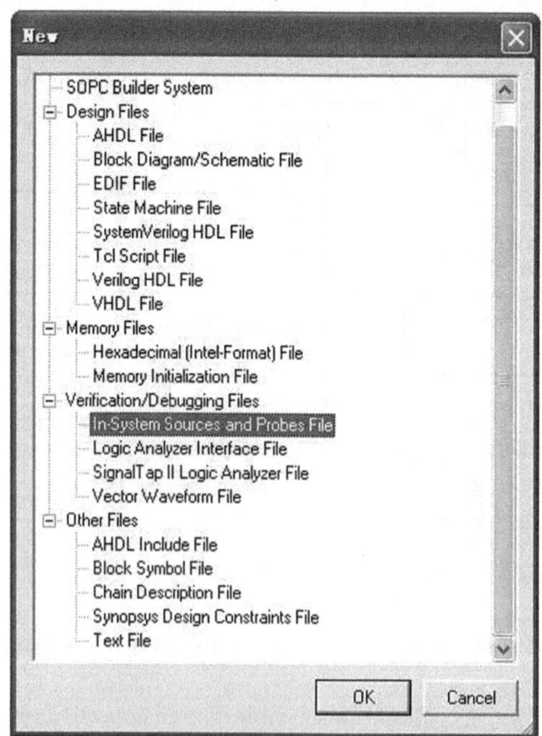

图 7-50　建立 In-System Sources and Probes

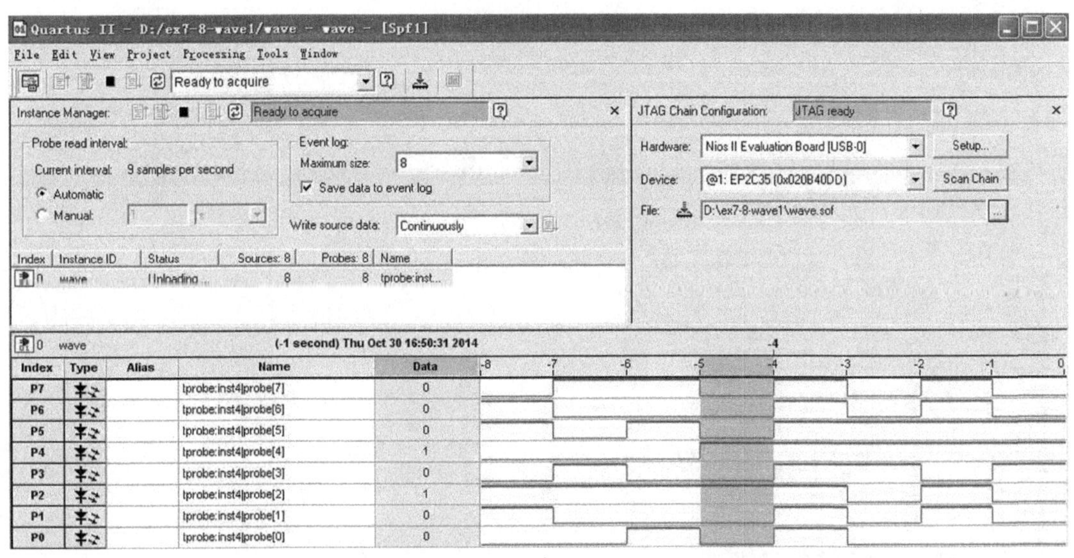

图 7-51 In-System Sources and Probes 正弦数据测试结果

7.9 Quartus Ⅱ连接 Modelsim 时序仿真

Quartus Ⅱ连接 Modelsim 时序仿真主要步骤：

(1)建立 Modelsim 工程；

(2)在 Project 窗口导入 wave. vho；

(3)编译 wave. vho 文件(其由 Quartus Ⅱ生成)；

(4)建立测试台文件。

以 wave. vho 为源生成并编辑 wave_tb. vhd 文件，仿真 wave_tb 文件，建立工程如图 7-52所示。

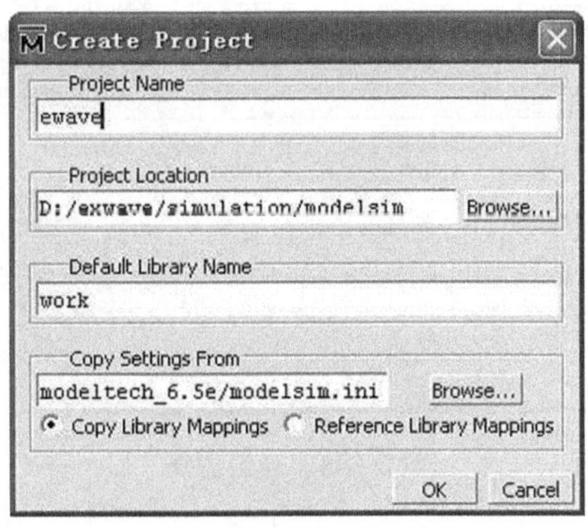

图 7-52 建立工程

Altera 仿真库放在 Modelsim 安装目录下的\cycloneii 中，要映射 cycloneii 库到 Modelsim 安装目录下的\cycloneii 中，建立的 Altera 仿真库如图 7-53 所示。

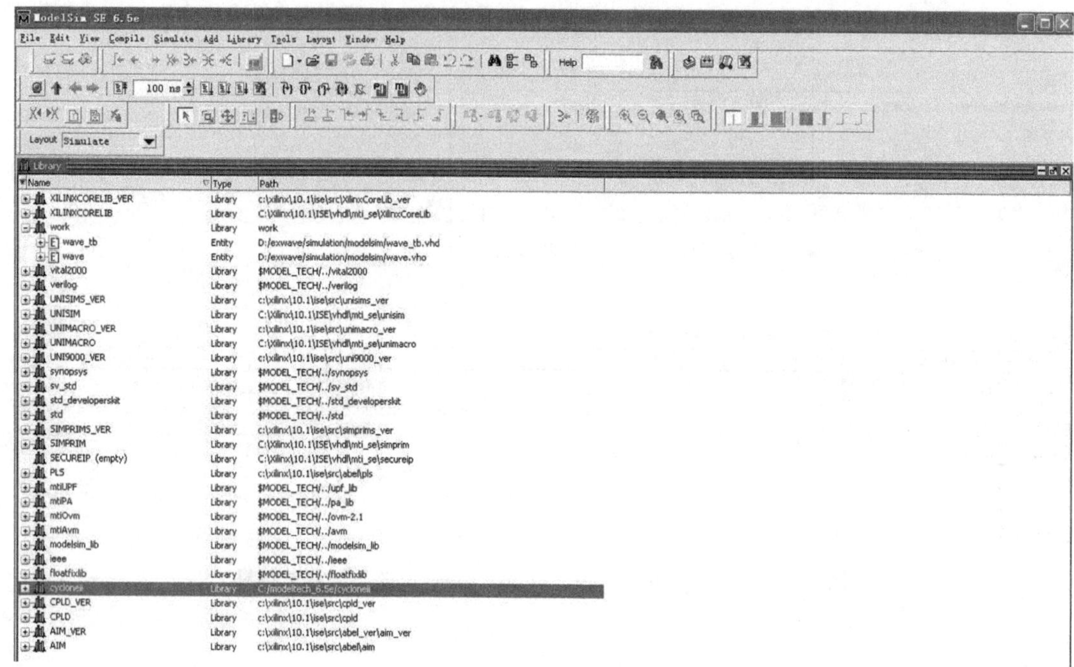

图 7 - 53　建立 Altera 仿真库

开始仿真的设置如图 7 - 54 所示。

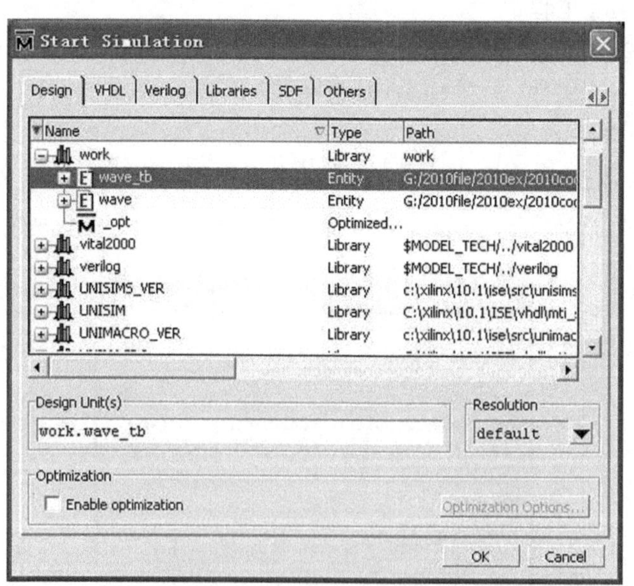

图 7 - 54　开始仿真

选中 SDF 选项，导入延时文件 *.sdo(后仿真)，在 Transcript 窗键入

```
Vsim>add wave *
Vsim>run 140 us
```

选择 dout 信号，选择 Format→Analog，Radix→Unsigned。

可以用 Analog(Custom)改变数据范围等,观察到正弦波 Modelsim 时序仿真结果。

7.10 SRAM 设计与仿真

SRAM 的功能与模块设计如表 7-4 和图 7-55 所示。

表 7-4 Sram 功能

cs_b	oe_b	we_b	mode	i/o pin
H	X	X	不选	High-Z
L	H	H	不输出使能	High-Z
L	L	H	读	data Out
L	X	L	写	data In

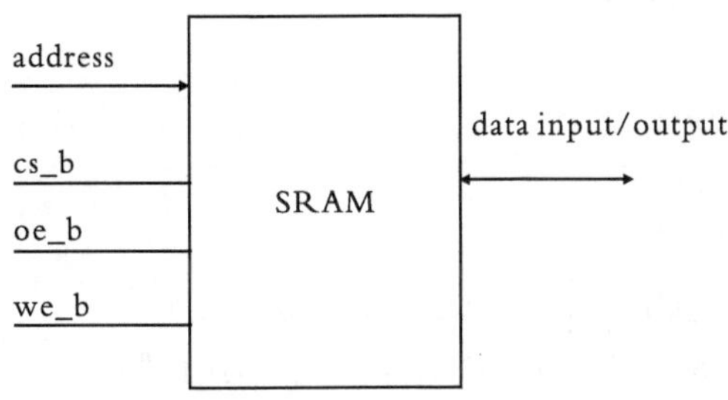

图 7-55 SRAM 模块设计

示例 7-9 单口单地址双向的 SRAM 设计

```
Library Ieee;
Use Ieee.Std_Logic_1164.All;
Use Ieee.Numeric_Std.All;

Entity ram6116 Is
  Port(cs_b,we_b,oe_b:In Std_Logic;
       address: In Unsigned(7 Downto 0);
       io:Inout Unsigned (7 Downto 0));                --双向接口
End Entity ram6116;

Architecture simple_ram Of ram6116 Is
Type ramtype Is Array(0 to 255) Of Unsigned(7 Downto 0);     --2 维数组
Signal ram1: ramtype:=(Others=>(Others=>'0'));
Begin
  io<="ZZZZZZZZ" When(cs_b='1' or we_b='0' or oe_b='1')
  Else ram1(To_Integer(address));                      --read from RAM
```

```
  Process(we_b,cs_b) Is
  Begin
    If cs_b='0' And Rising_Edge(we_b) Then
    --ram1(To_Integer(address'delayed))<=io;                     --写存储器
      ram1(To_Integer(address))<=io;                              --写存储器
    End If;
  End Process;
End Architecture simple_ram;
```

示例 7-9a　Ram6116 测试基准 1

```
Library Ieee   ;
Use Ieee.Numeric_Std.All    ;
Use Ieee.Std_Logic_1164.All    ;

Entity ram6116_tb   Is
End Entity ram6116_tb;

Architecture ram6116_tb_arch Of ram6116_tb Is
Signal io     :  Unsigned (7 Downto 0) :="10101010";                --"AA"
Signal oe_b    :Std_Logic :='0';
Signal we_b    :Std_Logic :='0';
Signal address:Unsigned (7 Downto 0):="00000000"    ;               --"00"
Signal cs_b    :Std_Logic    :='1';
Signal clk,clk1:Std_Logic :='1';
Signal c,c1:Integer Range 0 To 30:=0;

  Component ram6116 Is
    Port (io   : Inout Unsigned (7 Downto 0) ;
           oe_b: In Std_Logic ;
           we_b: In Std_Logic ;
           address: In Unsigned (7 Downto 0) ;
           cs_b: In Std_Logic );
  End Component ram6116;

Begin
  DUT: ram6116
  Port Map (io    => io    ,
            oe_b    => oe_b   ,we_b    => we_b    ,
            address => address    , cs_b    => cs_b ) ;
```

```
  cs_b<='0' After 10 ns;                                    --片选,'0'有效
  clk <= Not clk After 10 ns;
  clk1<= Not clk1 After 20 ns;
  Process(clk,we_b) Is
  Begin
    If (clk'Event And clk='1')  Then
      If (c<10) Then                                         --写 5 个数
         c<=c+1;
         we_b<= Not we_b ;
      Elsif c>=10 Then
         we_b<='1';
      End If;
    End If;
  End Process;

  Process(clk1) Is
  Begin
    If (clk1'Event And clk1='1')  Then
      If (c1<10) Then
        c1<=c1+1;
        io<=io+1;
        address<=address+1;
      Elsif c1>=10 And address>0 Then
        io<= "ZZZZZZZZ" ;
        address<=address-1;
      End If;
    End If;
  End Process;
  oe_b<='0' After 20 ns;
End Architecture ram6116_tb_arch;
```

SRAM 仿真结果如图 7-56 所示。

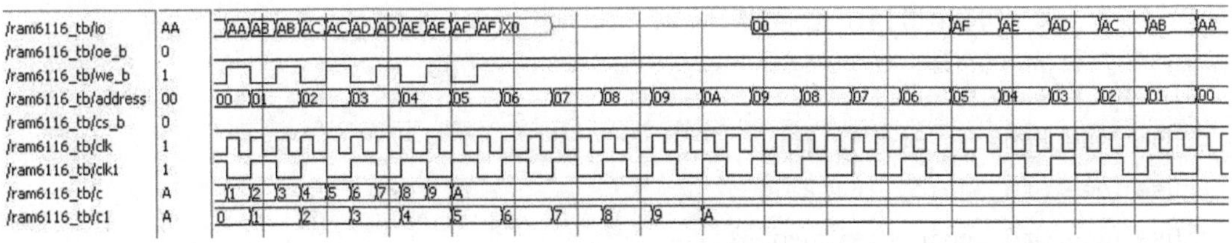

图 7-56　SRAM 仿真结果

在00～05地址写AA,AB,AC,AD,AE,AF;从05～01地址读出AF,AE,AD,AC,AB,AA;如堆栈在工作。

SRAM模块设计应用测试原理如图7-57所示。SRAM模块设计应用测试流程如图7-58所示。

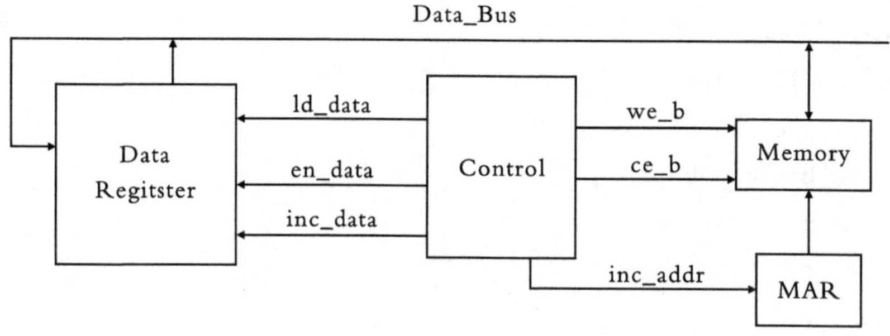

图7-57 SRAM模块设计应用测试

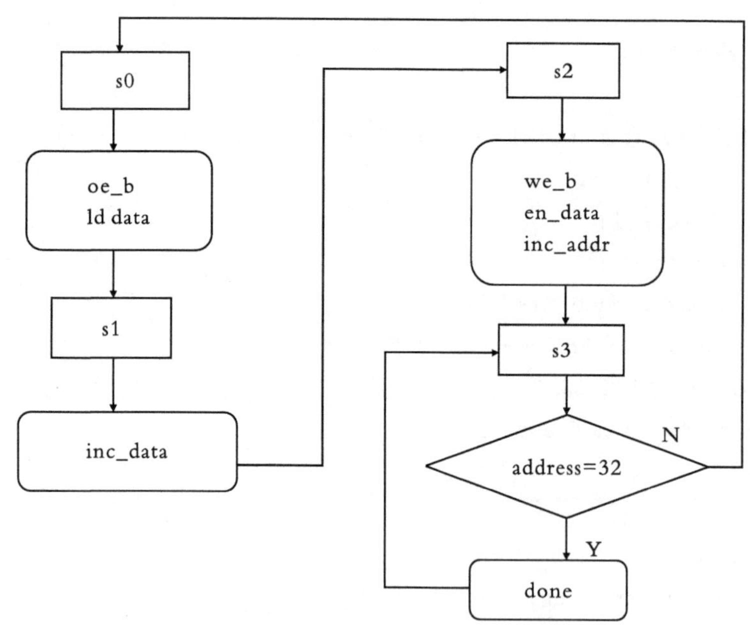

图7-58 SRAM模块设计应用测试流程

示例7-9b Ram6116测试基准2

```
Library Ieee;
Use Ieee.Numeric_Std.All;
Use Ieee.Std_Logic_1164.All;
Use Ieee.Std_Logic_Unsigned.All;

Entity ram6116_system Is
```

```
End Entity ram6116_system;

Architecture ramtest Of ram6116_system Is
  Component ram6116 Is
    Port(cs_b,we_b,oe_b:In Std_Logic;
         address:In Unsigned (7 Downto 0);
         io:Inout Unsigned (7 Downto 0));
  End Component ram6116;

Signal state,next_state:Integer Range 0 To 3;
Signal inc_addr,inc_data,ld_data,cs_b,done:Std_Logic:='0';
Signal clk,oe_b:Std_Logic:='1';
Signal en_data,we_b:Std_Logic:='0';
Signal data:Unsigned (7 Downto 0):="00000000";
Signal address:Unsigned (7 Downto 0):="00000000";
Signal io:Unsigned(7 Downto 0):="01010101";
Begin
  ram1:ram6116 Port Map(cs_b,we_b,oe_b,address,io);

  control:Process(state,address) Is
  Begin
    Case state Is
      When 0=> we_b<='1'; oe_b<='1';ld_data<='1';inc_data<='0';
                en_data<='0';
                next_state<=1;
      When 1=>we_b<='1'; oe_b<='1';ld_data<='0'; inc_data<='1';
               en_data<='0';inc_addr<='0';
               next_state<=2;
      When 2=>we_b<='0'; oe_b<='0';inc_addr<='1';en_data<='1';
               next_state<=3;
      When 3=> we_b<='1'; oe_b<='1'; en_data<='0';
                done<='1';
                next_state<=0;
        If(address ="00100000")Then
        Else next_state<=0;
        End If;
    End Case;
  End Process control;
  register_update:Process(clk) Is
```

```
  Begin
    If Rising_Edge(clk) Then state<=next_state;
      If(ld_data='1') Then data<=io ; End If;
      If(inc_data='1' And  oe_b='0') Then data<=data+1; End If;
      If(inc_addr='1' And we_b='0') Then address<=address+1 After 20ns;
        If(address ="00100000") Then  address<="00000000";End If;
      End If;
    End If;
    clk<= Not clk After 100ns;
  End Process register_update;
  io<=data When en_data<='1' Else "ZZZZZZZZ";
End Architecture ramtest;
```

示例 7－9 程序说明：在读存储器时，io 作为输入时，其作为输出的功能应该用“ZZZZZZZZ”封住。

将数据写入了 SRAM，通过点击 View→Memory 可以看到写入存储器的数据，如图 7－59 所示，同时将数据从 io 端口读出，如图 7－60 所示。

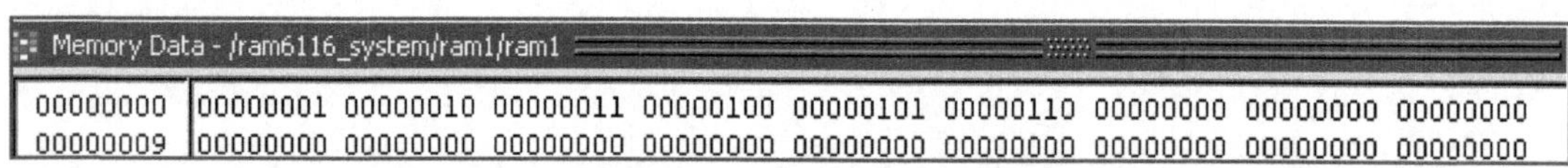

图 7－59　SRAM 仿真写入存储器的数据

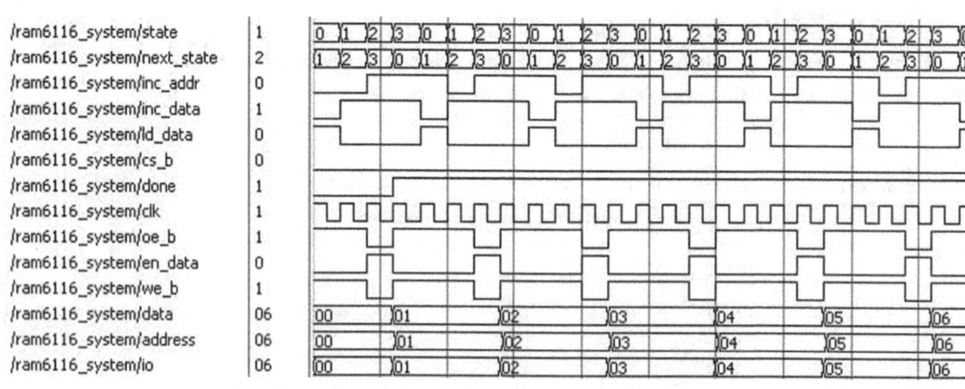

图 7－60　写入存储器的数据从 io 端口读出

示例 7－10　双口双地址单向 SRAM 设计

```
Library Ieee;
Use Ieee.Std_Logic_1164.All;
Use Ieee.Std_Logic_Arith.All;
Use Ieee.Std_Logic_Unsigned.All;
```

```
Entity sramd Is
  Generic(Width : Integer:=8;                  --共存储 64 位,位宽 8
          Depth : Integer:=8;                  --存储深度
          Addr : Integer:=3);                  --3 位地址
  Port(data : In Std_Logic_Vector(Width-1 Downto 0);
       q : Out Std_Logic_Vector(Width-1 Downto 0);
       clk : In Std_Logic;
       nwe,nre : In Std_Logic;
       waddress: In Std_Logic_Vector(Addr-1 Downto 0);        --写地址
       raddress: In Std_Logic_Vector(Addr-1 Downto 0));       --读地址
End Entity sramd;

Architecture behav Of sramd Is
Type mem Is Array (0 To Depth-1) Of
    Std_Logic_Vector(Width-1 Downto 0);
Signal ramtmp : mem;
Begin
  pwrite: Process (clk) Is
  Begin
    If (clk'Event And clk='1') Then
      If (nwe ='0') Then
        ramtmp (Conv_Integer (waddress)) <= data;
      End If;
    End If;
  End Process pwrite;

  pread: Process (clk) Is
  Begin
    If (clk'Event And clk='1') Then
      If (nre = '0') Then
        q <= ramtmp(Conv_Integer (raddress));
      End If;
    End If;
  End Process pread;
End Architecture behav;
```

双端口 RAM 设计如图 7 - 61 和图 7 - 62 所示。

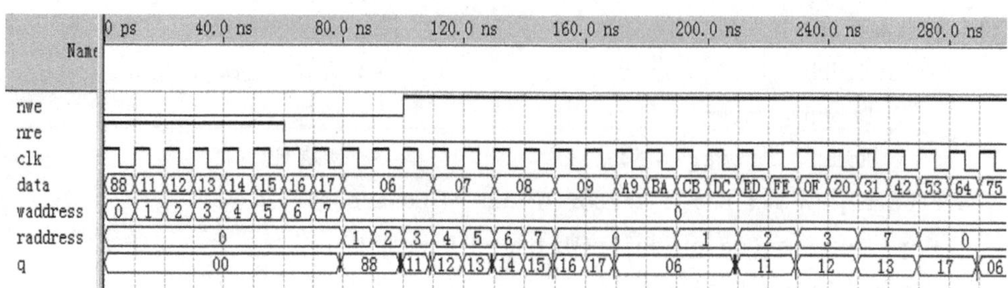

图 7-61　双口双地址单向 SRAM 仿真 1

说明:在写信号没有完成时,也可以读。在地址 7,0 写时,从地址 0,1 开始的读也开始了。第一个 0 地址的读为 88,因为第二个 0 的写还没有开始,第二个 0 地址的读就是 06。换句话说,在写的同时可以读。

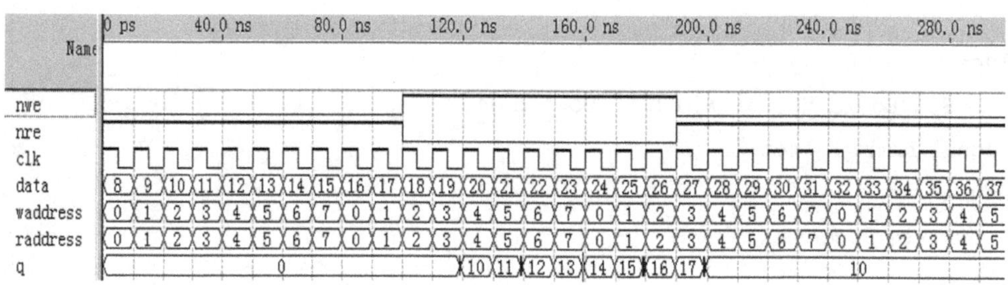

图 7-62　双口双地址单向 SRAM 仿真 2

说明:从地址 0 开始,写入数据 8,9,10,11,12 等。从地址 2 开始,读出了数据 10,11,12 等。

注意:地址数 n 与 Depth 的关系:2^n=Depth。

示例 7-11　双口单地址单向 SRAM 设计

```
Library Ieee;
Use Ieee.Std_Logic_1164.All;
Use Ieee.Std_Logic_Unsigned.All;

Entity sram1 Is
  Port(clk,cs,wr: In Std_Logic;
       adr: In Std_Logic_Vector(3 Downto 0);
       din: In Std_Logic_Vector(7 Downto 0);
       dout: Out Std_Logic_Vector(7 Downto 0));
End Entity sram1;

Architecture behav Of sram1 Is
Subtype word Is Std_Logic_Vector(7 Downto 0);            --数组自定义
```

```
Type memory Is Array(0 To 15) Of word;                          --2 维数组自定义
Signal sram: memory;                                            --2 维数组信号声明
Signal adr_In: Integer Range 0 To 15;
Begin
  adr_in<=Conv_Integer(adr);                                    --标准逻辑矢量转为整数
  pwrite: Process(wr,cs,adr_in,din,sram,clk) Is
  Begin
    If(clk'Event And clk='1') Then
      If(cs='0' And wr='0') Then
        sram(adr_in)<=din;                                      --数据写入 sram
      End If;
    End If;
  End Process pwrite;

  pread: Process(wr,cs,adr_in,sram,clk) Is
  Begin
    If(clk'Event And clk='1') Then
      If(wr='0' And wr='0') Then
        dout<=sram(adr_in);                                     --数据读入 sram
      Else
        dout<=(Others=>'Z');
      End If;
    End If;
  End Process pread;
End Architecture behav;
```

双口单地址单向 SRAM 仿真如图 7-63 所示。

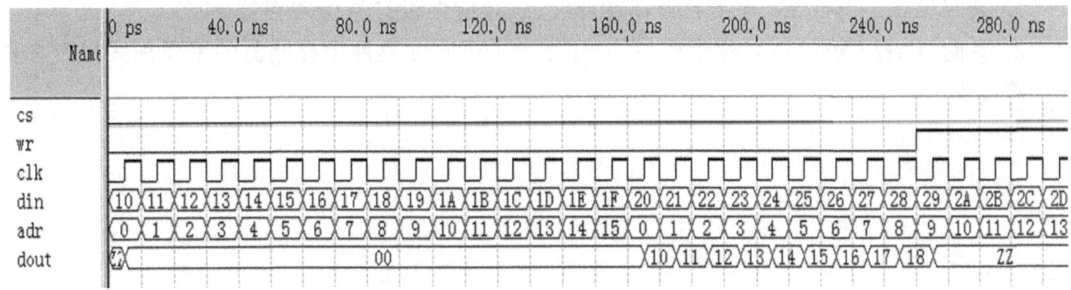

图 7-63　双口单地址单向 SRAM 仿真

程序说明：在存储器设计与应用中往往要用到 2 位数组，在 VHDL 中 2 位数组要采用自定义类型的方式来说明。要根据应用需求来决定'Z'的使用，因为一些 FPGA 内部没有高祖态。

本章示例 7－9b 是很有意思的，它将存储器应用通过状态机来设计，使得存储器可以按照一定的节拍工作，前面介绍的运算也可以按照节拍，这样就容易做系统整合了。

习 题

7－1　设计采用 1 位加法器实现的 32 位串行加法器，设计框图如图 7－64 所示。

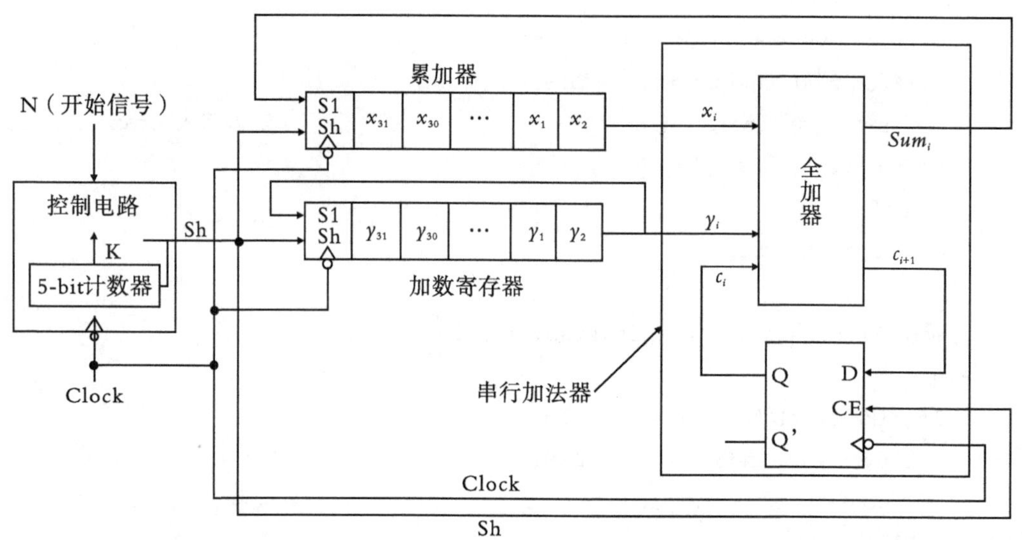

图 7－64　1 位加法器实现 32 位串行加法器

7－2　查阅资料设计增加多种波形发生器的波形种类。

7－3　查阅资料实现同步与异步的 FIFO 设计与仿真。

扩展学习与总结

1. 查阅 Altra Cyclone 器件手册，采用嵌入式存储器设计移位寄存器。

2. 查阅 Altra Cyclone 器件手册，进一步了解 Cyclone 器件中存储器的工作模式以及如何外接存储器。

8　通信模块设计

本章的主要内容与方法：

(1)数字相关器设计。

(2)巴克码生成与检测。

(3)扰码与解扰码。

(4)基于 DDS 的调制解调。

(5)串并转换。

(6)校验与纠错编解码。

(7)传输编解码。

8.1　采用流水线技术设计高速数字相关器

在数字通信系统中，常用一个特定的序列作为数据开始的标志，称为帧同步字。在数字传输的过程中，发送端要在发送数据之前插入帧同步字；接收机需要在已解调的数据流中搜寻帧同步字，以确定帧的位置和帧定时信息。帧同步字一般为一系列连续的码元，在接收端需要对这一系列连续的码元进行检测，如果与预先确定的帧同步字吻合，则说明接收端与发送端的数据是保持同步的，开始接收，否则不能进行接收。

数字相关器常用于帧同步检测、扩频接收机、误码校正以及模式匹配等领域。

8.1.1　数字相关器原理

数字相关器检测等长度的两个数字序列间相等的位数，比较等长度的两个数字序列之间有多少位相同，多少位不同，即序列间的相关运算。

一位相关器，即异或门，其结果可以表示两个一位数据的相关程度。异或为 0 表示数据位相同；异或为 1 表示数据位不同。多位数字相关器可以由多个一位相关器级联构成。N 位数字相关器的运算可以分解为以下两个步骤：

(1)对应位进行异或运算，得到 N 个一位相关运算结果；

(2)统计 N 位相关运算结果中 0 或 1 的数目，得到 N 位数字中相同位和不同位的数目。

数字相关器用于帧同步检测时，当数字相关器接收到一组数据时，在时钟的上升沿对帧同步字进行检测。如果帧同步字是一个连续 16 位的码元，数字相关器在进行检测的过程中，只有当连续检测到 16 位的码元与预先设定的帧同步字完全相同时，才由输出端输出信号表示帧同步；否则，任何一位出现不相等，数字相关器又将重新开始进行检测，直到出现连续的 16 位码元与预先设定的 16 位码元完全相等时才进行输出。

8.1.2　数字相关器的设计

在 16 位并行数字相关器中，由于实现 16 位并行相关器需要的乘积项、或门过多，因此为降低耗用资源，可以分解为 4 个 4 位相关器，然后用两级加法器相加得到全部 16 位的相关结果，其结构图如图 8-1 所示。如果直接实现该电路，整个运算至少要经过三级门延时。随着相关数目的增加，速度还将进一步降低。为提高速度，采用“流水线技术”进行设计，模块中对每一步运算结果都进行锁存，按照时钟节拍逐级完成运算的全过程。虽然每组输入值需要经过三个节拍后才能得到运算结果，但是，每个节拍都有一组新值输入到第一级运算电路，每级运算电路上都有一组数据同时进行运算，所以总的来讲，每步运算花费的时间只有一个时钟周期，从而使系统工作速度基本等于时钟工作频率。

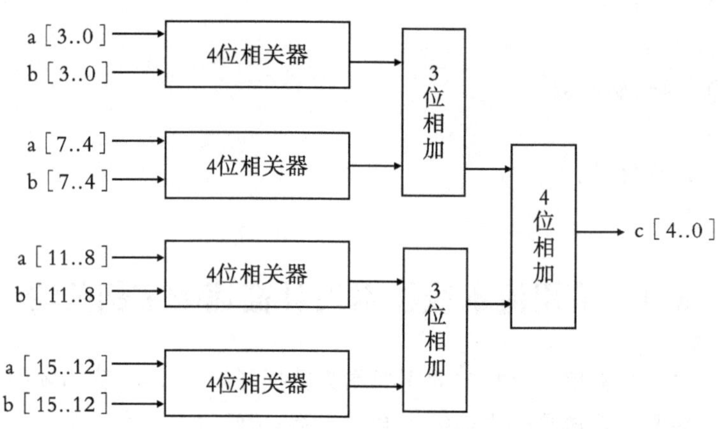

图 8-1 并行 16 位数字相关器

在使用 VHDL 进行高速数字相关器设计时，主要实现 4 位相关器和多位加法器模块的设计。其中元件有 4 位相关器模块、3 位加法器模块和 4 位加法器模块。

示例 8-1 并行 16 位数字相关器代码

```
--并行 16 位数字相关器顶层设计 correlator.vhd
Library Ieee;
Use Ieee.Std_Logic_1164.All;
Use Work.All;

Entity correlator Is
  Port(output:Out Std_Logic_Vector(4 Downto 0);          --比较结果
       input:In Std_Logic_Vector(15 Downto 0);
       ctrl:In Std_Logic;
       clk:In Std_Logic);                                 -- 同步时钟
End Entity correlator;

Architecture correlator_beba Of correlator Is
Signal a,b: Std_Logic_Vector(15 Downto 0);                --a 基数;b 和 a 比较
Signal sum1,sum2,sum3,sum4: Std_Logic_Vector(2 Downto 0);
```

```
Signal temp1,temp2  :Std_Logic_Vector(3 Downto 0);

Begin
  Process(clk,Input) Is
  Begin
    If (clk'Event And clk='1')Then
      If ctrl='0'Then  a<=input;
      Else b<=input;
      End if;
    End If;
  End Process;

  c0: detect Port Map(sum1,a(3 Downto 0),b(3 Downto 0),clk);
                                                    --检测[3..0]位
  c1: detect Port Map(sum2,a(7 Downto 4),b(7 Downto 4),clk);
                                                    --检测[7..4]位
  c2: detect Port Map(sum3,a(11 Downto 8),b(11 Downto 8),clk);
                                                    --检测[11..8]位
  c3: detect Port Map(sum4,a(15 Downto 12),b(15 Downto 12),clk);
                                                    --检测[15..12]位
  c4: addn Generic Map(N=>3)Port Map(temp1,sum1,sum2,clk);
                                                    --3 位加法求部分和
  c5: addn Generic Map(N=>3)Port Map(temp2,sum3,sum4,clk);
  c6: addn Generic Map(N=>4)Port Map(output,temp1,temp2,clk);
                                                    --4 位加法求和
End Architecture correlator_beba;

--4Bit correlator  4 位相关模块实体说明 detect.vhd
Library Ieee;
Use Ieee.Std_Logic_1164.All;
Entity detect Is                                    --4 位相关模块实体声明
  Port(sum: Out Std_Logic_Vector(2 Downto 0);       --比较结果
       a,b : In Std_Logic_Vector(3 Downto 0);       --两组数据
       clk : In Std_Logic );                        --同步时钟
End Entity detect;

--4 位相关模块方法 1,结构体 detect_one
Architecture detect_one Of detect Is
Signal ab: Std_Logic_Vector(3 Downto 0);
```

```
Begin
  ab<=a Xor b;

  Process(ab) Is
  Begin
    Case ab Is                              --利用 Case 语句完成 4 个 1 位相关结果的统计
      When "0000"=>sum<="100";                               --4;四位都相同
      When "0001"|"0010"|"0100"|"1000"=>sum<="011";          --3;只有 1 位不同
      When "0011"|"0101"|"0110"|"1001"|"1010"|"1100"
           =>sum<="010";                                     --2;两位不同
      When "0111"|"1011"|"1101"|"1110"=>sum<="001";          --1;三位不同
      When "1111"=>sum<="000";                               --0;都不同
      When Others=>Null;
    End Case;
  End Process;
End Architecture detect_one;

--4 位相关模块方法 2,结构体 detect_two
Architecture detect_two Of detect Is
Signal ab : Std_Logic_Vector(3 Downto 0);
Begin
  ab<=a Xor b;                                               --判断 a,b 是否相同

  Process(clk) Is
  Begin
    If clk'Event And clk='1' Then
      If ab="1111" Then sum<="000";                --列出各种组合,输出相应相关值
      Elsif ab="0111" Or ab="1011" Or ab="1101" Or
            ab="1110" Then sum<="001";
      Elsif ab="0001" Or ab="0010" Or ab="0100" Or
            ab="1000" Then sum<="011";
      Elsif ab="0000" Then sum<="100";
      Else sum<="010";
      End If;
    End If;
  End Process;
End Architecture detect_two;

--n 位加法器 addn.vhd
```

```
Library Ieee;
Use Ieee. Std_Logic_1164. All;
Use Ieee. Std_Logic_Unsigned. All;
Entity addn Is                                     --n 位加法器实体说明
  Generic(n:Positive);
  Port(add: Out Std_Logic_Vector(N Downto 0);      --和
       a,b : In Std_Logic_Vector(N-1 Downto 0);    --两组数据,加数,被加数
       clk : In Std_Logic );                       -- 同步时钟
End Entity addn;

Architecture addn_beba Of addn Is
Signal aa,bb: Std_Logic_Vector(N Downto 0);
Begin
  aa<='0'&a;                                       --扩展,便于加法进位
  bb<='0'&b;

  Process(clk,a,b) Is
  Begin
    If (clk'Event And clk='1') Then
      add<=aa+bb;                                  --前面进行了扩展,进位不会溢出
    End If;
  End Process;
End Architecture addn_beba;
```

示例 8-2　顺序 16 位数字相关器代码

```
Library Ieee;
Use Ieee. Std_Logic_1164. All;
Use Ieee. Std_Logic_Unsigned. All;
Use Work. All;

Entity addm Is
  Generic(N: Positive:=4);
  Port(a,b: In Std_Logic_Vector(15 Downto 0);
       clk,en: In Std_Logic;
       c: Out Std_Logic_Vector(N-1 Downto 0));
End Entity addm;

Architecture rtl Of addm Is
Type darray Is Array(0 To 3) Of Std_Logic_Vector(3 Downto 0);
```

```
Signal sum: darray;
Signal c0,c1: Std_Logic_Vector(3 downto 0);
Signal qs: Std_Logic_Vector(1 Downto 0);

  Function detectf(a,b:Std_Logic_Vector) Return Std_Logic_Vector Is
  Variable ab,sum: Std_Logic_Vector(N-1 Downto 0);
  Begin
    ab:=a Xor b;

    Case ab Is                         --利用 Case 语句完成 4 个 1 位相关结果的统计
      When "0000"=>sum:="0100";                              --4;四位都相同
      When "0001"|"0010"|"0100"|"1000"=>sum:="0011";         --3;只有 1 位不同
      When "0011"|"0101"|"0110"|"1001"|"1010"|"1100"=>sum:="0010";
                                                             --2;两位不同
      When"0111"|"1011"|"1101"|"1110"=>sum:="0001";          --1;三位不同
      When"1111"=>sum:="0000";                               --0;都不同
      When Others=>Null;
    End Case;
    Return sum;
  End Function detectf;

Alias x3: Std_Logic_Vector(3 Downto 0) Is a(15 Downto 12);   --别名
Alias x2: Std_Logic_Vector(3 Downto 0) Is a(11 Downto 8);
Alias x1: Std_Logic_Vector(3 Downto 0) Is a(7 Downto 4);
Alias x0: Std_Logic_Vector(3 Downto 0) Is a(3 Downto 0);
Alias y3: Std_Logic_Vector(3 Downto 0) Is b(15 Downto 12);
Alias y2: Std_Logic_Vector(3 Downto 0) Is b(11 Downto 8);
Alias y1: Std_Logic_Vector(3 Downto 0) Is b(7 Downto 4);
Alias y0: Std_Logic_Vector(3 Downto 0) Is b(3 Downto 0);
Begin

  Process(clk) Is                                            --4 计数
  Begin
    If(clk'Event And clk='1') Then
      a<=a1; b<=b1;
      If(qs="11") Then qs<="00";
      Elsif(qs<"11") Then
        qs<=qs+1;
      End If;
```

```
    End If;
  End Process;

Alias y0: Std_Logic_Vector(3 Downto 0) Is b(3 Downto 0);
Begin

  Process(clk) Is
  Begin
    If(clk'Event And clk='1') Then
      a<=a1; b<=b1;
      If(qs="11") Then qs<="00";
      Elsif(qs<"11") Then
        qs<=qs+1;
      End If;
    End If;
  End Process;

  gen_16: clkgen3 Generic Map(15)                              --分频元件例化
    Port Map (i_clk_50m =>clk,o_clk=>o_clk);

  Process(o_clk) Is
  Begin
    If(o_clk'Event And o_clk='1') Then
      c0<=sum(0)+sum(1);
      c1<=sum(2)+sum(3);
      c<=c0+c1;
    End If;
  End Process;

  Process(clk,qs,x0,x1,x2,x3,y0,y1,y2,y3) Is                   --计数值编码
  Begin
    If(clk'Event and clk='1) Then
      Case qs Is
        When "00"=>sum(0)<=detectf(x0,y0);
        When "01"=>sum(1)<=detectf(x1,y1);
        When "10"=>sum(2)<=detectf(x2,y2);
        When "11"=>sum(3)<=detectf(x3,y3);
        When Others=>sum(3)<="1000";
      End Case;
```

```
    End If;
  End Process;
End Architecture rtl;
```

程序说明：

(1)示例 8-1 并行程序，采用了并行语句生成(Generate)语句调用元件。

(2)示例 8-2 顺序程序，采用了函数形式调用元件，为了体现顺序，采用计数及编码分为了 4 个节拍，将相关的加法带入时钟进程中，形成流水线。

8.2　巴克码生成与检测

设计内容与要求：

(1)采用 2 种方法生成巴克码(计数编码法，计数查表法)。

(2)采用 3 种方式检测巴克码(移位分频逻辑运算法，相关运算法，状态机法)。

(3)巴克码生成与巴克码检测联调。

生成并检测巴克码的方法，如图 8-2 所示。

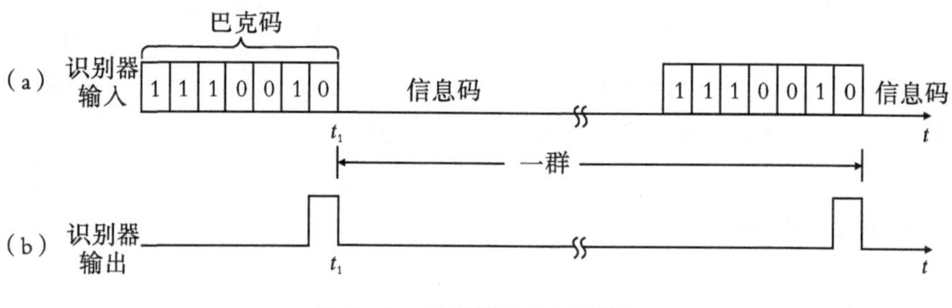

图 8-2　巴克码生成与检测

8.2.1　巴克码生成原理

在数字同步通信中，发送端需要在信息的起止时刻插入具有特殊意义的码，称为群同步码。群同步的特殊码组的基本要求是：具有尖锐单峰特性的自相关函数，便于与信息码区别，码长适当，以保证传输效率。巴克码是符合上述要求的特殊码组之一，常用于群同步。

在数字通信系统中，同步是非常关键的。由于信号的远距离传输，不可避免存在信号延时、干扰、非线性失真、收发两端的时钟偏差等。为保证数字传输信号的有效性，必须进行同步。根据同步的作用，可以分为：载波同步、位同步、帧同步、网同步。

数字通信中，信号流的最小单元是码元，若干码元构成一个帧，若干个帧再构成一个复帧，……。在接收端，必须分辨出每个帧的起始和接收，否则，无法正确恢复信息。这种同步，被称为帧同步(又称群同步)。帧同步可以用多种方法来实现，在此仅列举一种：连贯插入法，即在每一帧的开头连续插入一个特殊码组，比如巴克码，收端检测到该特殊码组的存在，就意味着帧开始了。

巴克码是一个有限长的数字序列。一个 n 位巴克码序列 $\{x_i\}$，其中 $1 \leqslant i \leqslant n$，$x_i$ 取值为 $+1$ 或者 -1，其局部自相关函数满足

$$R(j)=\sum_{i=1}^{n-j}x_i x_{x+j}=\begin{cases}n, & j=0\\ 0,\pm 1, & 0<j<n\\ 0, & j\geqslant n\end{cases} \tag{8-1}$$

即当 $j=0$ 时，巴克码的局部自相关函数达到峰值；当其他 j 值时，$R(j)$ 在 ± 1 附近波动，可以用作帧同步的特殊码组。

符合上述自相关特性的码组是存在的，比如 $\{+1,+1,+1,-1,-1,+1,-1\}$ 就是 7 位巴克码序列。当 $j=0$ 时，$R(j)=\sum_{i=1}^{7}x_i^2=7$，达到峰值；当 $j=1$ 时，$R(j)=1$；当 $j=3,5,7$ 时，$R(j)=0$；当 $j=2,4,6$ 时，$R(j)=-1$。

一个 n 位长的码组 $\{x_1, x_2, x_3,\cdots,x_n\}$，其中 x_i 的取值为 $+1$ 或 -1，若它的局部相关函数为

$$R(j)=\sum_{i=1}^{n-j}x_i x_{i+1}=\begin{cases}\pm 1,\\ 0, j\geqslant n\end{cases} \tag{8-2}$$

则称这种码组为巴克码，其中 j 表示错开的位数。目前已找到的所有巴克码组如表 8-1 所示。其中，+、- 号表示 x_i 的取值为 +1、-1，分别对应二进制码的“1”或“0”。

表 8-1　7 位巴克码组自相关函数计算

自相关函数	$R(0)$	$R(1)$	$R(2)$	$R(3)$
码组 x_i	+++--+-	+++--+-	+++--+-	+++--+-
码组 x_i+1	+++--+-	-+++--+	+-+++--	-+-+++-
相关结果	+++++++	-++-+--	+-+---+	-+---++
相关计算	1+1+1+1+1+1+1	1+1-1+1-1-1	-1+1-1-1+1	-1-1+1+1
算数结果	7	0	-1	0

如表 8-1 方法，可计算出：$j=0, R(j)=7$；$j=1,3,5$ 时，$R(j)=0$；$j=2,4,6$ 时，$R(j)=-1$；根据定义，$R(7)=0$。

根据这些值，利用偶函数性质，可以作出 7 位巴克码的 $R(j)$ 与 j 的关系曲线，其自相关函数在 $j=0$ 时具有尖锐的单峰特性。这一特性正是连贯式插入群同步码组的主要要求之一。

根据以上讨论的原理给出巴克码的检出模型。其关键是，若需要在数字信号流中检出巴克码组，只要检测序列的自相关函数即可。

选择 7 位巴克码来进行设计，序列为 $\{+1,+1,+1,-1,-1,+1,-1\}$。

8.2.2　巴克码检测原理

代码序列检测器是一种同步时序电路，用于搜索、检测输入的二进制代码串中是否出现指定的代码序列。以 7 位巴克码为例，用 7 级移位寄存器、相加器和判决器就可以组成一个巴克码识别器。当输入码元的“1”进入某移位寄存器时，该移位寄存器的 1 端输出电平为 +1，0 端输出电平为 -1；反之，当进入“0”码时，该移位寄存器的 0 端输出电平为 +1，1 端输出电平为 -1。各移位寄存器输出端的接法与巴克码的规律一致，这样识别器实际上是对输入的巴克码进行相关运算。当一帧信号到来时，首先进入识别器的是群同步码组，只有当 7 位巴克码在

某一时刻,正好已全部进入 7 位寄存器时,7 位移位寄存器输出端都输出+1,相加后得最大输出+7,其余情况相加结果均小于+7。若判别器的判决门限电平定为+6,那么就在 7 位巴克码的最后一位 0 进入识别器时,识别器输出一个同步脉冲表示一群的开头。

示例 8-3 巴克码生成与检测

巴克码生成与检测顶层设计如图 8-3 所示。

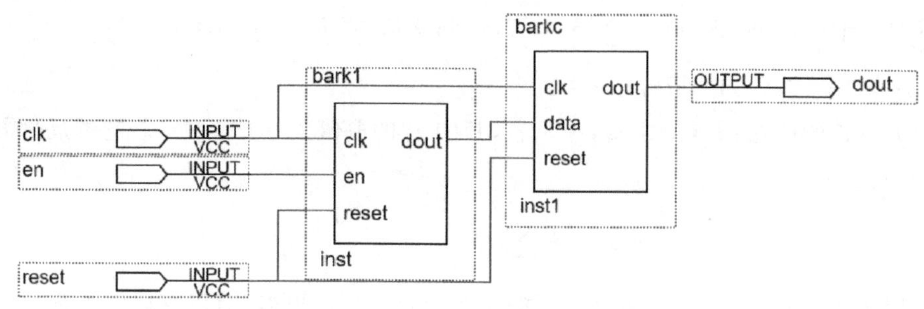

图 8-3 巴克码生成与检测顶层设计

将图 8-3 原理图转换为 VHDL,在 QuartusⅡ中自动生成 VHDL 顶层描述,操作如下:File→Create/Update→Create HDL Design for Current File

```
--示例 8-3 程序,巴克码生成与检出,barkt.vhd
Library Ieee ;
Use Ieee.Std_Logic_1164.All;
Use Ieee.Std_Logic_Unsigned.All;
Use Work.All;

Entity barkt Is                                    --巴克码生成与检出顶层实体
  Port(clk,en, reset : In Std_Logic;
       dout : Out Std_Logic);
End Entity barkt;

Architecture bdf_type Of barkt Is
Signal SYNTHESIZED_WIRE_0 :  Std_Logic;
Begin

  b2v_inst : bark1                                 --巴克码生成元件例化
  Port Map(clk => clk, reset => reset,en =>en,
           dout => SYNTHESIZED_WIRE_0);

  b2v_inst1 : barkc                                --巴克码检出元件例化
  Port Map(clk => clk,
```

```
          data => SYNTHESIZED_WIRE_0,
          reset => reset, dout => dout);
End Architecture bdf_type;
```

示例 8-3a 巴克码生成程序

```
Library Ieee ;
Use Ieee.Std_Logic_1164.All;
Use Ieee.Std_Logic_Unsigned.All;

Entity bark1 Is                                      --一个实体,两个结构体
  Port(clk,en,reset: In Std_Logic;
       dout: Out Std_Logic);
End Entity bark1;

Architecture bark_a Of bark1 Is                      --巴克码生成方法1
Signal count : Std_Logic_Vector(2 Downto 0);
Signal tmp1: Std_Logic;
Begin
  dout<=tmp1;

  Process(clk,reset) Is
  Begin
    If reset='0' Then
      count<=(Others=>'0');
    Elsif Rising_Edge(clk) Then                      --时钟上升沿
      If en='1' Then
        If count="110" Then count<="000";
        Else count<=count+1;
        End if;
        Case count Is
          When "000"=>tmp1<='1';
          When "001"=>tmp1<='1';
          When "010"=>tmp1<='1';
          When "011"=>tmp1<='0';
          When "100"=>tmp1<='0';
          When "101"=>tmp1<='1';
          When "110"=>tmp1<='0';
          When Others=>tmp1<='0';
        End Case;
```

```
        End If;
      End If;
    End Process;
End Architecture bark_a;

Architecture bark_b Of bark1 Is                                   --巴克码生成方法 2
Constant barkcode: Std_Logic_Vector(6 Downto 0):="0100111";
Signal count: Integer Range  0 To 6;
Begin

    Process(clk,reset) Is
    Begin
      If reset='0' Then count<=0;
      Elsif Rising_Edge(clk) Then
        If en='1' Then
          If count=6 Then count<=0;
          Else count<=count+1;
          End If;
          Dout<= barkcode(count);
        Else count<=0;
        End If;
      End If;
    End Process;
End Architecture bark_b;
```

示例 8-3b 巴克码检测程序

```
Library Ieee ;
Use Ieee.Std_Logic_1164.All;
Use Ieee.Std_Logic_Unsigned.All;

Entity barkc Is
    Port(clk,data,reset: In Std_Logic;
         dout: Out Std_Logic);                          --同步脉冲
End Entity barkc;

Architecture barkca Of barkc Is                         --移入数据后逻辑运算比对
Signal temp: Std_Logic_Vector(6 Downto 0);
Signal tmp1,clk1: Std_Logic;
Signal coun: Std_Logic_Vector(2 Downto 0);
```

```
Begin
  dout<=tmpl;

  Process(clk,reset) Is
  Begin
    If reset='0' Then  temp<=(Others=>'0');
    Elsif Rising_Edge(clk) Then  temp(0)<=data;
      For i In 1 To 6 Loop
        temp(i)<=temp(i-1);                       --移位寄存器
      End Loop;
    End If;
  End Process;

  Process(clk,reset) Is
  Begin
    If(reset='0') Then
       coun<="000";
    Elsif(clk'Event And clk='1') Then
      If(coun="101") Then coun<="000";
      Else coun<=coun+'1';
      End If;
      clk1<=coun(2);
    End If;
  End Process;

  Process(clk1) Is
  Begin
    If ( (Not temp(0)) And temp(1) And (Not temp(2))And (Not temp(3))
        And temp(4) And temp(5) And  temp(6))='1' Then
        tmpl<='1';
    Else  tmpl<='0';
    End If;
  End Process;
End Architecture barkca;

Architecture barkcb Of barkc Is                    --相关运算法
Constant barkcode: Std_Logic_Vector(6 Downto 0):="1110010";
Signal temp: Std_Logic_Vector(6 Downto 0);
Begin
```

```
  Process(clk,reset) Is
  Variable sum:Integer Range -1 to +7;
  Variable reg:  Std_Logic_Vector(6 Downto 0);
  Begin
    sum:=0;
    If reset='0'Then
      temp<=(Others=>'0');
              dout<='0';
    Elsif clk'Event And clk='1'Then
      temp(0)<=data;
      For i In 1 To 6 Loop
        temp(i)<=temp(i-1);                          --左移移位寄存器
      End Loop;
      reg:=barkcode Xor temp;
    End If;
    For i In 1 To 6 Loop
    If (reg(i)='1') Then sum:=sum-1;
      Else sum:=sum+1;
      End If;
    End Loop;
    If sum<6 Then dout <='0';
    Else dout <='1';
    End If;
  End Process;
End Architecture barkcb ;

--电路需记忆初始状态,1,11,111,1110,11100,111001,1110010,8个状态
Architecture barkcc Of barkc Is                       --状态机方式
Type state_type Is (s1,s2,s3,s4,s5,s6,s7,s8);         --自定义枚举数据类型
Signal state: state_type;                             --自定义数据类型信号
Begin

  Demo_State:Process(clk,reset) Is                    --单进程Moore状态机
  Begin
    If reset='0'Then   state<=s1;                     --状态初值设定
    --dout<=(Others=>'0');
    Elsif clk'Event And clk='1'Then
      Case state Is
```

```
        When s1=>If data='1'Then state<=s2; End If;           --状态转换判定
        dout<='0';                                           --状态输出与数据输入无关
        When s2=>If data='1'Then state<=s3;                  --符合记忆顺序转换
                 Elsif data='0'Then state<=s1; End If;
                                                             --不符合记忆改变顺序
        dout<='0';
        When s3=>If data='1'Then state<=s4;
                 Elsif data='0'Then state<=s1; End If;
        dout<='0';
        When s4=>If data='0'Then state<=s5;
                 Elsif data='1'Then state<=s4;End If;
        dout<='0';
        When s5=>If data='0'Then state<=s6;
                 Elsif data='1'Then state<=s2; End If;
        dout<='0';
        When s6=>If data='1'Then state<=s7;
                 Elsif data='0'Then state<=s1; End If;
        dout<='0';
        When s7=>If data='0'Then state<=s8;
                 Elsif data='1'Then state<=s3; End If;
        dout<='1';
        When s8=>If data='0'Then state<=s1;
                 Elsif data='1'Then state<=s2;End If;
        dout<='0';
        When Others=> state<=s1;
      End Case;
    End If;
  End Process;
End Architecture barkcc;
```

8.3 扰码与解扰码

8.3.1 扰码与解扰码简介

用NRZ码进行基带信号传输的缺点是，其频谱会因数据出现连“1”、连“0”而包含大的低频成分，不适应信道的传输特性，也不利于从中提取出时钟信息。解决办法之一是采用扰码技术，使信号受到随机化处理，变为伪随机序列，又称为“数据随机化”和“能量扩散”处理。

扰码虽然“扰乱”了原有数据的本来规律，但因为是特意“扰乱”的，在接收端很容易去加扰，恢复成原数据流。

实现加扰和解码，需要产生伪随机二进制序列，再与输入数据逐个比特作运算。这个伪随机二进制序列也称为 m 序列，m 序列与传输数据流的 NRZ 码进行 Xor 扰码运算后，数据“1”和“0”的连续游程都很短，且出现的概率基本相同。

8.3.2 m 序列生成

产生伪随机序列（Pseudo Random Sequence Generator，PRSG）的电路可分为线性反馈移位寄存器（Linear Feedback Shift Registers，LFSR）和非线性反馈移位寄存器。

m 序列是最长线性反馈移位寄存器序列的简称。m 序列是一伪随机序列，具有与随机噪声类似的尖锐自相关特性，但它不是真正随机的，而是按一定的规律形式周期性地变化。由于 m 序列容易产生，规律性强，有许多优良的特性，在扩频通信和码分多址系统中最早获得应用。m 序列产生电路通用形式如图 8-4 所示。

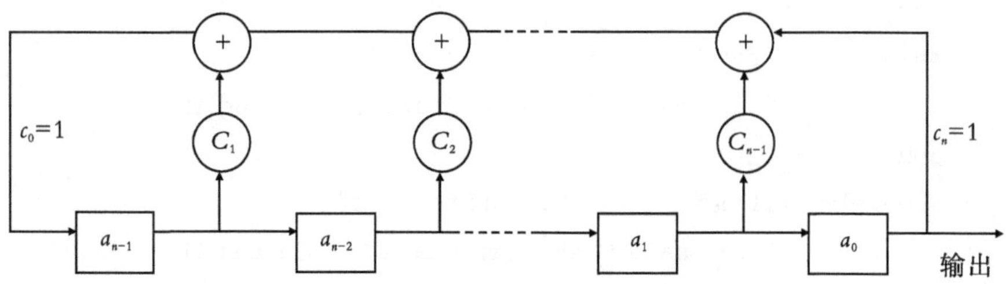

图 8-4　m 序列产生电路通用形式

一般，一个 n 级反馈移存器可能产生的 m 序列的最长周期为 2^n-1。如图 8-4 所示，a_{n-i} $(i=1,2,\cdots,n)$ 是各移位寄存器的状态，$C_i(i=1,2,\cdots,n)$ 对应各移存器的反馈系数，$C_i=1$ 表示该级移存器参与反馈，$C_i=0$ 表示该级移存器不参与反馈，C_0 和 C_n 不能等于 0，这是因为 $C_0=0$ 意味着移存器无反馈，而 $C_n=0$ 则意味着反馈移存器蜕化为 $n-1$ 级或更少级的反馈移存器。反馈函数为

$$a_k = C_1 a_{k-1} + C_2 a_{k-2} + \cdots + C_n a_{k-n}(\text{模 2 加}) \tag{8-3}$$

m 序列的特征多项式或称本征多项式也可以为公式(8-4)所示：

$$f(x) = C_{n-1}x^{n-1} + C_{n-2}x^{n-2} + \cdots + C_1 x + C_0 = f(x) = \sum_{i=1}^{n} C_i X^i \tag{8-4}$$

$C_0, C_1, \cdots, C_{n-1}$ 是相应码元的数值(0 或 1)。X 表示单位延时，X 向左移一位就是 X^2，向右移一位就是 X^{-1}。本原多项式可以通过 MATLAB 求解，例

```
n=4;
x=gfprimfd(n,'all');      %求出 n=4 的所有本原多项式的系数序列
for i=1,size(x);  %将系数序列写成解析式，循环语句是依次写出所有的本原多项式
gfpretty(x(2,:))
gfpretty(x(5,:))
end
```

MATLAB 计算结果：

1　0　1　0　0　1　　$1+x^3+x^5$

1　0　0　1　0　1　　$1+x^2+x^5$

1 1 1 1 0 1

1 1 1 0 1 1

1 1 0 1 1 1

1 0 1 1 1 1

特征多项式的系数 C_i 为 1 或者 0，X_i 为寄存器的延时，由 n 级移位寄存器通过模 2 加能出现各种不同状态形成的 m 序列，共有 2^n 个状态，除去全“0”状态外还剩下 2^n-1 种状态。

若 m 序列的本征多项式为 $f(x)=1+x^2+x^5$ 则实现它的电路如图 8-5 所示。

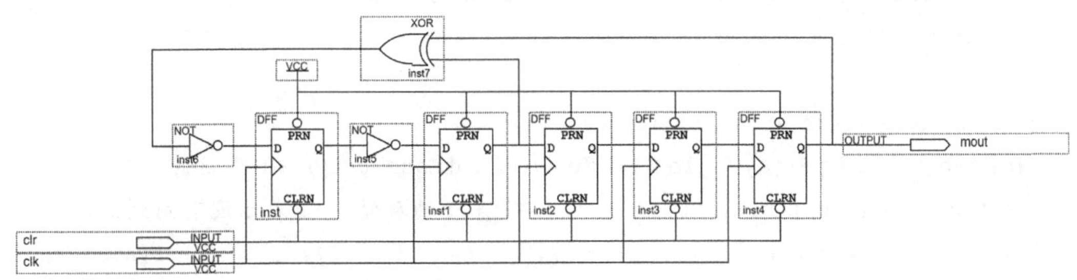

图 8-5 m 序列 $f(x)=1+x^2+x^5$ 电路

参见 6.3.1 节 LFSR 计数器与二进制计数器设计中 LFSR 计数器。图 8-5 这个设计中的反相器是防止电路发生死锁状态而添加的。

8.3.3 有关加扰与解扰的设计

扰码器与解扰器原理：扰码器是在发送端使用移位寄存器产生 m 序列，然后信息序列与 m 序列作模二加，其输出即为加扰的随机序列；解扰器是在接收机端使用相同的扰码序列与收到的被扰信息模二加，将原信息得到恢复。扰码器与解扰器的设计原理如图 8-6 所示。

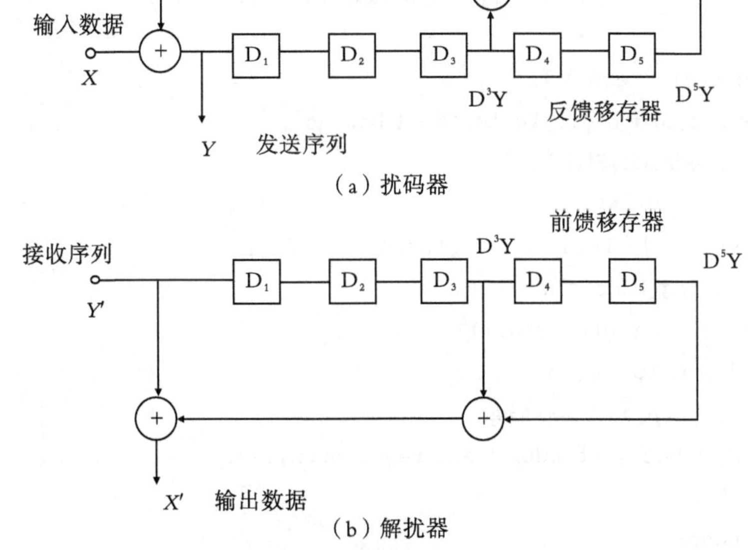

(a) 扰码器

(b) 解扰器

图 8-6 扰码器与解扰器的设计原理

示例 8 - 4　线性反馈移位寄存器

```
Library Ieee;
Use Ieee.Std_Logic_1164.All;
Entity nlfsr Is
Generic(N:Integer Range 1 To 36:=8);                          --LFSR的范围在1～36之间
  Port(reset,clk: In Std_Logic;
       dout : Out Std_Logic_Vector(N-1 Downto 0));
End Entity nlfsr ;

Architecture nlfsr_Arch Of nlfsr Is
Type tap_table Is Array (1 To 36,1 To 4) Of Integer Range -1 To 36;
Constant taps:tap_table:=(          --有≤4的反馈,-1表示没有用到的反馈链接
  (0, -1, -1, -1), (1, 0, -1, -1), (1, 0, -1, -1),            --1,2,3
  (1, 0, -1, -1), (2, 0, -1, -1), (1, 0, -1, -1),             --3,4,5
  (0, -1, -1, -1), (6, 0, -1 ,-1), (4, 0, -1, -1),            --7,8,9
  (3, 0, -1, -1), (2, 0, -1, -1), (7, 4, 3, -1),              --10,11,12
  (4, -1, -1, -1), (12, 11, 1, -1), (1, 0, -1, -1),           --13,14,15
  (5, 0, -1, -1), (3, 0 ,-1, -1), (7, 0, -1, -1),             --16,17,18
  (6, 5, 1, -1), (3, 0, -1, -1), (2, 0, -1, -1),              --19,20,21
  (1, 0, -1, -1), (5, 3, 2, -1), (4, 3, 1, -1),               --22,23,24
  (8, 7, 1, -1), (8, 7, 1, -1), (3, 0, -1, -1),               --25,26,27
  (2, 0, -1, -1), (1, 0, -1, -1), (16,15, -1, -1),            --28,29,30
  (3, -1, -1, -1), (28, 27, 1, -1), (13, 0, -1, -1),          --31,32,33
  (15, 14, 1, -1), (2, 0, -1, -1), (11, 0, -1, -1));          --34,35,36
Begin
  p0:Process(clk,reset) Is
  Variable reg:Std_Logic_Vector(N-1 Downto 0);
  Variable feedback:Std_Logic;
  Begin
    If reset = '1' Then reg:=(Others=>'1');
    Elsif Rising_Edge(clk) Then
      feedback:=reg(taps(N,1));
      For i In 2 To 4 Loop
        If taps(n,i)>=0 Then
          feedback:=feedback Xor reg(taps(n,i));
        End If;
      End Loop;
      reg:=feedback & reg(n-1 Downto 1);
```

```
    End If;
    dout <=reg;
  End Process p0;
End Architecture nlfsr_arch;
```

程序说明:这种数值表示是不好的编程习惯,当条件不满足时,算法会崩溃,因此后面各级采用异或,taps 表中的－1 项忽略。

8.4　基于 DDS 的调制解调

设计要求:

(1)利用 FPGA 实现 AD9850 的主要功能。

(2)实现基于 DDS 的调制解调。

数字直接频率综合技术(Direct Digital Fraquency Synthesis,DDFS,简称 DDS)的基本原理是从相位概念出发,以数控振荡器的方式产生频率,通过相位可控制的正弦波直接合成所需要波形的技术。从时域角度,即利用 Nyquist 时域采样定理,在时域中进行频率合成,通过查表法产生波形,DDS 的频率及步进容易控制,合成的频率取决于累加器及查找表的速度。

基于 DDS 的电路一般包括基准时钟、频率累加器、相位累加器、幅度/相位转换电路、D/A 转换器。频率累加器对输入信号进行累加运算,产生频率控制数据。相位累加器由 N 位全加器和 N 位累加寄存器级联而成,对代表频率的二进制码进行累加运算,是典型的反馈电路,产生累加结果。幅度/相位转换电路实质上是一个波形寄存器或是 ROM,以供查表使用,读出的数据送入 D/A 转换器和低通滤波器。图 8－7 给出了商用 DDS AD9850 的原理框图。图 8－8为相位累加器位宽与采样点的关系图。

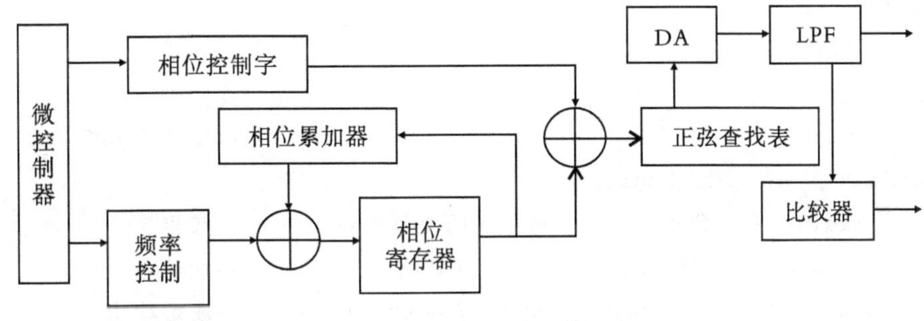

图 8－7　AD9850 DDS 原理框图

DDS 输出的频率为:$f_o=Kf_c/2^N$,其中,$K=\Delta PHASE$;

DDS 输出频率的分辨率为:$\Delta f_{min}=1\times f_c/2^N$。

调制说明:

(1)BPSK(二进制相位键控,Binary Phase-Shift Keying):2 电平调相,或称 PSK,用 0,1 两个数码分别对载波的相位进行调制。载波有两个不同的相位如 0,180。

(2)QPSK(四进制相位键控):4 电平调相,用 00,01,11,10,4 个数码分别对载波的相位进行调制。载波有 4 个不同的相位如:π/2 系统,180,0,90,270;π/4 系统,225,45,135,315。

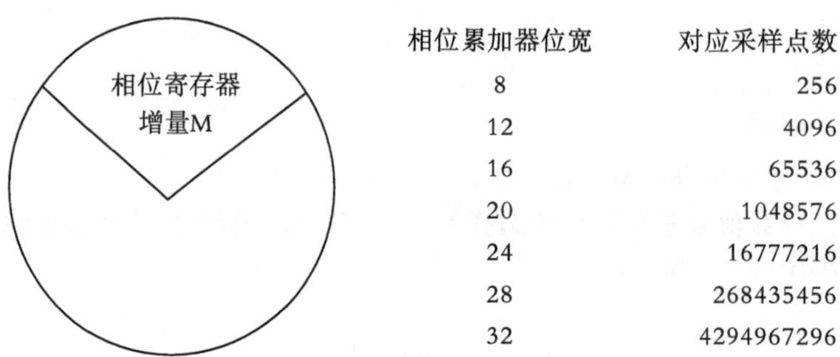

相位累加器位宽	对应采样点数
8	256
12	4096
16	65536
20	1048576
24	16777216
28	268435456
32	4294967296

图 8-8　相位累加器位宽与采样点的关系

(3)OPSK(八进制相位键控)：8电平调相，用000,001,011,010,…8个数码分别对载波的相位进行调制，载波有8个不同的相位。

示例 8-4　DDS 的模型与仿真

单纯频率可调的 DDS 电路加上 yadd 加法模块即成为相位可调的 DDS 电路，如图 8-9 所示。

示例 8-4a　相位频率可调的 DDS 电路模型程序

如图 8-9 所示，DDS 是由调用系统模块 D 触发器、流水线加法器、基本 ROM 的正弦波信号发生器，以及 16 位转 8 位电路组成的。实验证明这种方式实现的性能好于自行编写的 VHDL 文件的效果。以下程序的目的主要是为了对图 8-9 作说明。

```
Library Ieee;                                              -- DDS 的模型 VHDL 描述
Use Ieee.Std_Logic_1164.All;
Library Work;
Use Work.All;

Entity DDS Is                                              --实体名 DDS
  Port(clk :  In  Std_Logic;
       freqin : In  Std_Logic_Vector(15 Downto 0);         --频率字
       pword :  In  Std_Logic_Vector(7 Downto 0);          --相位字
       dout :  Out  Std_Logic_Vector(7 Downto 0));         --波形输出
End Entity DDS;

Architecture bdf_type Of DDS Is

Signal SYNTHESIZED_WIRE_0 :  Std_Logic_Vector(15 Downto 0);
Signal SYNTHESIZED_WIRE_1 :  Std_Logic_Vector(15 Downto 0);
Signal SYNTHESIZED_WIRE_2 :  Std_Logic_Vector(7 Downto 0);
Signal SYNTHESIZED_WIRE_3 :  Std_Logic_Vector(7 Downto 0);
```

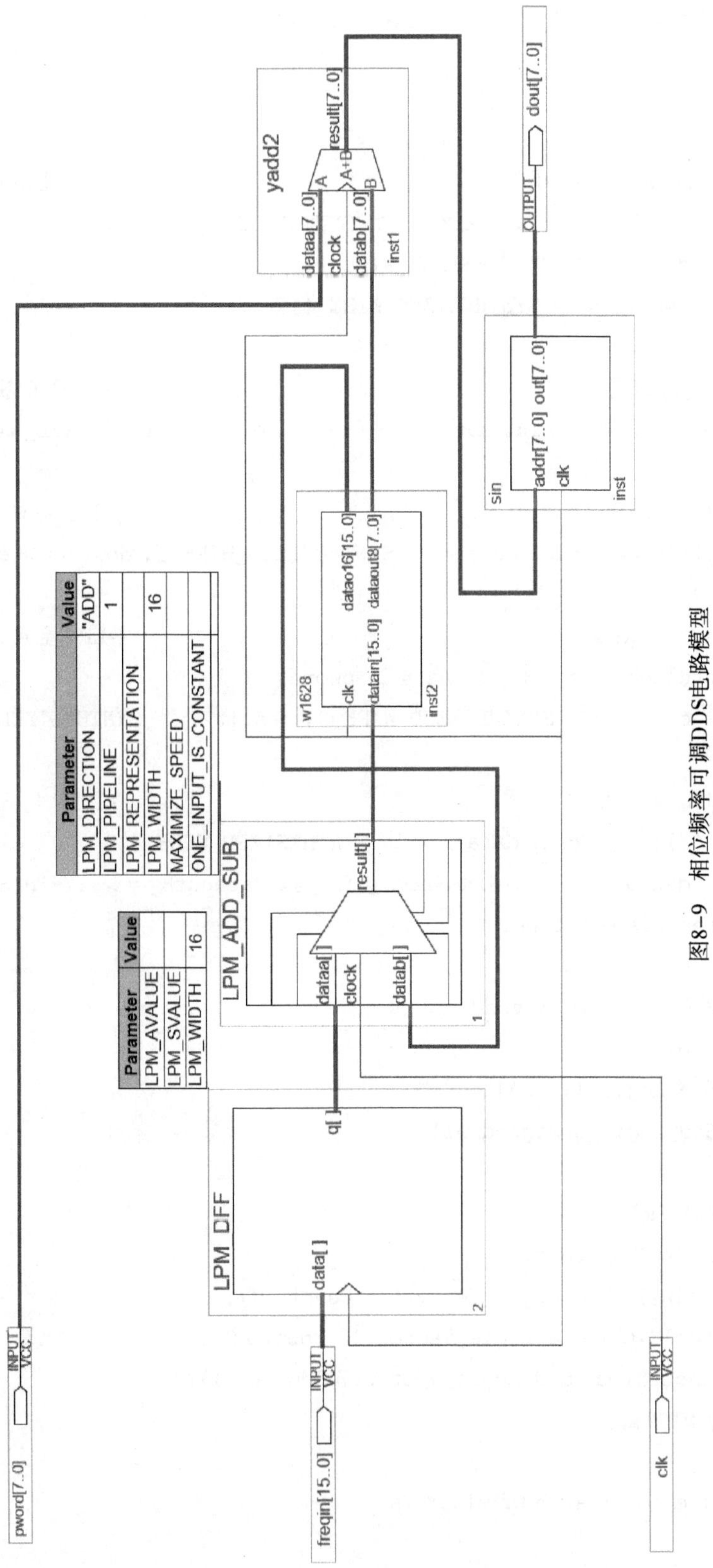

图8-9 相位频率可调DDS电路模型

```
Signal SYNTHESIZED_WIRE_4 :  Std_Logic_Vector(15 Downto 0);

Begin

  b2v_1 : lpm_add_sub_0                              --LPM 生成的 16 们加法器
  Port Map(clock => clk, dataa => SYNTHESIZED_WIRE_0,
           datab => SYNTHESIZED_WIRE_1,
           result => SYNTHESIZED_WIRE_4);

  b2v_2 : lpm_dff_1                                  --16 倍 D 触发器
  Port Map(clock => clk, data => freqin, q => SYNTHESIZED_WIRE_0);

  b2v_inst : sin                                     --256 字正弦发生器
  Port Map(clk => clk, addr => SYNTHESIZED_WIRE_2, dout => dout);

  b2v_inst1 : yadd2                                  --LPM 生成的 8 位加法器
  Port Map(clock => clk, dataa => pword,
           datab => SYNTHESIZED_WIRE_3, result => SYNTHESIZED_WIRE_2);

  b2v_inst3 :w1628                                   --16 位转 8 位的位宽变换
  Port Map(clk => clk, datain => SYNTHESIZED_WIRE_4,
           datao16 => SYNTHESIZED_WIRE_1, dataout8 => SYNTHESIZED_WIRE_3);
End Architecture bdf_type;

--16 位转 8 位电路,w1628.vhd
Library Ieee;
Use Ieee.Std_Logic_1164.All;
Use Ieee.Std_Logic_Unsigned.All;

Entity w1628 Is
  Port(clk:In Std_Logic;
       datain:In Std_Logic_Vector(15 Downto 0);
       datao16:Out Std_Logic_Vector(15 Downto 0);
       dataout8:Out Std_Logic_Vector(7 Downto 0));
End Entity w1628;

Architecture w1628_arch Of w1628 Is
Begin
  wl:Process(clk) Is
```

```
  Variable  s:Std_Logic_Vector(7 Downto 0);
  Begin
    If(clk'Event And clk='1')Then
       s:=datain(15 Downto 8) ;
    End If;
    dataout8<=s;
    datao16<= datain;
  End Process wl;
End Architecture w1628_arch;
```

示例 8-4b DDS 的仿真

```
--DDS 测试程序 dds_tb.vhd
Library Ieee  ;
Library Cycloneii  ;                                    --声明 Cycloneii 库
Use Cycloneii.Cycloneii_Components.All ;
--调用 Cycloneii_Components 包
Use Ieee.Std_Logic_1164.All  ;
Use Ieee.Std_Logic_Unsigned.All  ;
Use Ieee.Std_Logic_Arith.All  ;

Entity dds_tb  Is
End Entity dds_tb;

Architecture dds_tb_arch Of dds_tb Is
Signal pword: Std_Logic_Vector (7 Downto 0):="00000000" ;
                                                        --相位控制字初值
Signal dout  :Std_Logic_Vector (7 Downto 0)  ;
Signal freqin: Std_Logic_Vector (15 Downto 0):="0000111111111111";
                                                        --频率控制字初值
Signal clk ,clk1  :  Std_Logic:='0'  ;
Signal cb   : Bit_Vector( 15 Downto 0);
Signal coun: Integer Range 0 To 10:=0;

  Component DDS Is                                      --元件 DDS 声明
    Port (pword  : In Std_Logic_Vector (7 Downto 0) ;
          dout : Out Std_Logic_Vector (7 Downto 0) ;
          freqin: In Std_Logic_Vector (15 Downto 0) ;
          clk   :In Std_Logic );
  End Component DDS;
```

```
Begin
  DUT  : DDS                                                    --例化 DDS 元件
  Port Map (pword    => pword   , dout    => dout   ,
            freqin => freqin   , clk    => clk ) ;
  clk <=Not clk After 10 ns;
  clk1<=Not clk1 After 12800 ns;

  Process(clk1) Is
  Begin
    If clk1 'Event And clk1='1' Then
      If coun<10 Then
        coun<=coun+1;
        freqin<=To_Stdlogicvector(To_Bitvector(freqin) srl 1);
      --通过右移位产生频率控制字,srl 可以采用 Bit,Boolean 类型
      End If;
    End If;
  End Process;
End Architecture dds_tb_arch;
```

程序说明：

(1)Cycloneii 库要先编译,加入到工程中。

(2)图 8-10 是相位控制字设定为常数,改变频率字得到的仿真结果。

(3)移位操作符有 6 种,sll, srl, sla,sra,rol,ror。sll, srl 移位补 0;sla,sra 移位补首位;rol,ror 循环移位。移位操作符的操作类型是 Bit,Boolean 类型。

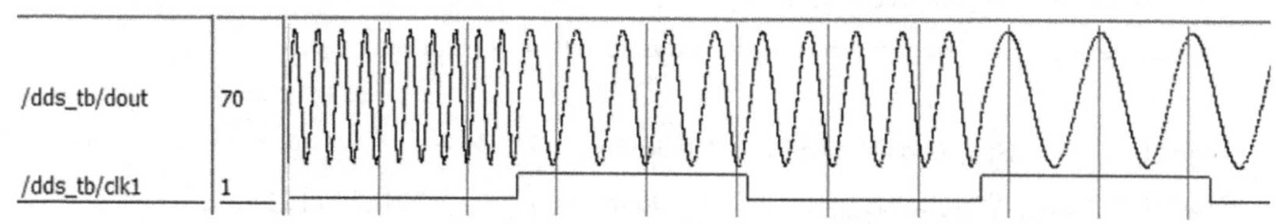

图 8-10　DDS 电路 Modelsim 仿真步进效果

8.4.1　DDS 步进方波的实现

FPGA DDS 步进方波的实现与 AD9850 的实现机制相同,可以通过在 DDS 输出的步进正弦波之后接比较器来实现。比较器的电路实现主要有三种方法：

(1)74LS85 集成数值比较器;

(2)数字逻辑运算从低位进行扩展;

(3)直接调用 LPM 模块实现。

前两种方法这里不介绍。

采用 LPM 模块的方法简便易行。具体选用 LPM_COMPARE 模块,参数设置如图 8-11

所示。可以看到模块的各种比较结果可以复选。

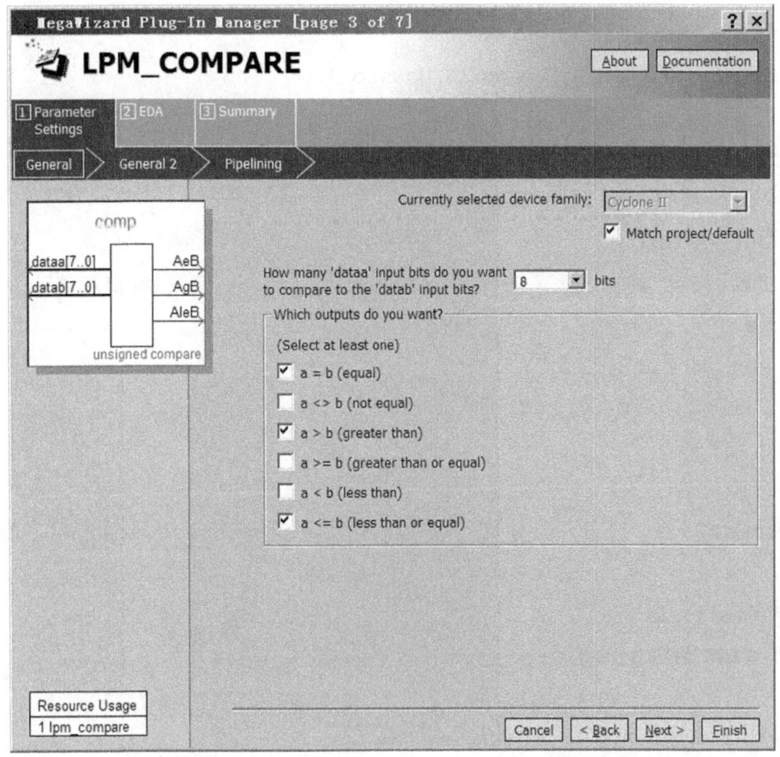

图 8-11　LPM_COMPARE 模块的参数设置

对于数值在 0 与 127 之间的正弦数据，找到正弦的零点，比如数值 64，做大于等于比较，结果为真时，输出高电平；结果为假时，输出低电平，这样就获得了与正弦 DDS 步进频率一样的步进正弦波。

我们还可以通过数控分频来实现方波的步进。

示例 8-4c　可实现步进方波的数控分频器

```
Library Ieee;
Use Ieee.Std_Logic_1164.All;
Use Ieee.Std_Logic_Unsigned.All;

Entity dctrl Is
  Port (clk : In Std_Logic;
        din : In Std_Logic_Vector(7 Downto 0);
        dout : Out Std_Logic);
End Entity dctrl;

Architecture dctrl_Arch Of dctrl Is
```

```
Signal full:Std_Logic;
Begin
  Process(clk,din) Is
  Variable cnt:Std_Logic_Vector(7 Downto 0);
  Begin
    If(clk'Event And clk='1') Then
      If(cnt="11111111") Then
        cnt:=din;
        full<='1';                              --溢出标志
      Else
        cnt := cnt + '1';
        full <= '0';
      End If;
    End If;
  End Process;

  Process(full) Is
  Variable cnt2:Std_Logic;
  Begin
    If(full'Event And full='1') Then
      cnt2 := Not cnt2;
      If(cnt2 = '1') Then
        dout <= '1';
      Else
        dout <= '0';
      End If;
    End If;
  End Process;
End Architecture dctrl_arch;
```

示例 8－5　基于 DDS 的调制

基于 DDS 的调制总图如图 8－12 所示。

```
--将总图转换为 VHDL,msktop.vhd
Library Ieee;
Use Ieee.Std_Logic_1164.All;
Use Work.All;

Entity msktop Is                                --调制实验顶层文件
  Port(clr :  In   Std_Logic;
       clk :  In   Std_Logic;                   --时钟
```

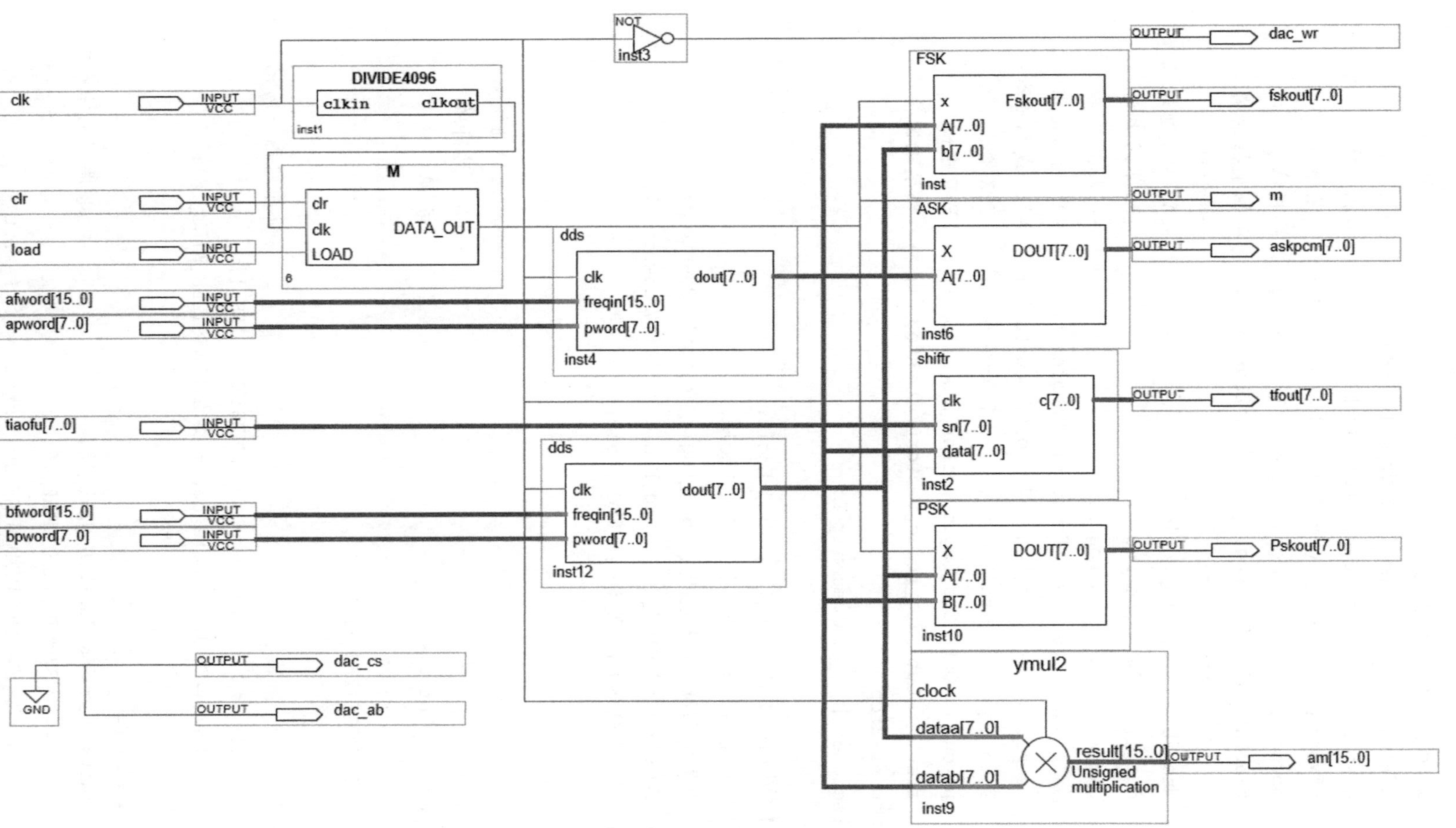

图8-12　基于DDS的调制总图

```
        load : In  Std_Logic;                                   --加载信号
        afword : In  Std_Logic_Vector(15 Downto 0);             --频率字
        apword : In  Std_Logic_Vector(7 Downto 0);              --相位字
        bfword : In  Std_Logic_Vector(15 Downto 0);             --频率字
        bpword : In  Std_Logic_Vector(7 Downto 0);              --相位字
        tiaofu : In  Std_Logic_Vector(7 Downto 0);              --幅度控制端
        m : Out  Std_Logic;                                     --m序列
        dac_ab : Out  Std_Logic;                                --AD7302 通道选择
        dac_cs : Out  Std_Logic;                                --AD7302 片选
        dac_wr : Out  Std_Logic;                                --AD7302 写控制
        am : Out  Std_Logic_Vector(15 Downto 0);                --AM 调幅输出
        askpcm : Out  Std_Logic_Vector(7 Downto 0);             --ASK 或 PCM 输出
        fskout : Out  Std_Logic_Vector(7 Downto 0);             --FSK 输出
        pskout : Out  Std_Logic_Vector(7 Downto 0);             --PSK 输出
        tfout : Out  Std_Logic_Vector(7 Downto 0));             --幅度控制输出
End Entity msktop;
Architecture bdf_type Of msktop Is

Signal SYNTHESIZED_WIRE_17 :  Std_Logic;
Signal SYNTHESIZED_WIRE_18 :  Std_Logic;
Signal SYNTHESIZED_WIRE_19 :  Std_Logic_Vector(7 Downto 0);
Signal SYNTHESIZED_WIRE_20 :  Std_Logic_Vector(7 Downto 0);
Signal SYNTHESIZED_WIRE_16 :  Std_Logic;

Begin
  dac_ab <= '0';                                                --DA 选择通道 a
  dac_cs <= '0';                                                --DA 片选有效
  dac_wr <= Not(clk);                                           --写 DA

  m <= SYNTHESIZED_WIRE_18;

  b2v_6 : m                                                     --m 序列元件例化
  Port Map(clr => clr, clk => SYNTHESIZED_WIRE_17,
           load => load, data_out => SYNTHESIZED_WIRE_18);

  b2v_inst : fsk
  Port Map(x => SYNTHESIZED_WIRE_18,                            --fsk 调制元件例化
           a => SYNTHESIZED_WIRE_19,
           b => SYNTHESIZED_WIRE_20, fskout => fskout);
```

```
    b2v_inst1 : divide4096                              --分频模块(DAC 需要的)例化
    Port Map(clkin => clk,
            clkout => SYNTHESIZED_WIRE_17);

    b2v_inst10 : psk                                    --psk 调制元件例化
    Port Map(x => SYNTHESIZED_WIRE_18,
            a => SYNTHESIZED_WIRE_20,
            b => SYNTHESIZED_WIRE_19, dout => Pskout);

    b2v_inst12 : dds                                    --dds 元件例化
    Port Map(clk => SYNTHESIZED_WIRE_17,
            freqin => bfword, pword => bpword,
            dout => SYNTHESIZED_WIRE_20);

    b2v_inst2 : shiftr                                  --右移移位寄存器例化
    Port Map(clk => SYNTHESIZED_WIRE_17,
            data => SYNTHESIZED_WIRE_19,
            sn => tiaofu, c => tfout);

    b2v_inst4 : dds                                     --dds 元件例化
    Port Map(clk => SYNTHESIZED_WIRE_17,
            freqin => afword, pword => apword,
            dout => SYNTHESIZED_WIRE_19);

    b2v_inst6 : ask                                     --ask 调制元件例化
    Port Map(x => SYNTHESIZED_WIRE_18,
            a => SYNTHESIZED_WIRE_19,
            dout => askpcm);

    b2v_inst9 : ymul2                                   --乘法器元件例化
    Port Map(clock => SYNTHESIZED_WIRE_17,
            dataa => SYNTHESIZED_WIRE_20,
            datab => SYNTHESIZED_WIRE_19, result => am);
End Architecture bdf_type;
```

程序说明:在设计时可以采用原理图的方式,作为设计存档一定要同时将设计中采用原理图的文件转成 VHDL 文件,这不仅使存档更可靠,便于注释,而且以后设计移植、设计复用等更方便。

示例 8－6　基于 DDS 的调制测试程序

```
Library Cycloneii   ;
Library Ieee   ;
Use Cycloneii. Cycloneii_Components. All   ;
Use Ieee. Std_Logic_1164. All   ;

Entity msktop_tb   Is
End Entity msktop_tb;

Architecture msktop_tb_arch Of msktop_tb Is
Signal dac_wr   :   Std_Logic :='1'   ;
Signal dac_ab   :   Std_Logic :='1'   ;
Signal dac_cs   :   Std_Logic :='1'   ;
Signal clk    :  Std_Logic :='0'   ;
Signal clr    :  Std_Logic   :='1';
Signal load   :   Std_Logic   :='1';
Signal tiaofu:   Std_Logic_Vector (7 Downto 0) :="11111111"   ;
Signal afword:   Std_Logic_Vector (15 Downto 0):="0000000011111111" ;
Signal bfword:   Std_Logic_Vector (15 Downto 0):="0000000011111111" ;
Signal apword:   Std_Logic_Vector (7 Downto 0) :="00000000"   ;
Signal bpword:   Std_Logic_Vector (7 Downto 0) :="10000000" ;
Signal am    :  Std_Logic_Vector (15 Downto 0)   ;
Signal tfout :  Std_Logic_Vector (7 Downto 0)   ;
Signal pskout:   Std_Logic_Vector (7 Downto 0)   ;
Signal fskout:   Std_Logic_Vector (7 Downto 0)   ;
Signal askpcm:   Std_Logic_Vector (7 Downto 0)   ;
Signal m    :     Std_Logic   ;

  Component msktop   Is
    Port ( dac_cs,dac_wr,dac_ab : Out Std_Logic ;
          clr,load,clk: In Std_Logic ;
          afword,bfword: In Std_Logic_Vector (15 Downto 0) ;
          apword,bpword: In Std_Logic_Vector (7 Downto 0) ;
          tiaofu: In Std_Logic_Vector (7 Downto 0) ;
          am: Out Std_Logic_Vector (15 Downto 0) ;
          tfout : Out Std_Logic_Vector (7 Downto 0) ;
          m: Out Std_Logic ;
          pskout,fskout,askpcm: Out Std_Logic_Vector (7 Downto 0) );
```

```
  End Component msktop;
Begin
  DUT  : msktop
  Port Map ( dac_wr   => dac_wr, dac_ab   => dac_ab,
             dac_cs   => dac_cs, clr   => clr,  clk   => clk,
             load  => load,      m   => m,
             afword   => afword, bfword => bfword,
             apword   => apword, bpword => bpword,
             am   => am,
             tfout => tfout, tiaofu   => tiaofu,
             pskout   => pskout,fskout => fskout,
             askpcm   => askpcm) ;
  clr <='0' After 200 ns;
  load<='0' After 200 ns;
  clk <=Not clk After 10 ns;
End Architecture msktop_tb_arch;
```

程序说明：

(1)FSK,ASK 可以采用相位控制字相同,频率控制字不同的方式仿真。

(2)PSK 可以采用相位控制字不同,频率控制字不相同的方式仿真。

(3)脉冲编码调制 PCM 从 askout 引出,脉冲编码调制与 ASK 调制不同的是方波与正弦波的频率关系不同。

(4)程序中 AM 调制采用乘法器产生了 16 位数,对于 8 位 DA 是不行的。

8.4.2 ASK 调制与 PCM 调制

数字信号对载波振幅调制称为振幅键控即 ASK(Amplitude-Shift Keying)。其表达式为

$$\varphi_{\mathrm{ASK}}(t)=\begin{cases}A\cos\omega_0 t,\text{“1”}\\0,\text{“0”}\end{cases} \tag{8-5}$$

ASK 调制与 PCM 调制在程序上只是分频关系不一样,代码是一样的。

ASK 调制 VHDL 程序的结构体如下：

```
Architecture ask_arch Of ask Is
Constant b:Std_Logic_Vector(7 Downto 0):="10000000";


Begin
  askout<=a When x='1'                    --DDS1 输出
           Else b;                        --常数输出
End Architecture ask_arch;
```

PCM(Pulse-code modulation)脉冲编码调制是实现语音信号数字化的一种方法。PCM 主要包括抽样、量化、编码。这里我们设计实现的是抽样,调制以后的波形如图 8-13(a)所

示。语音信号用正弦波表示，抽样信号用了与 ASK 调制同样的 m 序列。ASK 调制波形如图 8-13(b)所示。

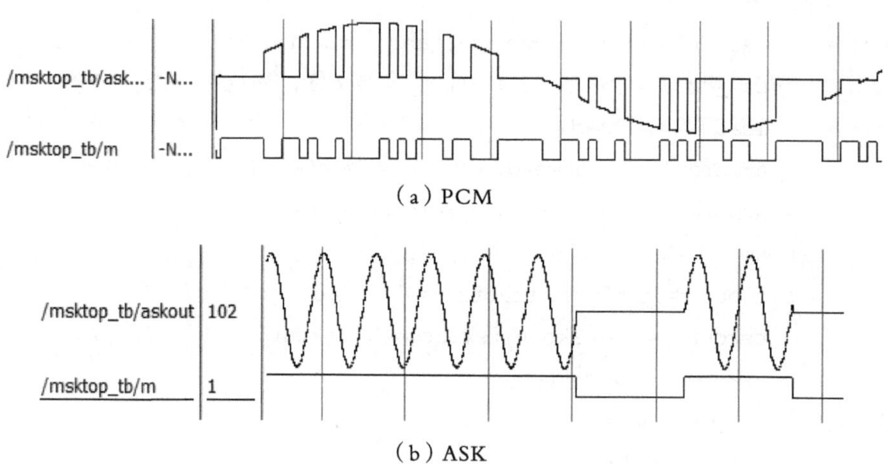

(a) PCM

(b) ASK

图 8-13　ASK 与 PCM 脉冲编码调制波形

8.4.3　FSK 调制与解调

数字信号对载波频率调制称为频移键控即 FSK(Frequency-Shift Keying)。

频移键控(FSK)是用不同频率的载波来传送数字信号，用数字基带信号控制载波信号的频率。其表达式为

$$\varphi_{\text{FSK}}(t)=\begin{cases}A\cos\omega_0 t,\text{“1”}\\A\cos\omega_1 t,\text{“0”}\end{cases} \tag{8-6}$$

FSK 调制 VHDL 程序的结构体：

```
Architecture fsk_arch Of fsk Is
Begin
  fskout<=a When x='1'                    --DDS1 的输出
          Else b;                         --DDS2 的输出
End Architecture fsk_arch;
```

FSK Modelsim 仿真如图 8-14(a)所示，FSK 嵌入式逻辑分析仪测试结果如图 8-16(b)所示，与示波器上观察完全一致，可以看到在 m 序列信号发生‘0’与‘1’的转换时，FSK 输出发生了如公式(8-6)所示的频率变化。

FSK 信号解调控制模块原理图，如图 8-15 所示。

将正弦波信号过零检测转换成与之对应的方波信号，再经过与解调控制模块时钟的分频相比较进行计数，然后经过判决后决定输出高电平或低电平。

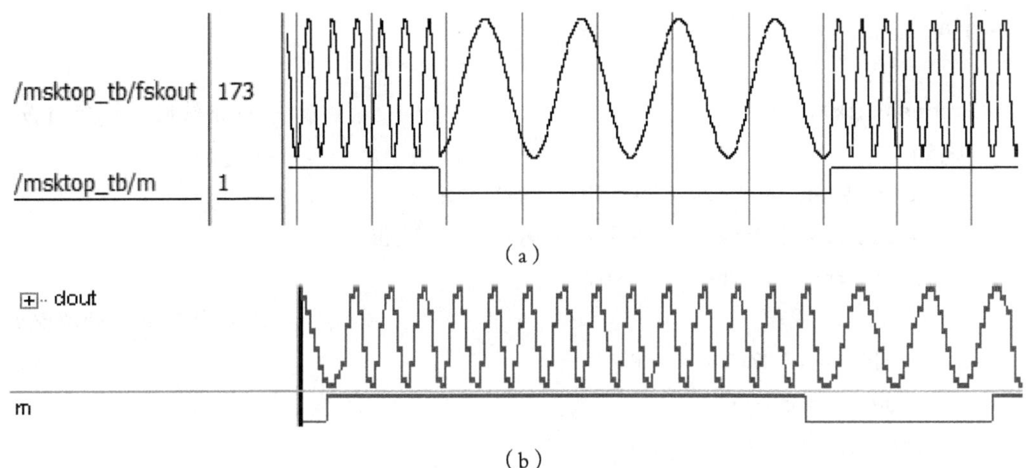

图 8-14 FSK 仿真与测试结果

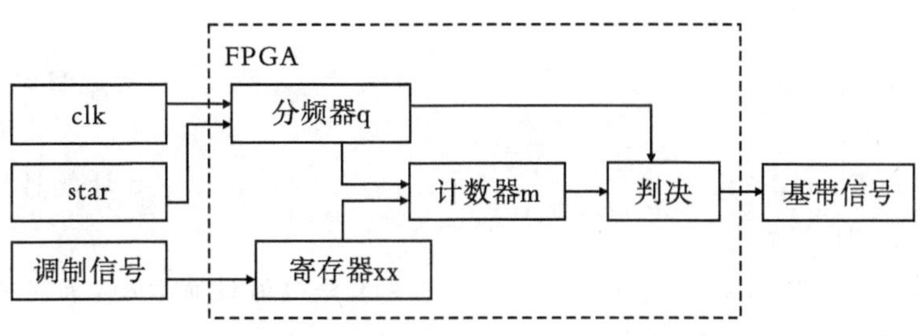

图 8-15 FSK 解调控制模块原理设计

示例 8-7 FSK 解调控制模块程序

```
--先将 FSK 调制信号经过过 0 比较,再采用此程序
Library Ieee;
Use Ieee.Std_Logic_Arith.All;
Use Ieee.Std_Logic_1164.All;
Use Ieee.Std_Logic_Unsigned.All;

Entity pl_fsk2 Is
  Port(clk, start:In Std_Logic;                    --系统时钟, 同步信号
       x    :In Std_Logic;                          --调制信号
       y    :Out Std_Logic);                        --基带信号
End Entity pl_fsk2;

Architecture behav Of pl_fsk2 Is
Signal q: Integer Range 0 To 17;                    --分频计数器
Signal xx: Std_Logic;                               --寄存器
Signal m: Integer Range 0 To 6;                     --计数器
```

```
Begin

  Process(clk) Is                                   --对系统时钟进行 q 分频
  Begin
    If clk'Event And clk='1' Then xx<=x;
    --在 clk 信号上升沿时,x 信号对中间信号 xx 赋值
      If start='0' Then q<=0;
                                                    --If 语句完成 q 的循环计数
      Elsif q=17 Then q<=0;
      Else q<=q+1;
      End If;
    End If;
  End Process;

  Process(xx,q) Is                                  --此进程完成 FSK 解调
  Begin
    If q=17 Then m<=0;                              --m 计数器清零
    Elsif q=16 Then
      If m<=4 Then y<='0';             --If 语句通过 m 大小,来判决 y 输出的电平
      Else y<='1';
      End If;
    Elsif  xx'Event And xx='1' Then m<=m+1;         --计 xx 信号的脉冲个数
    End If;
  End Process;
End Architecture behav;
```

8.4.4 BPSK 调制

数字信号对载波相位调制称为相移键控 PSK(Phase-Shift Keying)。

数字相位调制(相移键控)是用数字基带信号控制载波的相位,使载波的相位发生跳变的一种调制方式。其表达式为

$$\varphi_{\mathrm{PSK}}(t)=\begin{cases}A\cos\omega_0 t, & \text{“1”}\\ A\cos(\omega_0 t+\pi), & \text{“0”}\end{cases} \tag{8-7}$$

PSK Modelsim 仿真结果如图 8-16 所示,可以看到在 m 序列信号发生‘0’与‘1’的转换时,PSK 输出发生了如公式(8-7)所示的 180°相位变化。

由于 PSK 系统抗噪声性能优于 ASK 和 FSK,而且频带利用率较高,所以,在中、高速数字通信中被广泛采用。

与 FSK 信号给入类似,PSK 用伪随机码给出信号序列,设信元长度与正弦波周期一致,设计二选一电路。当信元为‘1’时,输出初始相位为 0 的正弦波;当信元为‘0’时,输出初始相位为π的正弦波。PSK 结构体程序与 FSK 相同,由 DDS 控制字决定调制方式。

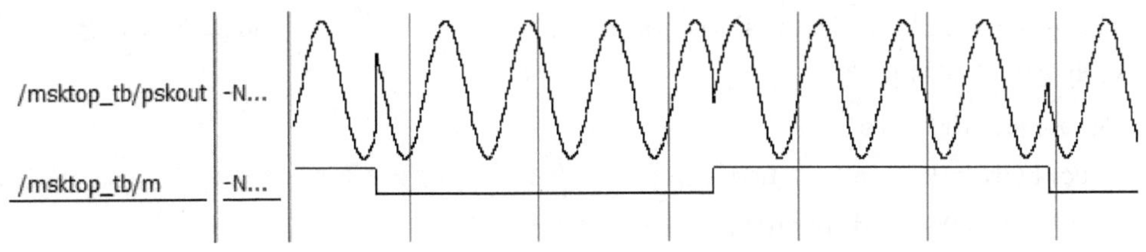

图 8-16　PSK Modelsim 仿真结果

8.5　移位寄存器及其典型应用

8.5.1　移位寄存器

示例 8-8　四位移位寄存器

```
Library Ieee;
Use Ieee.Std_Logic_1164.All;

Entity Shift Is                                    --实体 Shift 有 2 个结构体
  Port(a, clk : In Std_Logic;
       b :  Out Std_logic);
End Entity shift;

Architecture gen_shift Of shift Is                 --第一个结构体
--结构化循环调用描述
  Component dff   Is                               --D 触发器元件说明
    Port(d, clk : In Std_Logic;
         q : Out Std_Logic);
  End Component dff ;

Signal z : Std_Logic_Vector( 0 To 3 );
Begin
  z(0) <= a;

  g1 : For i In 0 To 2 Generate
    dffx : dff Port Map( z(i), clk, z(i + 1));
  End Generate g1;
  b <= z(3);
End Architecture  gen_shift;
```

```
Architecture rtl_shift Of shift Is                    --第二个结构体
--结构化直接调用描述
  Component dff  Is
    Port(d, clk : In Std_Logic;
         q : Out Std_logic);
  End Component dff ;

Signal z : Std_Logic_Vector( 0 To 4 );
Begin
  z(0) <= a;
  dff1: dff Port Map( z(0), clk, z(1) );
  dff2: dff Port Map( z(1), clk, z(2) );
  dff3: dff Port Map( z(2), clk, z(3) );
  b <= z(3);
End Architecture rtl_shift ;
```

示例 8-9　任意位移位寄存器

```
Library Ieee;
Use Ieee.Std_Logic_1164.All;
Entity shift Is
  Generic (len : Integer:=8);                         --8 位
  Port(a, clk : In Std_Logic;
       b :Out Std_Logic);
End Entity shift;

Architecture if_shift Of shift Is
  Component dff Is
    Port(d, clk : In Std_Logic;
         q :Out Std_Logic);
  End Component dff;

Signal z : Std_Logic_Vector( 1 To (len -1) );
Begin
  g1 :For i In 0 To (len -1) Generate
    g2 :If i = 0 Generate                             --第一级
      dffx : dff Port Map( a, clk, z(i + 1));
    End Generate g2;

    g3 :If i = (len -1) Generate                      --最末级
```

```
      dffx : dff Port Map( z(i), clk, b );
    End Generate g3;

    g4 :If (i > 0) And i < (len -1) Generate          --中间级
      dffx : dff Port Map( z(i), clk, Z(i + 1) );
    End Generate g4;

  End Generate g1;
End Architecture if_shift;
--可以自行编写 D 触发器的 VHDL,也可以采用 Altera 库中的 dff。
```

示例 8-10 任意位循环左右移位寄存器

```
Entity shiftlr Is
Generic(N: Positive: = 8);                            --指定移位寄存器为 8 位
  Port(lshift: In Boolean: = True;                    -- lshift 为 True 左移
       d: In Bit_Vector( N Downto 1);                 --数据输入
       qout: Out Bit_Vector( N Downto 1);             --数据输出
       clk,ld,sh: In Bit);                            --时钟,加载,移位
End Entity shiftlr;

Architecture shiftlr_arch Of shiftlr Is
Signal q,shifter: Bit_Vector(N Downto 1);
Signal shiftin: Bit;
Begin
  qout<=q;
  genrs: If Not lshift Generate                       --右移
    shiftin <=q(1);
    shifter<=shiftin&q(N Downto 2);
  End Generate genrs;

  genls: If lshift Generate                           --左移
    shiftin <=q(N);
    shifter<=q(N-1 Downto 1)&shiftin;
  End Generate genls;

  Process(clk,q) Is
  Begin
    If (clk'Event And clk='1') Then                   --上升沿触发
      If ld='1' Then                                  --加载
```

```
        q<=d;
      Elsif sh='1'Then
        q<=shifter;                              --移位寄存
      End If;
    End If;
  End Process;
End Architecture shiftlr_arch;
```

8.5.2 移位寄存器的应用——串并变换

串并变换在电子系统设计中是经常用到的,它与移位寄存器的利用密切相关。

所谓串入/并出移位寄存器,即输入数据是一个接一个的依次进入,输出时则是整组一次送出。并入/串出即输入数据整组一次进入,输出时则是一个接一个的依次送出。

这里介绍八位串入/并出的移位寄存器的设计方法和并入/串出的移位寄存器的设计方法。

示例 8-11 八位并入/串出程序

```
Library Ieee;
Use Ieee.Std_Logic_1164.All;
Use Ieee.Std_Logic_Unsigned.All;
Entity piso Is
  Port(data_in: In Std_Logic_Vector(7 Downto 0);
       clk,nload,reset: In Std_Logic;
       data_out: Out Std_Logic);
End Entity piso;

Architecture piso_arch Of piso Is
Signal q: Std_Logic_Vector(7 Downto 0):=(Others=>'0');
Signal coun: Std_Logic_Vector(2 Downto 0):=(Others=>'0');
Begin

  Process(nload,clk) Is
  Begin
    If nload='0'Then  q<=data_in;          --并行加载数据
    End If;
  End Process;

  Process(clk,reset) Is
  Begin
    If reset='0'Then coun<="000";
```

```
    Elsif(clk'Event And clk='1') Then
      If(coun="111") Then  coun<="000";
      Else  coun<=coun+'1';
      End If;
    End If;
  End Process;

  Process(coun,clk) Is
  Begin
    If clk'Event And clk='1' Then
      If coun="000" Then data_out<=q(3);        --串行输出
      Elsif coun="001" Then data_out<=q(2);
      Elsif coun="010" Then data_out<=q(1);
      Elsif coun="011" Then data_out<=q(0);
      Elsif coun="100" Then data_out<=q(7);
      Elsif coun="101" Then data_out<=q(6);
      Elsif coun="110" Then data_out<=q(5);
      Elsif coun="111" Then data_out<=q(4);
      --Elsif Others=> Null;
      End If;
    End If;
  End Process;
End Architecture piso_arch;
```

示例 8-12 八位串入/并出程序

```
Library Ieee;
Use Ieee.Std_Logic_1164.All;
Use Ieee.Std_Logic_Unsigned.All;
Entity  sipo  Is
  Port(data_in: In Std_Logic;
       clk,reset: In Std_Logic;
       data_out: Out Std_Logic_Vector(7 Downto 0));
End Entity sipo ;

Architecture sipo_arch Of sipo Is
Signal q: Std_Logic_Vector(7 Downto 0);
Signal coun: Std_Logic_Vector(2 Downto 0);
Signal clk1: Std_Logic;
```

```
Begin
  p1: Process(clk) Is
  Begin
    If clk'Event And clk='1' Then
      q(0)<= data_in;                          --串行输入
      For i In 1 To 7 Loop
        q(i)<=q(i-1);
      End Loop;
    End If;
  End Process p1;

  p2: Process(clk,reset) Is
  Begin
    If(reset='0') Then  coun<="000";
    Elsif(clk'Event And clk='1') Then
      If(coun="111") Then  coun<="000";
      Else  coun<=coun+'1';
      End If;
      clk1<=coun(2);
    End If;
  End Process p2;

  p3: Process(clk1) Is
  Begin
    If(clk1'Event And clk1='1') Then
    data_out<=q;                               --并行输出
    End If;
  End Process p3;
End Architecture sipo_arch;
--注意:串并转换中的频率变化
```

示例 8-13 N bit 串并转换器

```
Entity spconv Is
  Generic(N:Integer:=8);                       --8 位
  Port(clk,data:In Bit;
       convert: Out Bit_Vector(N-1 Downto 0));
End Entity spconv;

Architecture spconv_arch Of spconv Is
```

```
Signal s: Bit_Vector(convert'Range);         --数组 convert 的范围属性,N-1 to 0
Begin

  g: For i In convert'Range Generate
    g1:If (i<convert'Left) Generate          --数组 convert 的左边界属性,N-1
      Process Is
      Begin
        Wait Until (clk'Event And clk ='1');  --时钟上升沿
        s(i+1) <=s(i);
      End Process;
    End Generate g1;
    g2:If (i= convert'Right) Generate            --数组 convert 的右边界属性
      Process Is
      Begin
        Wait Until (clk'Event And clk ='1');
        s(i) <=data;
      End Process;
    End Generate g2;
    convert(i)<= s(i);
  End Generate g;
End Architecture spconv_arch;
```

程序说明：N bit 串并转换就是 N bit 移位寄存器，数据在第 N 个时钟周期并行读出。N bit 串并转换仿真图如图 8-17 所示。

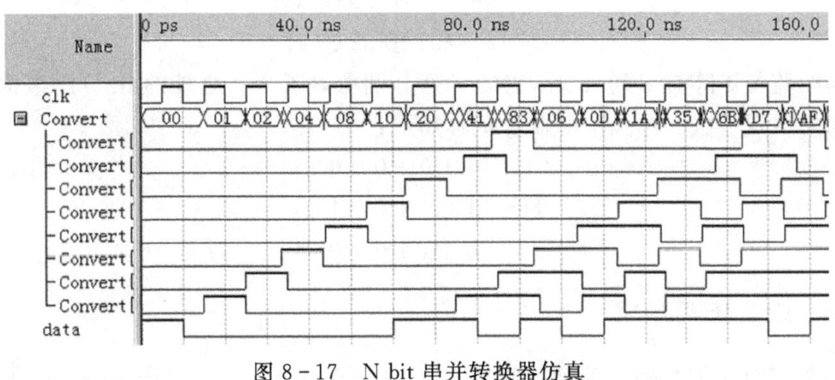

图 8-17　N bit 串并转换器仿真

8.6　校验与纠错编解码设计

8.6.1　汉明(Hamming)编解码简介

在信源编码数据的基础上增加一些冗余码元(又称监督码或检验码)，使监督码元与信息码元之间建立一种确定的关系，称为差错控制编码或纠错编码。

在接收端，根据监督码元与信息码元之间已知的特定关系，可实现检错和纠错，完成此任务的过程称之为误码控制译码(解码)。

差错控制编码的基本操作方法是：在发送端被传输的信息序列上附加一些监督码元，这些多余的码元与信息之间以某种确定的规则建立校验关系。接收端按照既定的规则检验信息码元与监督码元之间的关系，一旦传输过程中发生差错，则信息码元与监督码元之间的校验关系将受到破坏，从而可以发现错误，乃至纠正错误。

Hamming 码是一种纠正单个错误的线性分组码。分组码一般用(n,k)表示。码长为：$n=2^m-1$；信息码位：$k=2^m-m-1$；监督码位：$r=n-k$；这里 m 为大于等于 2 的正整数，给定 m 后，既可构造出具体的汉明码(n,k)。要检出 e 个错误，要求最小码距为 $d \geq e+1$；要纠正 t 个错误，要求最小码距为 $d \geq 2t+1$；要检出 e 个错误纠正 t 个错误，要求最小码距为 $d \geq e+t+1, e>t$。

Hamming 码的特点：最小码距 d=3；纠错能力 t=1。

8.6.2 汉明编码原理(8,4)

信息码为 d0 d1 d2 d3，生成汉明编码 d0 d1 d2 d3 p0 p1 p2 p3 p4，有

$$p0=a0 \oplus a1 \oplus a2$$
$$p1=a0 \oplus a1 \oplus a3$$
$$p2=a0 \oplus a2 \oplus a3$$
$$p3=a1 \oplus a2 \oplus a3$$

即

$$a4=a0 \oplus a1 \oplus a2$$
$$a5=a0 \oplus a1 \oplus a3$$
$$a6=a0 \oplus a2 \oplus a3$$
$$a7=a1 \oplus a2 \oplus a3$$

汉明码的监督矩阵有 n 列 m 行，它的 n 列分别由除了全 0 之外的 m 位码组构成，每个码组只在某列中出现一次。系统中的监督矩阵如下：

$$H=\begin{vmatrix} 1\,1\,0\,1\,0\,0\,0\,1 \\ 1\,0\,1\,1\,0\,0\,1\,0 \\ 0\,1\,1\,1\,0\,1\,0\,0 \end{vmatrix}$$

8.6.3 汉明译码原理

s0 对序列进行奇偶校验。校验方法是：$s0= a0 \oplus a1 \oplus a2 \oplus a3 \oplus a4 \oplus a5 \oplus a6 \oplus a7$；1 的个数为奇数 s0=1，为偶数 s0=0。

下列三个式子是监督关系表达式：

$$s1=a0 \oplus a1 \oplus a3 \oplus a5$$
$$s2=a0 \oplus a2 \oplus a3 \oplus a6$$
$$s3=a1 \oplus a2 \oplus a3 \oplus a7$$

根据表达式，得到表 8－2。

表 8-2 监督关系与错码位置表

s1s2s3	s3s2s1	错码位置	s1s2s3	s3s2s1	错码位置
1 1 0	0 1 1	a0	0 0 0	0 0 0	a4
1 0 1	1 0 1	a1	1 0 0	0 0 1	a5
0 1 1	1 1 0	a2	0 1 0	0 1 0	a6
1 1 1	1 1 1	a3	0 0 1	1 0 0	a7

译码说明:汉明译码的方法,可以采用计算校正因子,然后确定错误图样并加以纠正。当(s3s2s1)= "000"|"001"|"010"|"100" 时,对应 a4～a7 位。s3s2s1="011"说明 a0 位出错,将 a0 取反。如此类推即可完成译码。程序设定了三种状态,如表 8-3 所示。

表 8-3 校正因子与错误图样

校正因子	e	d	sec	
s="0000"	1	0	0	no errors
s0=1 s3～1= "000"|"001"|"010"|"100" s3～1= "110" "101" "011" "111"	0	0	1	single bit error
s0=0 And s(3 Downto 1) /= "000"	0	1	0	double error

示例 8-14 汉明编码程序

```
-- 汉明(Hamming) 编码
-- 用并行语句的 4 位 Hamming 编码
-- The output vector is connected to the individual parity bits
-- using an aggregate assignment.

Library Ieee;
Use Ieee.Std_Logic_1164.All;
Entity hamenc Is
  Port(datain:In Bit_Vector(0 To 3);                --d0 d1 d2 d3
       hamout : In Bit_Vector(0 To 7));             --d0 d1 d2 d3 p0 p1 p2 p3 p4
End Entity hamenc;

Architecture hamenc_arch Of hamenc Is
Signal p0, p1, p2, p4 : Bit;                        --check bits
Begin
  --产生校验位
  p0 <= (datain(0) Xor datain(1)) Xor datain(2);
```

```
  p1 <= (datain(0) Xor datain(1)) Xor datain(3);
  p2 <= (datain(0) Xor datain(2)) Xor datain(3);
  p4 <= (datain(1) Xor datain(2)) Xor datain(3);
  --连接到输出端
  hamout(4 To 7) <= p0&p1&p2&p4;
  hamout(0 To 3) <= datain(0 To 3);
End Architecture hamenc_arch;
```

示例 8-15　汉明解码程序

```
-- Hamming Decoder
-- This Hamming decoder accepts an 8-bit Hamming code
--(produced by the encoder above) And perForms
-- single error correction And double error detection.

Library Ieee;
Use Ieee. Std_Logic_1164. All;
Entity hamdec Is
  Port(hamin : In Bit_Vector(0 To 7);          --d0 d1 d2 d3 p0 p1 p2 p3 p4
       dataout : Out Bit_Vector(0 To 3);        --d0 d1 d2 d3
       sec, ded, ne : Out Bit);                 --diagnostic outputs
End Entity hamdec;

Architecture hamdec_arch Of hamdec Is
Begin

  Process(hamin) Is
  Variable s : Bit_Vector(3 Downto 0);
  Begin
    --Generate s Bits
    s(0):= (((((((hamin(0) Xor hamin(1)) Xor hamin(2)) Xor hamin(3))
            Xor hamin(4)) Xor hamin(5)) Xor hamin(6)) Xor hamin(7));
    s(1):= (((hamin(0) Xor hamin(1)) Xor hamin(3)) Xor hamin(5));
    s(2):= (((hamin(0) Xor hamin(2)) Xor hamin(3)) Xor hamin(6));
    s(3):= (((hamin(1) Xor hamin(2)) Xor hamin(3)) Xor hamin(7));
    If (s = "0000") Then                                   --no errors
      ne <= '1';
      ded <= '0';
      sec <= '0';
      dataout(0 To 3) <= hamin(0 To 3);
```

```
    Elsif (s(0) = '1') Then                                --single bit error
      ne <= '0';
      ded <= '0';
      sec <= '1';
      Case s(3 Downto 1) Is
        When "000"|"001"|"010"|"100" =>
          dataout(0 To 3) <= hamin(0 To 3);                -- parity errors
        When "011" => dataout(0) <= Not hamin(0);
          dataout(1 To 3) <= hamin(1 To 3);
        When "101" => dataout(1) <= Not hamin(1);
          dataout(0) <= hamin(0);
          dataout(2 To 3) <= hamin(2 To 3);
        When "110" => dataout(2) <= Not hamin(2);
          dataout(3) <= hamin(3);
          dataout(0 To 1) <= hamin(0 To 1);
        When "111" => dataout(3) <= Not hamin(3);
          dataout(0 To 2) <= hamin(0 To 2);
      End Case;
                                                           --double error
    Elsif (s(0) = '0') And (s(3 Downto 1) /= "000") Then
      ne <= '0';
      ded <= '1';
      sec <= '0';
      dataout(0 To 3) <= "0000";
    End If;
  End Process;
End Architecture hamdec_arch;
```

8.7 传输码型的生成

要开发和实现高效、高可靠性、高稳定性、高质量的通信系统，选择一种好的数据传输编译码方式是非常重要的。

各种各样的通信协议都依赖某种形式的编码方案。

传输码（又称为线路码）的结构取决于实际信道特性和系统工作的条件。在较为复杂的基带传输系统中，传输码的结构应具有下列主要特性：

(1)能从其相应的基带信号中获取定时信息；

(2)相应的基带信号无直流成分和只有很小的低频成分；

(3)不受信息源统计特性的影响，即能适应信息源的变化；

(4)尽可能地提高传输码型的传输效率；

(5)具有内在的检错能力。

……

满足或部分满足以上特性的传输码型种类繁多,主要有:CMI 码、AMI、HDB3、曼彻斯特(Manchester)等。

8.7.1　曼彻斯特(Manchester)编译码设计

1. 曼彻斯特码

曼彻斯特码(Manchester code)又称裂相码、双向码,一种用电平跳变来表示 1 或 0 的编码,其变化规则很简单,即每个码元均用两个不同相位的电平信号表示,每个码是一个周期的方波,但 0 码和 1 码的相位正好相反。其对应关系为

$$0 \rightarrow 01, 1 \rightarrow 10$$

曼彻斯特编码是自同步的编码方式,时钟同步信号隐藏在数据波形中。在曼彻斯特编码中,每一位的中间有一跳变。位中间的跳变既作时钟信号,又作数据信号;从高到低跳变表示"1",从低到高跳变表示"0"。

还有差分曼彻斯特编码,每位中间的跳变仅提供时钟定时,而用每位开始时有无跳变表示"0"或"1",有跳变为"0",无跳变为"1"。

曼彻斯特编码特点如下:

(1)传输流的速率是原始数据流的两倍,要占用较宽的频带。

(2)信号恢复简单,只要找到信号的边缘进行异步提取即可。

(3)10 Mb/s 以太网(Ethernet)采用曼彻斯特码。

2. 曼彻斯特编译码电路设计

曼彻斯特编码电路可用 Moore 状态机来实现。曼彻斯特编码状态如图 8-18 所示。

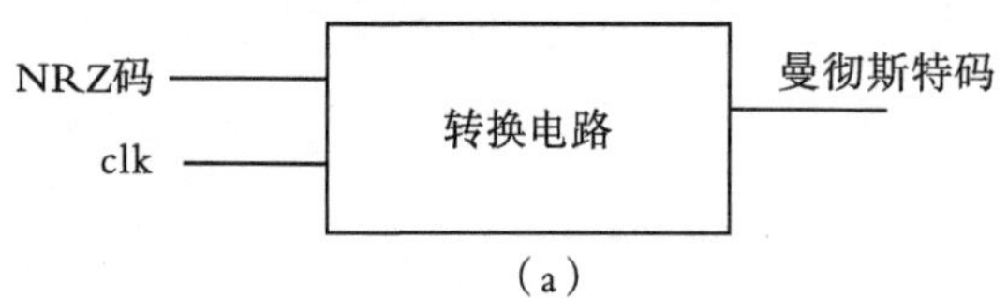

(a)

当前状态	下一个状态		当前输出
s0	s1	s3	0
s1	s2	—	0
s2	s1	s3	1
s3	—	s0	1

(b)

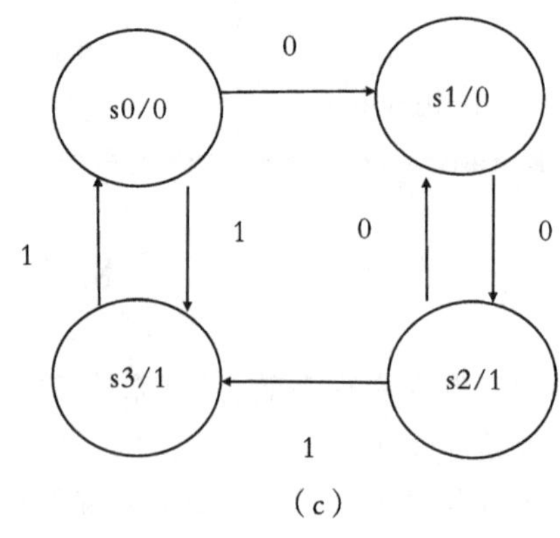

(c)

图 8-18　曼彻斯特编码状态

示例 8-16a　曼彻斯特码编码

```
-- mcode.vhd 其中含有 3 种曼彻斯特码编码方法
```

```
Library Ieee;
Use Ieee.Std_Logic_1164.All;

Entity  mcode  Is                                          --定义实体名为 mcode
  Port(clk,reset, datain: In Std_Logic;                    --时钟、复位、基带输入
       dataout : OutStd_Logic);                            --编码输出
End  Entity  mcode;

Architecture mcode_arch1 Of mcode Is                       --曼彻斯特码编码方法 1
Type state_type Is (s0, s1, s2, s3);
--自定义类型 state_type，有 4 个状态值 s0, s1, s2, s3
Signal state   : state_type;                               --定义一个信号
Begin
  Reg:Process (clk, reset) Is                          --2 进程 Moore 状态机的时序进程
  Begin
    If reset = '1' Then  state <= s0;                      --异步复位，状态初始化
    Elsif (Rising_Edge(clk)) Then
      Case state Is                                        --Case 选择语句
        When s0=>                                          --状态转换
        If datain = '0' Then
          state <= s1;
        Else state <= s3;
        End If;
        When s1=>
        If datain = '0' Then
          state <= s2;
        Else state <= s2;                                --无此 Else 会产生无效锁存器
        End If;
        When s2=>
        If datain = '1' Then
          state <= s3;
        Else state <= s1;
        End If;
        When s3 =>
        If datain = '1' Then
          state <= s0;
        Else state <= s0;                                --无此 Else 会产生无效锁存器
        End If;
      End Case;
```

```
    End If;
  End Process reg;

  com:Process (state) Is                                    --Moore 状态机
  Begin
    Case state Is
      When s0 => dataout <= '0';
      When s1 => dataout <= '0';
      When s2 => dataout <= '1';
      When s3 => dataout <= '1';
    End Case;
  End Process com;
End Architecture mcode_arch1;

Architecture mcode_arch2 Of mcode Is                        --曼彻斯特码编码方法 2
Signal con:Std_Logic_Vector(1 Downto 0):=(Others=>'0'));
Signal flag:Std_Logic:='0';
Begin

  Process(clk) Is
  Begin
    If clk'Event And clk='1' Then
      If flag='0' Then                                      --锁存器
        If datain='1' Then con<="10";
        Else con<="01";
        End If;
      End If;

      If flag='1' Then                                      --锁存器
        dataout<=con(1);
        flag<=Not flag;
      Else
        dataout<=con(0);
        flag<=Not flag;
      End If;
    End If;
  End Process;
End Architecture mcode_arch2;
```

```
Architecture mcode_arch3 Of mcode Is                    --曼彻斯特码编码方法 3
Begin
  dataout<=datain Xor clk;
End Architecture mcode_arch3;                           --功能仿真正确

Configuration mcode_cfg Of mcode Is                 --配置语句说明,附在主体程序之后
  For mcode_arch1                                       --指定实体对应结构体
  End For;
End Configuration mcode_cfg;                            --配置语句说明结束
```

示例 8－16b　曼彻斯特码译码

```
-- mdecode.vhd 曼彻斯特码译码程序,其中含 2 种译码方法
Library Ieee;
Use Ieee.Std_Logic_1164.All;

Entity mdecode Is
  Port(clk,datain:In Std_Logic;                         --时钟、基带码输入
       datad:Out Std_Logic);                            --编码输出
End Entity mdecode;

Architecture mdecode_arch1 Of mdecode Is                --译码方法 1
Signal con:Std_Logic_Vector(1 Downto 0);
Signal flag:Std_Logic_Vector(1 Downto 0);
Signal syn:Std_Logic;
Begin

  Process(clk) Is
  Begin
    If clk'Event And clk='1' Then
      con<=con(0) & datain;
    End If;
  End Process;

  Process(clk) Is
  Begin
    If clk'Event And clk='1' Then
      If con="11" Or con="00" Then
        flag<="11";
      Elsif syn='0' Then
```

```
        flag<='0'& flag(0);
      Else
        flag<=flag(1)&(Not flag(0));
      End If;
    End If;
  End Process;

  Process(clk) Is
  Begin
    If clk'Event And clk='1'Then
      If flag="11" Then
        Case con Is
          When "01"=>
            datad <='0';
            syn<='0';
          When "10"=>
            datad <='1';
            syn<='0';
          When Others=> syn<='0';
        End Case;
      End If;
    End If;
  End Process;
End Architecture mdecode_arch1 ;

Architecture mdecode_arch2 Of mdecode Is              --译码方法 2
Signal clk1:Std_Logic;
Begin
  Process(clk) Is
  Begin
    If(clk'Event And clk='1')Then                     --上升沿触发
      clk1<=Not clk1;                                 --二分频
    End If;
  End Process;
  datad <= Not(clk1 Xor datain);                      --译码
End Architecture mdecode_arch2;
```

示例 8-16c　曼彻斯特码编译码联调

```
--mendecod.vhd
Library Ieee;
Use Ieee.Std_Logic_1164.All;
Use Work.All;
Entity mendecode Is
  Port(clk,reset, datain: In  Std_Logic;            --时钟、复位、基带输入
       datac :  Out  Std_Logic;                     --曼彻斯特码编码
       datad :  Out  Std_Logic);                    --译码出入,恢复基带信号
End Entity mendecode;

Architecture mendecode_arch Of mendecode Is
Signal wire_0 :  Std_Logic;
Begin
  datac <=wire_0;

  inst1 : mcode                                     --编码元件例化
  Port Map(clk => clk, datain => datain,
           reset => reset, dataout => wire_0);
  inst3 : mdecode                                   --译码元件例化
  Port Map(clk => clk,
           datain => wire_0, datad => datad);
End Architecture mendecode_arch;
```

程序说明:采用异或的方式编译码看似容易,在设计时对电路时序同步等要求非常高,要用通过锁相环同步的信号,有时还要通过布局布线约束,甚至加比特矫正才可能得到正确的设计结果。曼彻斯特码编译码验证电路仿真结果如图 8-19 所示。

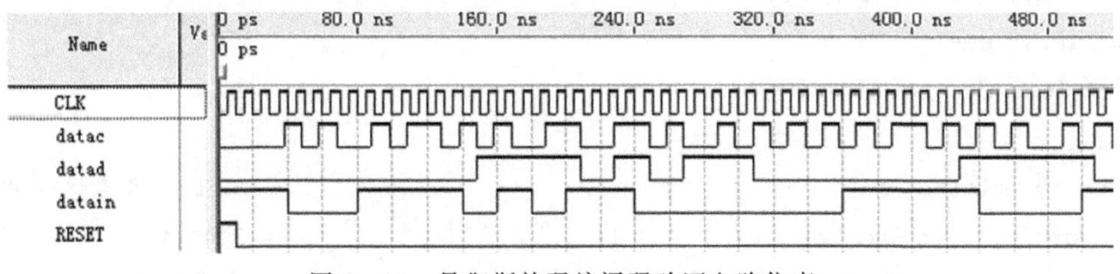

图 8-19　曼彻斯特码编译码验证电路仿真

8.7.2　传号反转码(CMI)编解码设计

CMI(Coded Mark Inversion)码又称为传号反转码,是一种 1B2B 码(一位信息码,二位码元),是二电平不归零码。CCITT 建议 CMI 码作为 PCM 脉冲编码调制器四次群的接口码型。

CMI 码的编码规则如表 8-4 所示。

表 8-4 CMI 码的编码规则

输入码字	输出结果
0	01
1	00/11 交替

因为 CMI 码在正常情况下，10 不可能出现，连续的 00 和 11 也不可能出现，从而不会连续出现 4 个以上的 0 码或 1 码，这种相关性可以用来检测因信道而产生的部分错误。这种码的传输速率为编码前的原信号速率的两倍，要占用较宽的频带。

CMI 编码原理：在 CMI 编码中，输入码字为 0 直接输出 01 码型。对于输入为 1 的码字，其输出 CMI 码字存在两种结果 00 或 11 码，因而对输入 1 的状态必须记忆。

CMI 解码原理示意图如图 8-20 所示。

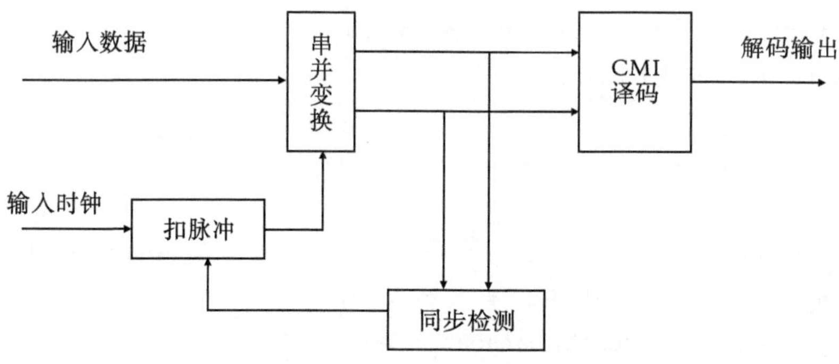

图 8-20 CMI 解码原理

CMI 译码电路由串并变换器、译码器、同步检测器、扣脉冲电路等部分组成。

CMI 解码端，需进行同步。同步过程的设计可根据码字的状态进行，因为在输入码字中不存在 10 码型，如果出现 10 码，则必须调整同步状态。

示例 8-17a 传号反转码 CMI 编码

```
-- CMI 编码程序 cmicode.vhd
Library Ieee;
Use Ieee.Std_Logic_1164.All;

Entity cmicode Is                                              -- CMI 编码实体说明
  Port(clk: In Std_Logic;                                      --时钟输入
        databin:In Std_Logic;                                  --基带信号输入
        datacout: Out Std_Logic_Vector(1 Downto 0));           --CMI 码输出
End Entity cmicode;

Architecture cmicode_arch Of cmicode Is
Signal flag: Std_Logic;
```

```
--标志位,用于基带信号为1时的判断,进而决定编码为"00"还是"11"
Begin
  Process(clk,databin) Is
  Begin
    If clk'event And clk='1' Then                      --上升沿触发
      If databin='1' And flag='1' Then                 --基带信号为1时
        datacout<="00";
        flag<='0';                                     --改变标志位的值
      Elsif databin='1' And flag='0' Then              --基带信号为1时
        datacout<="11";
        flag<='1';                                     --改变标志位的值
      Else                                             --基带信号为0时
        datacout<="01";
      End If;
    End If;
  End Process;
End Architecture cmicode_arch;                         -- CMI编码描述结束
```

示例 8-17b 传号反转码 CMI 译码

```
-- CMI译码程序 cmidcode.vhd
Library Ieee;
Use Ieee.Std_Logic_1164.All;

Entity cmidcode Is                                     -- CMI译码实体说明
  Port(clk: In Std_Logic;                              --时钟输入
       datacin: In Std_Logic_Vector(1 Downto 0);       --CMI码输入
       databout: Out Std_Logic);                       --基带信号输出
End Entity cmidcode;

Architecture cmidcode_arch Of cmidcode Is
Begin
  Process(clk,datacin) Is
  Begin
    If clk'event And clk='1' Then                      --上升沿触发
      If datacin="00" Or datacin="11" Then
        databout<='1';
      Else
        databout<='0';
      End If;
```

```
      End If;
    End Process;
End Architecture cmidcode_arch;                      -- CMI 译码描述结束
```

习 题

8-1 试采用 DDS 设计实现 DPSK，OPSK,QPSK 调制。

8-2 采用单 DDS 实现 ASK、PSK、FSK 调制,查阅资料进行调制解调的联调。

8-3 查阅资料实现 CRC 编解码。

8-4 设计一个串行数据检测器,要求是连续 4 个或 4 个以上的 1 时输出为 1,其他输入情况下为 0。编写测试模块并给出仿真波形。

8-5 设计一个序列发生器。要求根据输入的 8 位并行数据输出串行数据,如果输入数据在 0～127 之间则输出一位 0,如果输入数据在 128～255 之间则输出一位 1,同步时钟触发;串行输入生成的序列码,再转换成并行码与最初给出的并行码进行比较,相同则输出 1,即形成一个封闭系统测试,编写测试模块给出仿真波形。

8-6 连续给出包含多组巴克码序列的 7 位数据流,输出错误数组的个数,并用 LED 显示出来。

8-7 在外部开关的控制下改变产生不同的 m 序列。可以设置 m_sel,选择输出不同的 m 序列,具体如下:

(1)当 m_sel ="00"时,本原多项式为 13(八进制表示);

(2)当 m_sel ="01"时,本原多项式为 23(八进制表示);

(3)当 m_sel ="10"时,本原多项式为 103(八进制表示);

(4)当 m_sel ="11"时,本原多项式为 203(八进制表示)。

并以上述 m 序列构成扰码,并解扰码。

注意观察:如果不消除全 0 状态,观察 m 序列产生器的工作。

8-8 分别编写扰码和解扰码的 VHDL 程序,与巴克码生成检测实验结合,或自选其他实验联合调试,仿真和测试。

扩展学习与总结

设计 SPWM 波,并给出 2 个 SPWM 波的实际电路应用实现。

9　接口设计

本章的主要内容与方法：

(1)UART 通信设计。

(2)LCD 字符液晶显示控制设计。

(3)矩阵扫描键盘控制设计。

(4)可编程接口芯片 8255 设计。

9.1　UART/RS232 接口

1. UART(Universal Asynchronous Receiver Transmitter,通用异步收发器)说明

UART 广泛应于短距离串行传输。基本的 UART 通信采用两条信号线(RXD、TXD)来完成数据的全双工通信。UART 的传送速率用每秒钟传送数据位的数目来表示,称之为波特率。如:波特率 9600=9600 bps(位/秒)。一些设计模块如,WIFI 通信、蓝牙通信、USB 口通信、传感器等会提供 UART、SPI 或者 I^2C 转换接口。UART 的数据帧格式说明如下:

START	D0	D1	D2	D3	D4	D5	D6	D7	P	STOP
起始位	数 据 位(Data Bits)								校验位	停止位

(1)起始位(Start Bit):发送器通过发送起始位而后开始一个字符传送,起始位从 1 到 0 跳变使数据线处于逻辑 0 状态,提示接受器数据传输即将开始。

(2)数据位(Data Bits):数据位一般为 8 位一个字节的数据(也有 6 位、7 位的情况),低位(LSB)在前,高位(MSB)在后。

(3)校验位(Parity Bit):校验位用来判断接收的数据位有无错误,一般是奇偶校验。

(4)停止位(STOP):停止位用以标志一个字符传送的结束,它对应于逻辑 1 状态。

2. UART 的主要设计模块

发送器将准备输出的并行数据按照 UART 帧格式转为 TXD 信号串行输出;接收器接收 RXD 串行信号,并将其转化为并行数据;波特率发生器产生一个高于波特率 16 倍的本地时钟信号对输入 RXD 不断采样,使接收器与发送器保持同步。

示例 9-1　UART 顶层程序

```
--文件名:uart_top.vhd
--功能:顶层映射
--连接:直连方式
Library Ieee;
```

```
Use Ieee.Std_Logic_1164.All;

Use Ieee.Std_Logic_Unsigned.All;
Use Work.All;

Entity uart Is
  Port(clk50mhz,reset: In Std_Logic;                        --系统时钟,复位
       rxd: In Std_Logic;                                   --收信号
       txdbuft: In Std_Logic_Vector(7 Downto 0);            --待发输入数据缓冲
       rbuft: Out Std_Logic_Vector(7 Downto 0) ;            --接收数据缓冲
       txd: Out Std_Logic );                                --发信号
End Entity uart;

Architecture uart_arch Of uart Is
Signal b: Std_Logic;                                        --16 倍波特率时钟
Signal xmit_p: Std_Logic;                                   --发启动信号
Signal xbuf: Std_Logic_Vector(7 Downto 0);                  --待发送数据缓冲区
Signal txd_done_out: Std_Logic;                             --发就绪
Signal rev_buf: Std_Logic_Vector(7 Downto 0);               --接收数据缓冲区
Signal rev_ready: Std_Logic;                                --收就绪

Begin
  --波特率产生元件例化--
  uart_baud: gen_div Generic Map (163)                      --参见第 6 章分频部分
  --326 分频,326 * 16 * 9600= 50MHz,波特率为 9600
    Port Map(clk=>clk50mhz, reset=> reset, bclk=> b);

  --串口接收元件例化--
  uart_receive: uart_r
    Port Map(bclkr=> b, resetr=> reset, rxdr=>rxd,
             r_ready=>rev_ready, rbuf=>rev_buf);

  --串口发送模块例化--
  uart_transfer: uart_t
    Port Map(bclkt=> b, resett=> reset, xmit_cmd_p=>xmit_p,
             txdbuf=>xbuf, txd=>txd, txd_done=>txd_done_ out);

  --信号窄化模块例化--
  narr_rev_ready: narr_sig
```

```
  Port Map(sig_in=>rev_ready,                    --输入需窄化信号 rev_ready
          clk=> b, reset=> reset,
          narr_prd=>x"03",                       --窄化信号高电平持续 3 个 clk
          周期
          narr_sig_out=>xmit_p);                 --输出窄化后信号 xmit_p

  Process (rev_ready, reset, rev_buf,clk_b) Is
  Begin
    If reset='1' Then
      xbuf<="00000000";
    Else
      If Rising_Edge(rev_ready) Then             --接收完毕
        xbuf<=rev_buf;                           --装载数据,可送到 LED 等接口
      End If;
    End If;
  End Process;
End Architecture uart_arch;
```

示例 9-1a　发送器程序

/* 文件名:transfer.vhd。

功能:UART 发送器。

说明:系统由五个状态(x_idle,x_start,x_wait,x_shift,x_stop)和一个进程构成。

在空闲状态,传送线为逻辑'1',数据的传送以一个起始位开始,接着从低位开始传送的数据,然后是停止位 */

```
Library Ieee;
Use Ieee.Std_Logic_1164.All;
Use Ieee.Std_Logic_Arith.All;
Use Ieee.Std_Logic_Unsigned.All;

Entity transfer Is
  Generic(framlent:Integer:=8);
  Port(bclkt,resett,xmit_cmd_p:In Std_Logic;      --定义输入输出信号
       txdbuf:In Std_Logic_Vector(7 Downto 0):="11001010";
       txd:Out Std_Logic;
       txd_done:Out Std_Logic);
End Entity transfer;

Architecture behavioral Of transfer Is
Type states Is(x_idle,x_start,x_wait,x_shift,x_stop);
```

```
--定义状态
Signal state:states:=x_idle;
Begin
  Process(bclkt,resett,xmit_cmd_p,txdbuf)Is                        --主控时序、组合进程
  Variable xcnt16:Std_Logic_Vector(4 Downto 0):="00000";              --定义中间变量
  Variable xbitcnt:Integer:=0;
  Variable txds:Std_Logic;
  Begin
    If resett='1'Then state<=x_idle;txd_done<='0';txds:='1';    --复位
    Elsif  Rising_Edge(bclkt) Then
      Case state Is
        When x_idle=>                         --状态 1,等待数据帧发送命令
          If xmit_cmd_p='1'Then state<=x_start;txd_done<='0';
          Else state<=x_idle;
          End If;
        When x_start=>                        --状态 2,发送信号至起始位
          If xcnt16>="01111" Then state<=x_wait;xcnt16:="00000";
          Else xcnt16:=xcnt16+1;txds:='0';state<=x_start;
          End If;
        When x_wait=>                         --状态 3,等待状态
          If xcnt16>="01110" Then
            If xbitcnt=framlent Then state<=x_stop;xbitcnt:=0;
            Else state<=x_shift;
            End If;
            xcnt16:="00000";
          Else xcnt16:=xcnt16+1;state<=x_wait;
          End If;
        When x_shift=>txds:=txdbuf(xbitcnt);   --状态 4,将待发数据进行并串转换
        xbitcnt:=xbitcnt+1;state<=x_wait;
        When x_stop=>                         --状态 5,停止位发送状态
          If xcnt16>="01111" Then
            If xmit_cmd_p='0'Then state<=x_idle;xcnt16:="00000";
            Else xcnt16:=xcnt16; state<=x_stop;
            End If;
            txd_done<='1';
          Else xcnt16:=xcnt16+1;txds:='1'; state<=x_stop;
          End If;
        When Others=>state<=x_idle;
      End Case;
```

```
    End If;
    txd<=txds;
  End Process;
End Architecture behavioral;
```

示例 9-1b 接收器程序

```
/* 文件名:receiver.vhd。
功能:UART 接收器。
说明:系统由五个状态(r_start,r_center,r_wait,r_sample,r_stop)和两个进程构
成 */
Library Ieee;
Use Ieee.Std_Logic_1164.All;
Use Ieee.Std_Logic_Arith.All;
Use Ieee.Std_Logic_Unsigned.All;

Entity receiver Is
  Generic(Framlenr:Integer:=8);
  Port(bclkr,resetr,rxdr:In Std_Logic;                    --定义输入输出信号
       r_ready:Out Std_Logic;
       rbuf:Out Std_Logic_Vector(7 Downto 0));
End Entity receiver;

Architecture behavioral Of receiver Is
Type states Is(r_start,r_center,r_Wait,r_sample,r_stop);
                                                          --定义状态
Signal state:states:=r_start;
Signal rxd_sync:Std_Logic;
Begin
  pro1:Process(rxdr) Is
  Begin
    If rxdr='0'Then rxd_sync<='0';
    Else rxd_sync<='1';
    End If;
  End Process pro1;

  pro2:Process(bclkr,resetr,rxd_sync) Is                  --主控时序、组合进程
  Variable count:Std_Logic_Vector(3 Downto 0);            --定义中间变量
  Variable rcnt:Integer:=0;
  Variable rbufs:Std_Logic_Vector(7 Downto 0);
```

```
  Begin
    If resetr='1' Then state<=r_start;count:="0000";        --复位
    Elsif Rising_Edge(bclkr) Then
      Case state Is
        When r_start=>                                 --状态 1,等待起始位
          If rxd_sync='0' Then
            state<=r_center;r_ready<='0';rcnt:=0;
          Else state<=r_start;r_ready<='0';
          End If;
        When r_center=>                                --状态 2,求出每位的中点
          If rxd_sync='0' Then
            If count="0100" Then
              state<=r_wait;count:="0000";
            Else count:=count+1;state<=r_center;
            End If;
          Else state<=r_start;
          End If;
        When r_Wait=>                                  --状态 3,等待状态
          If count>="1110" Then
            If rcnt=Framlenr Then state<=r_stop;
            Else state<=r_sample;
            End If;
            count:="0000";
          Else count:=count+1;state<=r_wait;
          End If;
        When r_sample=>                                --状态 4,数据位采样检测
          rbufs(rcnt):=rxd_sync;rcnt:=rcnt+1;state<=r_wait;
        When r_stop=>                                  --状态 5,输出帧接收完毕信号
          r_ready<='1';rbuf<=rbufs;state<=r_start;
        When Others=>state<=r_start;
      End Case;
    End If;
  End Process pro2;
End Architecture behavioral;
```

示例 9-1c 信号窄化处理

```
--信号窄化处理程序 narr_sig.vhd
Library Ieee;
Use Ieee.Std_Logic_1164.All;
```

```
Use Ieee.Std_Logic_Unsigned.All;

Entity narr_sig Is
Port (sig_In, clk, reset: In Std_Logic;
      narr_prd: In Std_Logic_Vector(7 Downto 0);
                                              --窄化信号高电平持续的clk周期数
      narr_sig_out: Out Std_Logic);
End Entity narr_sig;

Architecture behave Of narr_sig Is
Signal narr_prd_cnt: Std_Logic_Vector(7 Downto 0);        --窄化信号高电平计数
Signal stop_narr_flag: Std_Logic;                  --停止窄化标志。0,窄化;1,不窄化
Begin
  narr_sig: Process (sig_in, reset, clk, stop_narr_flag) Is
  Begin
    If reset='1' Then
      narr_prd_cnt<=x"00";
      stop_narr_flag<='1'; narr_sig_out<='0';
    Elsif sig_in='0' Then
      narr_prd_cnt<=x"00";
      stop_narr_flag<='0'; narr_sig_out<='0';
    Elsif stop_narr_flag='0' Then                --sig_in的上升沿启动窄化信号
      If Rising_Edge(clk) Then
        narr_prd_cnt<=narr_prd_cnt+'1';  --每个时钟上升沿计数加1
        narr_sig_out<='1';
        If narr_prd_cnt=narr_prd Then         --narr_prd=0时,narr_sig_out是0
          narr_prd_cnt<=x"00";
          stop_narr_flag<='1'; narr_sig_out<='0';
        End If;
      End If;
    End If;
  End Process narr_sig;
End Architecture behave;
```

9.2　字符 LCD 显示控制

示例 9－2　字符 LCD1602 控制

```
/*--------------------------------------------------------------------------
--程序 lcd.vhd,采用 44780 控制器
Library Ieee;                                                --库声明
Use Ieee.Std_Logic_1164.All;
Use Ieee.Std_Logic_Unsigned.All;
Use Ieee.Std_Logic_Arith.All;
--------------------------------------------------------------------------*/
Entity lcd1602 Is                                            --LCD1602 实体声明
    Port(reset:In Std_Logic;
        gclkp1, gclkp2:In Std_Logic;                         --50 MHz
        lcd_rs, lcd_rw, lcd_e:Out Std_Logic;
        lcd_data:Out Std_Logic_Vector(7 Downto 0));          --LCD 数据输出
End Entity lcd1602;
Architecture behavioral Of lcd1602 Is                        --结构体声明
--LCD1602 内置字库,虽直接给地址就可以调用字符,但用函数提高了易用性
    Function char_to_integer(indata:Character) Return Integer Is
    Variable result:Integer Range 0 To 16#7F#;
    Begin
        Case indata Is
            When ' ' =>result:=32; When '!' =>result:=33;
            When '"' =>result:=34; When '#' =>result:=35;
            When '$' =>result:=36; When '%' =>result:=37;
            When '&' =>result:=38; When ''' =>result:=39;
            When '(' =>result:=40; When ')' =>result:=41;
            When '*' =>result:=42; When '+' =>result:=43;
            When ',' =>result:=44; When '-' =>result:=45;
            When '.' =>result:=46; When '/' =>result:=47;
            When '0' =>result:=48; When '1' =>result:=49;
            When '2' =>result:=50; When '3' =>result:=51;
            When '4' =>result:=52; When '5' =>result:=53;
            When '6' =>result:=54; When '7' =>result:=55;
            When '8' =>result:=56; When '9' =>result:=57;
            When ':' =>result:=58; When ';' =>result:=59;
            When '<' =>result:=60; When '=' =>result:=61;
```

```
        When '>' =>result:=62; When '?' =>result:=63;
        When '@' =>result:=64; When 'A' =>result:=65;
        When 'B' =>result:=66; When 'C' =>result:=67;
        When 'D' =>result:=68; When 'E' =>result:=69;
        When 'F' =>result:=70; When 'G' =>result:=71;
        When 'H' =>result:=72; When 'I' =>result:=73;
        When 'J' =>result:=74; When 'K' =>result:=75;
        When 'L' =>result:=76; When 'M' =>result:=77;
        When 'N' =>result:=78; When 'O' =>result:=79;
        When 'P' =>result:=80; When 'Q' =>result:=81;
        When 'R' =>result:=82; When 'S' =>result:=83;
        When 'T' =>result:=84; When 'U' =>result:=85;
        When 'V' =>result:=86; When 'W' =>result:=87;
        When 'X' =>result:=88; When 'Y' =>result:=89;
        When 'Z' =>result:=90; When '[' =>result:=91;
        When '\' =>result:=92; When ']' =>result:=93;
        When '^' =>result:=94; When '_' =>result:=95;
        When '`' =>result:=96; When 'a' =>result:=97;
        When 'b' =>result:=98; When 'c' =>result:=99;
        When 'd' =>result:=100; When 'e' =>result:=101;
        When 'f' =>result:=102; When 'g' =>result:=103;
        When 'h' =>result:=104; When 'i' =>result:=105;
        When 'j' =>result:=106; When 'k' =>result:=107;
        When 'l' =>result:=108; When 'm' =>result:=109;
        When 'n' =>result:=110; When 'o' =>result:=111;
        When 'p' =>result:=112; When 'q' =>result:=113;
        When 'r' =>result:=114; When 's' =>result:=115;
        When 't' =>result:=116; When 'u' =>result:=117;
        When 'v' =>result:=118; When 'w' =>result:=119;
        When 'x' =>result:=120; When 'y' =>result:=121;
        When 'z' =>result:=122; When '{' =>result:=123;
        When '|' =>result:=124; When '}' =>result:=125;
        When '~' =>result:=126; When Others=>result:=32;
    End Case;
    Return result;
End Function char_to_integer;
----------------------------------------------
--下面这段顺序编码状态值在 FPGA 中通常不采用,FPGA 中采用 one-hot 编码
--可以改写为枚举的状态,Quartus II 平台会自动生成状态编码值
```

```
Constant Idle:Std_Logic_Vector(3 Downto 0):="0000";
Constant Clear:Std_Logic_Vector(3 Downto 0):="0001";
Constant Returncursor:Std_Logic_Vector(3 Downto 0):="0010";
Constant Setmode:Std_Logic_Vector(3 Downto 0):="0011";
Constant Switchmode:Std_Logic_Vector(3 Downto 0):="0100";
Constant Shift:Std_Logic_Vector(3 Downto 0):="0101";
Constant Setfunction:Std_Logic_Vector(3 Downto 0):="0110";
Constant Setcgram:Std_Logic_Vector(3 Downto 0):="0111";
Constant Setddram:Std_Logic_Vector(3 Downto 0):="1000";
Constant Readflag:Std_Logic_Vector(3 Downto 0):="1001";
Constant Writeram:Std_Logic_Vector(3 Downto 0):="1010";
Constant Readram:Std_Logic_Vector(3 Downto 0):="1011";
---------------------------------------------
Signal state:Std_Logic_Vector(3 Downto 0);

Constant cur_inc:Std_Logic:='1';
Constant cur_dec:Std_Logic:='0';
Constant cur_shift:Std_Logic:='1';
Constant cur_noshift:Std_Logic:='0';
Constant open_display:Std_Logic:='1';
Constant open_cur:Std_Logic:='0';
Constant blank_cur:Std_Logic:='0';
Constant shift_display:Std_Logic:='1';
Constant shift_cur:Std_Logic:='0';
Constant right_shift:Std_Logic:='1';
Constant left_shift:Std_Logic:='0';
Constant datawidth8:Std_Logic:='1';
Constant datawidth4:Std_Logic:='0';
Constant twoline:Std_Logic:='1';
Constant oneline:Std_Logic:='0';
Constant font5x10:Std_Logic:='1';
Constant font5x7:Std_Logic:='0';
---------------------------------------------------
Type lcdram Is Array(0 To 31) Of Std_Logic_Vector(7 Downto 0);
Signal lcd_array:lcdram;                    --2 维信号

Signal data_in:Std_Logic_Vector(7 Downto 0);
Signal char_add:Std_Logic_Vector(5 Downto 0);
---------------------------------------------------
```

```
Signal clkint:Std_Logic;
Signal counter:Integer Range 0 To 127;
Signal flag:Std_Logic;
Signal div_counter:Integer Range 0 To 15;
Constant Divss:Integer:=15;

Signal period1us,period1ms,period1s:Std_Loqic;
Signal p1us,p1ms:Std_Logic;
Signal clk,fresh,freshplus:Std_Logic;
Begin

--Globle Clock Assignment
Globleclk:
Process(reset,gclkp1,period1us,period1ms,period1s,freshplus) Is
Variable count:Std_Logic_Vector(5 Downto 0);                    --1 MHz
Variable count1:Std_Logic_Vector(9 Downto 0);                   --1 kHz
Variable count2:Std_Logic_Vector(9 Downto 0);                   --1 Hz
Variable countt:Std_Logic_Vector(1 Downto 0);
Variable countr:Std_Logic_Vector(1 Downto 0);
Begin
    --gclkp:50 MHz
    --Period:1us(period1us<=gclkp1;)
    If(reset='0') Then count:=(Others=>'0');
    Elsif(gclkp1'Event And gclkp1='1') Then
        If(count>"110000") Then count:=(Others=>'0');           --1 us
        Else count:=count+1;
        End If;
    End If;
    period1us<=count(5);
    --Period:1 ms
    If(reset='0') Then count1:=(Others=>'0');
    Elsif(period1us'Event And period1us='1') Then
        If(count1>"1111100110")Then count1:=(Others=>'0');--1 ms
        Else count1:=count1+1;
        End If;
    End If;
    period1ms<=count1(9);
    clk<=count1(8);
    --Period:1 s
```

```
        If(reset='0') Then count2:=(Others=>'0');
        Elsif(period1ms'Event And period1ms='1') Then
          If(count2>"1111100110") Then count2:=(Others=>'0');  --1 s
          Else  count2:=count2+1;
          End If;
        End If;
        period1s<=count2(9);
        --Period:1 s
        If(reset='0') Then countt:=(Others=>'0');
        Elsif(period1s'Event And period1s='1') Then
            If(countt>"10") Then countt:=(Others=>'0');
            Else countt:=countt+1;
            End If;
        End If;
        freshplus<=countt(1);
```

/*限定位数的计数在计到进位时,进位溢出,就会循环计数,但当不希望循环计数时,就会像下面这样写 if…else 语句,结合标志信号方便做控制*/

```
        If(reset='0' Or freshplus='0') Then
            countr:=(Others=>'0');
        Elsif(Period 1ms'Event And period 1ms='1') Then
            If(countr<"11") Then countr:=countr+1;
            Else countr:="11";
            End If;
        End If;

        If(countr="10") Then fresh<='1';
        Else fresh<='0';
        End If;
    End Process Globleclk;

    Process(reset,clk,clkint) Is                                    --时钟进程
    Begin
        If(reset='0') Then clkint<='0';
        Elsif(clk'Event And clk='1')Then clkint<=Not clkint;
        End If;
    End Process;
    ---------------------------------------------
    Process(reset,clk) Is
    Variable temp:Std_Logic;
```

```
    Begin
        If(reset='0') Then temp:='0';
        Elsif(clk'Event And clk='0') Then temp:=Not temp;
        End If;
        lcd_e<=temp;
    End Process;
    ------------------------------
```

/＊LCD 控制器 447810 主要有 11 条指令：

指令 1：清显示，指令码 01H，光标复位到地址 00H 位置。

指令 2：光标复位，指令码 02H，光标返回到地址 00H，1.52 ms。

指令 3：光标和显示模式设置，约 40 μs。

指令 4：显示开关控制，约 40 μs。

指令 5：光标或显示移位，约 40 μs。

指令 6：功能设置命令，约 40 μs。

指令 7：字符发生器 RAM 地址设置，如，40H，约 40 μs。

指令 8：DDRAM 地址设置，如 80H，C0H，约 40 μs。

指令 9：读忙信号和光标地址，约 0 μs。

指令 10：写数据，约 40 μs。

指令 11：读数据，约 40 μs。

注：

(1)显示驱动程序先要对液晶模块功能进行初始化设置，约定显示格式等。

(2)指令要与 RS，R/W 信号配合，RS 和 R/W 是 00 为液晶设置，01 读忙，10 写数据，11 读数据。指令的后几位往往可以对 LCD 做不同的设置，44780 手册有说明。在本程序中分别用常数将前面位与后面位说明，即使不看手册，只要大概了解指令，就可猜出指令设置的含义，使程序可读性和可维护性好。

(3)注意显示字符时光标是自动移动的，无需人工干预。

(4)先要给输入显示位置的地址，如：0C0H，然后再输入要显示的字符代码。

(5)LCD 显示慢，在执行每条指令之前，一是先查忙标志，其为低电平不忙时，给指令，否则此指令失效。单片机编程常用这个办法，写指令前调用 DELAY 读忙和等待子程序；二是按照指令顺序及时序要求结合分频与状态机来写程序，在规定的状态给指令和数据，FPGA 编程常用这个办法。＊/

```
    lcd_rs<='1' When state=Writeram Or state=Readram Else '0';
    lcd_rw<='0' When state=Clear Or state=Returncursor
        Or state=Setmode Or state=Switchmode
        Or state=Shift Or state=Setfunction
        Or state=Setcgram
        Or state=Setddram
        Or state=Writeram Else '1';
    lcd_data<="00000001" When state=Clear       Else
```

```
        "00000010" When state=Returncursor  Else
        "000001"&cur_inc&cur_noshift
            When state=Setmode  Else
        "00001"&open_display & open_cur & blank_cur
            When state=Switchmode  Else
        "0001" & shift_cur&left_shift & "00"
            When state=Shift  Else
        "001" & datawidth8&twoline & font5x10&"00"
            When state=Setfunction  Else
        "01000000" When state=Setcgram  Else
        "10000000" When state=Setddram And counter=0  Else
        "11000000" When state=Setddram And counter/=0  Else
        data_in When state=Writeram
            Else"ZZZZZZZZ";
    char_add<=Conv_Std_Logic_Vector(counter,6)
            When state=Writeram And counter<32
            Else "000000";
    data_in<=lcd_array(Conv_Integer(char_add));
    ---------------------------------------------
    Process(reset,clkint,fresh) Is
    Begin
        If(reset='0') Then
            state<=Idle;
            counter<=0;
            flag<='0';
        Elsif(clkint'Event And clkint='1') Then
            Case state Is
                When Idle=>
                    If(flag='0') Then
                        state<=Setfunction;
                        flag<='1';
                        counter<=0;
                        div_counter<=0;
                    Else
                        If(fresh='1') Then flag<='0';
                        Else
                            If(div_counter<Divss) Then
                                div_counter<=div_counter+1;
                                state<=Idle;
```

```
                        Else
                            div_counter<=0;
                            state<=Shift;
                        --flag<='0';
                        End If;
                    End If;
                End If;
            ---------------------------------
            When Clear=>state<=Setmode;
            When Setmode=>state<=Writeram;
            When Returncursor=>state<=Writeram;
            When Switchmode =>state<=Clear;
            When Shift=>state<=Idle;
            When Setfunction=>state<=Switchmode;
            When Setcgram=>state<=Idle;
            When Setddram=>state<=Writeram;
            When Readflag=>state<=Idle;
            ---------------------------------
            When Writeram=>
                If(counter=15) Then
                    state<=Setddram;
                    counter<=counter+1;
                Elsif(counter/=15 And counter<32) Then
                    state<=Writeram;
                    counter<=counter+1;
                Else
                    state<=Shift;
                    counter<=0;
                End If;
            ---------------------------------
            When Readram=>state<=Idle;
            When Others=>state<=Idle;
        End Case;
    End If;
End Process;
-------------------------------------------------------------
Process(reset,fresh) Is                                  --LCD1602
显示内容
Begin
```

```
    lcd_array(0)<=Conv_Std_Logic_Vector(char_to_integer('W'),8);
    lcd_array(1)<=Conv_Std_Logic_Vector(char_to_integer('e'),8);
    lcd_array(2)<=Conv_Std_Logic_Vector(char_to_integer('l'),8);
    lcd_array(3)<=Conv_Std_Logic_Vector(char_to_integer('c'),8);
    lcd_array(4)<=Conv_Std_Logic_Vector(char_to_integer('o'),8);
    lcd_array(5)<=Conv_Std_Logic_Vector(char_to_integer('m'),8);
    lcd_array(6)<=Conv_Std_Logic_Vector(char_to_integer('e'),8);
    lcd_array(7)<=Conv_Std_Logic_Vector(char_to_integer(' '),8);
    lcd_array(8)<=Conv_Std_Logic_Vector(char_to_integer('t'),8);
    lcd_array(9)<=Conv_Std_Logic_Vector(char_to_integer('o'),8);
    lcd_array(10)<=Conv_Std_Logic_Vector(char_to_integer(' '),8);
    lcd_array(11)<=Conv_Std_Logic_Vector(char_to_integer(' '),8);
    lcd_array(12)<=Conv_Std_Logic_Vector(char_to_integer(' '),8);
    lcd_array(13)<=Conv_Std_Logic_Vector(char_to_integer(' '),8);
    lcd_array(14)<=Conv_Std_Logic_Vector(char_to_integer(' '),8);
    lcd_array(15)<=Conv_Std_Logic_Vector(char_to_integer(' '),8);
    lcd_array(16)<=Conv_Std_Logic_Vector(char_to_integer('E'),8);
    lcd_array(17)<=Conv_Std_Logic_Vector(char_to_integer('D'),8);
    lcd_array(18)<=Conv_Std_Logic_Vector(char_to_integer('A'),8);
    lcd_array(19)<=Conv_Std_Logic_Vector(char_to_integer(' '),8);
    lcd_array(20)<=Conv_Std_Logic_Vector(char_to_integer('a'),8);
    lcd_array(21)<=Conv_Std_Logic_Vector(char_to_integer('n'),8);
    lcd_array(22)<=Conv_Std_Logic_Vector(char_to_integer('d'),8);
    lcd_array(23)<=Conv_Std_Logic_Vector(char_to_integer(' '),8);
    lcd_array(24)<=Conv_Std_Logic_Vector(char_to_integer('F'),8);
    lcd_array(25)<=Conv_Std_Logic_Vector(char_to_integer('P'),8);
    lcd_array(26)<=Conv_Std_Logic_Vector(char_to_integer('G'),8);
    lcd_array(27)<=Conv_Std_Logic_Vector(char_to_integer('A'),8);
    lcd_array(28)<=Conv_Std_Logic_Vector(char_to_integer(' '),8);
    lcd_array(29)<=Conv_Std_Logic_Vector(char_to_integer('L'),8);
    lcd_array(30)<=Conv_Std_Logic_Vector(char_to_integer('a'),8);
    lcd_array(31)<=Conv_Std_Logic_Vector(char_to_integer('b'),8);
    End Process;
End Architecture behavioral;
```

9.3　4×4 矩阵扫描键盘与 LED 显示

设计目的与要求：

(1)了解 4×4 键盘扫描与 LED 显示设计原理。

(2)了解键盘取抖。

(3)设计一个 4×4 键盘矩阵与七段显示扫描程序。

(4)按下某一个键,在数码管上显示该键对应的键值。

设计原理介绍:

4×4 矩阵键盘与 LED 显示电路由如下模块组成:

· 时序产生电路;

· 键盘扫描电路;

· 弹跳抖动消除电路;

· 键盘译码电路;

· LED 显示电路。

键盘与 LED 显示对于 FPGA 而言是慢速器件,需要分频,扫描与抖动消除也需要分频。对于复杂电子系统设计定时与分频电路甚至需要专人来做。

1. 4×4 键盘矩阵

4×4 键盘,即 4 行 4 列键盘,按键有两种形式:直接式(线性)和矩阵式。

(1)直接式:一端 Vcc, 一端接 FPGA 的 I/O 接口。每个键占用一个 I/O 口,通常静态工作。

(2)矩阵式:矩阵有 M 行 N 列,有 M×N 个按键,占用 M+N 个 I/O 接口,通常动态工作。

4×4 直接式键盘,需 16 条 I/O 线;4×4 矩阵键盘,只需 8 个 I/O 接脚。

矩阵式接法又分共阴极(接法 A)如图 9-1 所示;共阳极接法(接法 B),需将图 9-1 中电阻接地改为接 Vcc,两者编码是取反关系,见表 9-1。

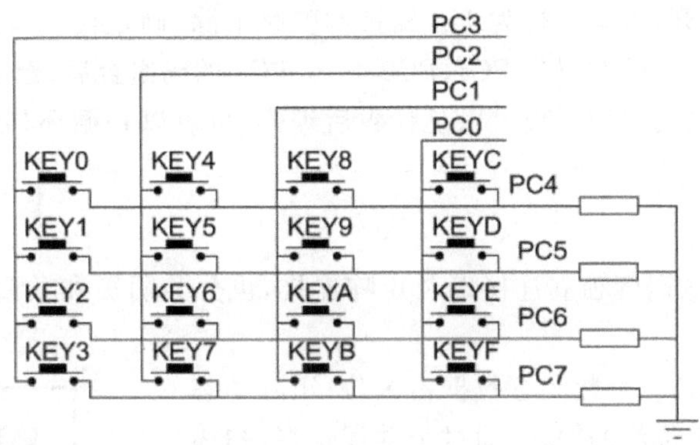

图 9-1 矩阵键盘接法 A

表 9-1 A 和 B 接法行列电平值与按键的对应关系

A 列(输出) PC3～PC0	A 行(输入) PC7～PC4	B 列(输出) PC3～PC0	B 行(输入) PC7～PC4	按键
1000	0001	0111	1110	0
1000	0010	0111	1101	1
1000	0100	0111	1011	2

续表

A 列（输出）PC3～PC0	A 行(输入) PC7～PC4	B 列（输出）PC3～PC0	B 行(输入) PC7～PC4	按键
1000	1000	0111	0111	3
0100	0001	1011	1110	4
0100	0010	1011	1101	5
0100	0100	1011	1011	6
0100	1000	1011	0111	7
0010	0001	1101	1110	8
0010	0010	1101	1101	9
0010	0100	1101	1011	A
0010	1000	1101	0111	B
0001	0001	1110	1110	C
0001	0010	1110	1101	D
0001	0100	1110	1011	E
0001	1000	1110	0111	F

因为无法预计什么时候有键按下，所以对列进行扫描，同时读取键盘行线的电平值。依序送 Hi 电位的扫描信号至列 PC0、PC1、PC2、PC3，信号为 1000、0100、0010、0001... 反复循环，一次一列为 Hi，每送出一列信号后，就从行 PC4、PC5、PC6、PC7 读入，若均为 Lo，代表没有按键按下，继续扫描下一列，若有一行为 Hi，经按键弹跳消除，确定代表有按键按下，停止扫描，再将 PC0、PC1、PC2、PC3 的数码与 PC4、PC5、PC6、PC7 的码组合后，经由软件译码，即可得到被按下的键值。具体的列扫描实现可以是计数后编码，也可以是循环移位，示例 9-3 采用的是前者。

2. 七段 LED 显示

LED 显示有每个数码管独立连接以及矩阵连接，也分共阴极和共阳极接法，工作也有静态与动态之分。

独立静态连接 LED 一个数码管数据占 8 位；矩阵动态连接各 8 位数据共用 8 位数据输入，通过公共端选择/扫描决定点亮的数码管。公共端选择/扫描与键盘列扫描方法一样。

动态显示可以依靠人眼的视觉暂留频率每秒为 24 帧的原理来实现类似静态的各个数码管的全显示。

七段 LED 显示数据(如图 9-2 所示)是(dot)gfedcba，dot 为小数点，所以七段 LED 显示数据是 8 位。

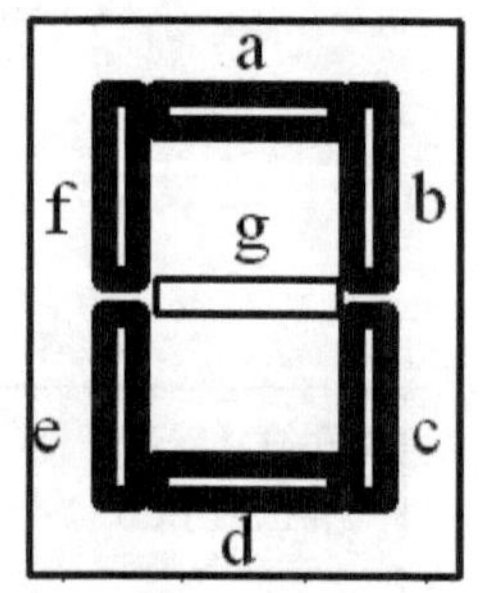

图 9-2　七段 LED 数码管段位编号

3. 键盘取抖

弹跳抖动一般在 ms 秒级，如图 9-3 所示。消除弹跳

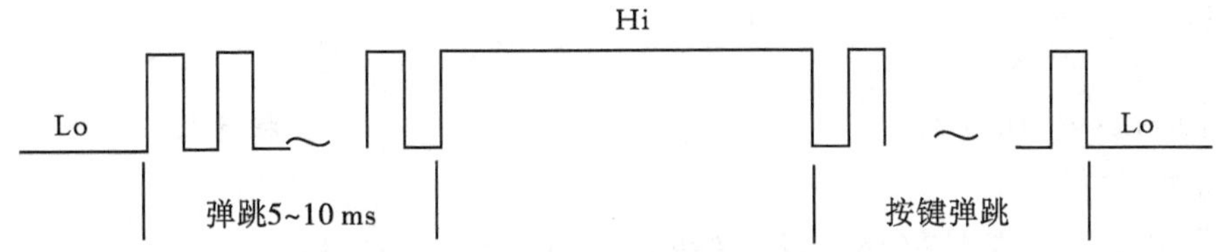

图 9-3 按键弹跳示意

抖动电路所使用脉冲信号的频率必须比其他电路使用的脉冲信号频率更高。

键盘的按键闭合与释放的瞬间会产生毛刺。系统会将其误认为是另一次输入，导致系统的误操作。

按键弹跳会影响到信号的检测，软件消除弹跳影响的方法是当读取有一位元为 Hi 时，延迟一段时间，此时间可设为 10ms，再读取一次，若值相同，代表确实有按键按下，否则，继续扫描下一列。同理，要检测按键是否真的放开时，也要经过一段延迟时间，当确定该位元为 Lo 时，即代表按键已放开。

将按下的键值处理后，需确定该按键已放开，才继续去检测有无下一按键被按下。也就是去进行下一列的扫描，否则，该按键所要执行的动作，可能会被处理很多次。

键盘去抖电路方法很多，最简单常用的是计数延迟的方法，参见示例 6－15。当某一键值保持一段时间不改变，认为是有效值，否则无效，重新计数。

示例 9-3 4×4 矩阵扫描键盘与 LED 显示

```
Library Ieee;
Use Ieee.Std_Logic_1164.All;
Use Ieee.Std_Logic_Unsigned.All;
Use Ieee.Std_Logic_Arith.All;

Entity keyscan1 Is
  Port(clk,clk1:In Std_Logic;                          --时钟,扫描时钟
       clr:In Std_Logic;
       kbcol1:In Std_Logic_Vector(3 Downto 0);         --列扫描信号
       kbrow:Out Std_Logic_Vector(3 Downto 0);         --行扫描信号
       segment:Out Std_Logic_Vector(6 Downto 0));      --7 段显示信号
End Entity keyscan1;

Architecture bev Of keyscan1 Is
Signal count:Std_Logic_Vector(1 Downto 0);
--Signal clk_div16:Std_Logic;
Signal d:Integer Range 16#0# To 16#f#;
Signal kbrow1:Std_Logic_Vector(3 Downto 0);
Signal keyv:Std_Logic_Vector(7 Downto 0);
```

```
Begin

  pa:Process(clk1) Is                                       --扫描计数
  Begin
    If clk'Event And clk='1' Then  count<=count+1;
    End If;
  End Process pa;

  pb:Process(clk1) Is
  Begin
    If(clk1'Event And clk1='1') Then
      Case count Is
        When "00"=>kbrow1<="1110";
        When "01"=>kbrow1<="1101";
        When "10"=>kbrow1<="1011";
        When "11"=>kbrow1<="0111";
        When Others=>Null;
      End Case;
      kbrow<=kbrow1;
    End If;
  End Process pb;

  pc:Process(clk1,kbcol1) Is
  Begin
    If(clk1'Event And clk1='1') Then
      keyv<=kbcol1 & kbrow1;
    End If;
  End Process pc;

  pd:Process(clk,clr) Is                                    --4×4 键值译码
  Begin
    If(clk'Event And clk='1') Then
      Case keyv Is
        When "11101110"=>d<=16#d#; When "11101101"=>d<=16#9#;
        When "11101011"=>d<=16#5#; When "11100111"=>d<=16#1#;
        When "11011110"=>d<=16#e#; When "11011101"=>d<=16#a#;
        When "11011011"=>d<=16#6#; When "11010111"=>d<=16#2#;
        When "10111110"=>d<=16#f#; When "10111101"=>d<=16#b#;
        When "10111011"=>d<=16#7#; When "10110111"=>d<=16#3#;
```

```
      When "01111110"=>d<=16#0#; When "01111101"=>d<=16#c#;
      When "01111011"=>d<=16#8#; When "01110111"=>d<=16#4#;
      When Others=>Null;
    End Case;
  End If;
End Process pd;

pe:Process(d,clk) Is
Begin
  If clr='0' Then segment<="0111111";                        --LED 7 段译码
  Elsif(clk1'Event And clk1='1') Then
    Case d Is
      When 16#0#=>segment(6 Downto 0)<="0111111";            --'0'
      When 16#1#=>segment(6 Downto 0)<="0000110";            --'1'
      When 16#2#=>segment(6 Downto 0)<="1011011";            --'2'
      When 16#3#=>segment(6 Downto 0)<="1001111";            --'3'
      When 16#4#=>segment(6 Downto 0)<="1100110";            --'4'
      When 16#5#=>segment(6 Downto 0)<="1101101";            --'5'
      When 16#6#=>segment(6 Downto 0)<="1111101";            --'6'
      When 16#7#=>segment(6 Downto 0)<="0000111";            --'7'
      When 16#8#=>segment(6 Downto 0)<="1111111";            --'8'
      When 16#9#=>segment(6 Downto 0)<="1101111";            --'9'
      When 16#a#=>segment(6 Downto 0)<="1110111";            --'A'
      When 16#b#=>segment(6 Downto 0)<="1111100";            --'B'
      When 16#c#=>segment(6 Downto 0)<="0111001";            --'C'
      When 16#d#=>segment(6 Downto 0)<="1011110";            --'D'
      When 16#e#=>segment(6 Downto 0)<="1111001";            --'E'
      When 16#f#=>segment(6 Downto 0)<="1110001";            --'F'
      When Others=>segment(6 Downto 0)<="0000000";           --' '
    End Case;
  End If;
End Process pe;
End Architecture bev;
```

9.4 可编程接口 8255 核设计

9.4.1 可编程接口 8255 芯片

Intel 8255 是配合 Intel 微处理器而设计的通用可编程并行 I/O 接口器件，它能把外围设备连到微型计算机总线上，通用性强、使用灵活，可以用程序来设置和改变芯片的工作方式，是一种典型的可编程并行接口芯片。8255 不仅与 Intel 公司的微处理器完全兼容，也与 89C51 单片机兼容，直接复位/置位特性便于实现控制性接口。

8255 共有三种模式：基本的输入/输出（方式 0）、有选通的输入/输出（方式 1）和双向总线输入/输出（方式 2）。

9.4.2 8255 核的内部结构

1. 端口与工作方式

8255 有 3 个与外设相连的 8 位数据端口，即端口 A、端口 B 和端口 C。编程人员可以通过软件将它们分别作为输入端口或输出端口，3 个端口在不同的工作方式下有不同的功能及特点，具体说明如下：

(1)A 口有一个 8 位数据输出缓冲/锁存器，一个 8 位数据输入缓冲/锁存器。

(2)B 口、C 口有一个 8 位数据输出缓冲/锁存器，一个 8 位数据输入缓冲器（无锁存器）。

8255 核由 3 种逻辑电路构成：缓冲/锁存器、组合逻辑电路和三态缓冲器。8255 核的内部结构如图 9 - 4 所示。

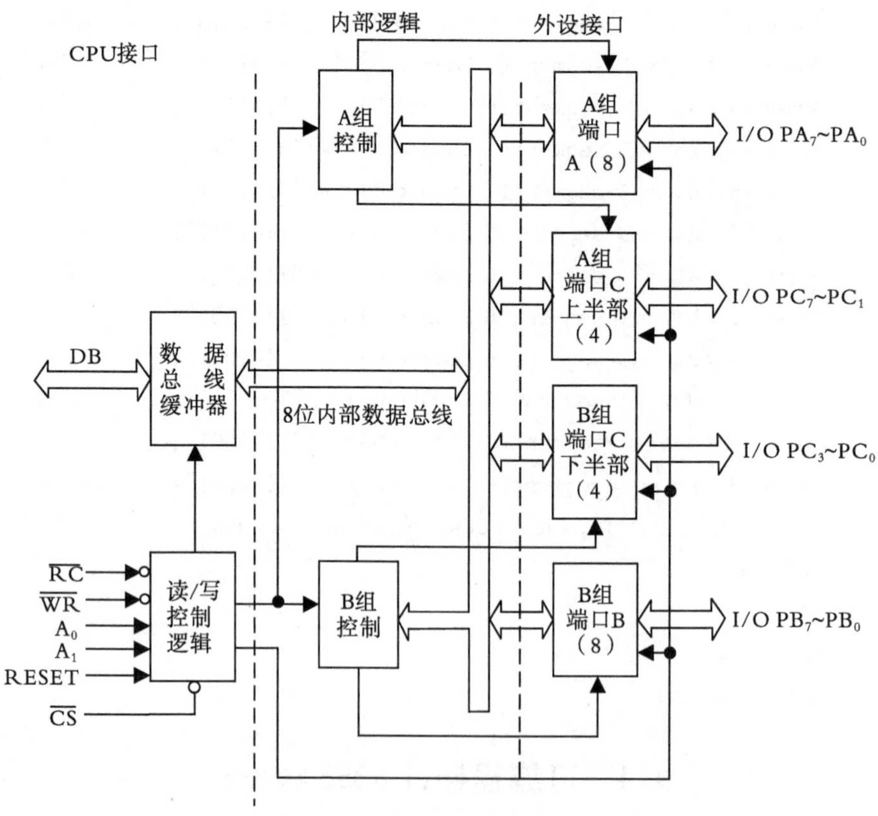

图 9 - 4 8255 核的内部结构

2. A 组和 B 组控制电路

A 组和 B 组这两组控制电路根据 CPU 的命令字控制 8255 A 工作方式。它们的控制寄存器先接收 CPU 送出的命令字，然后根据命令字分别决定两组的工作方式，也可根据 CPU 的命令字对端口 C 的每 1 位实现按位"复位"或"置位"。

A 组控制电路控制端口 A 和端口 C 的上半部($PC_7 \sim PC_4$)。

B 组控制电路控制端口 B 和端口 C 的下半部($PC_3 \sim PC_0$)。

3. 数据缓冲器

8255 具有双向 8 位缓冲器，用于传送 MCS－51 和 8255 间的控制字、状态字和数据字。

4. 读/写控制逻辑

读/写控制逻辑这部分电路可以接收 MCS－51 送来的读写命令和选口地址，用于控制对 8255 的读写。

9.4.3　8255 引脚与信号说明

1. 8255 引脚说明

(1)$D_0 \sim D_7$ 为双向数据总线，用来传递数据和控制字。

(2)RD 为读信号线，与其他信号线一起实现对 8255 接口的读操作。

(3)WR 为写信号线，与其他信号线一起实现对 8255 的写操作。

(4)CS 为片选信号线，低电平(有效)，低电平选中 8255，对 8255 进行操作。

(5)A0～A1 为端口地址选择信号线。8255 有 4 个端口，3 个为输入/输出端口，一个为控制寄存器端口。A_1A_0 选择端口：00 代表选择 A 口，01 代表选择 B 口，10 代表选择 C 口，11 代表选择控制寄存器。

(6)RESET 为复位信号输入，高电平有效。复位后，8255 的 A 口、B 口、C 口均被定义为输入。

(7)$PA_0 \sim PA_7$、$PB_0 \sim PB_7$、$PC_0 \sim PC_7$ 为 3 个输入/输出端口的引脚，其输入输出方向由软件来设定。

2. 8255 信号说明

(1)在构造体中定义了两条内部总线 int_bus_in、int_bus_out，8255 内部的 8 位数据都是通过这两条总线来进行输入或输出的。

(2)pa_latch、pb_latch 及 pc_latch 是 8255 芯片中 A 口、B 口及 C 口锁存器的输出。

(3)信号 ctrreg 是方式控制寄存器的输出。

(4)信号 st 为 C 口的输入/输出控制信号。当 st＝"01"时，C 口高 4 位为输出口，低 4 位为输入口；当 st＝"10"时，C 口高 4 位为输入口，低 4 位为输出口；当 st＝"11"时，C 口高低 4 为均为输入口。

(5)d 为输入或输出数据。

(6)ad 为端口地址选择信号，即 ad＝A_0&A_1。

示例 9－4　8255 程序

```
Library Ieee;
```

```
Use Ieee. Std_Logic_1164. All;
Use Ieee. Std_Logic_Arith. All;                                    --逻辑算数运算包
Use Ieee. Std_Logic_Unsigned. All;

Entity a8255 Is
    Port(reset,rd,wr,cs,a0,a1:In Std_Ulogic;
    --RESET 复位信号,RD 读信号,WR 写信号,CS 片选信号
    --A0、A1 端口地址选择信号
         pain:In Std_Logic_Vector( 7 Downto 0);
         paout:Out Std_Logic_Vector( 7 Downto 0);
         pbin:In Std_Logic_Vector( 7 Downto 0);
         pbout:Out Std_Logic_Vector( 7 Downto 0);
         pcl:Inout Std_Logic_Vector( 3 Downto 0);
         pch:Inout Std_Logic_Vector( 3 Downto 0);
         din:Out Std_Logic_Vector( 7 Downto 0));
    --dout:Out Std_Logic_Vector( 7 Downto 0));
End Entity a8255;

Architecture a8255_arch Of a8255 Is
Signal  internal_bus_out:Std_Logic_Vector( 7 Downto 0):="00000000";
--内部数据总线
Signal  internal_bus_in :Std_Logic_Vector( 7 Downto 0):="00000000";
Signal  st,ad,flag:Std_Logic_Vector( 1 Downto 0);
--st,C 口控制信号
Signal  ctrreg:Std_Logic_Vector( 7 Downto 0);
--方式控制字寄存器的输出
Signal  pa_latch  :Std_Logic_Vector( 7 Downto 0):="00000000";
--A 口锁存器的输出
Signal  pb_latch :Std_Logic_Vector( 7 Downto 0):="00000000";
--B 口锁存器的输出
Signal  pc_latch:Std_Logic_Vector( 7 Downto 0):="00000000";
--C 口锁存器的输出
Signal  d             :Std_Logic_Vector( 7 Downto 0):="00000000";

Begin
    ------------------------------------------------------------
    din <= internal_bus_out;
    paout<=pa_latch;
    p1: Process(rd,cs)   Is                                        --读进程
```

/＊当片选信号有效(cs=0)和读信号有效(rd=0)时,从A口或B口或C口读入外部设备提供的数据。＊/

```
        Begin
            st <= ctrreg(3) & ctrreg(0);
```

/＊C口输入/输出控制,ctrreg(3)C口高四位和ctrreg(0)C口低四位,1为输入,0为输出。＊/

```
            If( cs ='0' and rd ='0')Then
    --如果cs和rd信号有效,则8255读外部设备传送的数据
                If (a0 ='0' And a1 ='0' And ctrreg(4) ='1') Then
    --选中pa口且为输入口
                    internal_bus_in <=pain;
                Elsif(a0 ='1' And a1 ='0' And ctrreg(1) ='1')Then
    --选中pb口且为输入口
                    internal_bus_in <= pbin;
                Elsif(a0 ='0' And a1 ='1' And st ="01")Then
                    internal_bus_in(3 Downto 0)<=pcl(3 Downto 0);
    --选中C口
                Elsif(a0 ='0' And a1 ='1' And St ="10")Then
                    internal_bus_in(7 Downto 4)<=pch(3 Downto 0);
                Elsif(a0='0' and a1 ='1' And st ="11"
                    And ctrreg(7)='1')Then
    --pc口全为输入口,送pc口的8位数据给内部数据总线
                    internal_bus_in(3 Downto 0)<=pcl(3 Downto 0);
                    internal_bus_in(7 Downto 4)<=pch( 3 Downto 0);
                End If;
                --Else
                --internal_bus_in <= "ZZZZZZZZ";
    --若pa、pb、pc口均不是输入口,则送高阻态给总线
            End If;
            d <= internal_bus_in;              --把内部总线的数据传给D端口
        End Process p1;

        p2: Process(cs,wr,reset) Is            --写进程
        Variable ctrregf:      Std_Ulogic;     --选择标志寄存器,1位
        Variable bctrreg_v:    Std_Logic_Vector(3 Downto 0);
        --C口位控制字寄存器
        Begin

            If(reset='1')Then
```

/＊复位信号有效时,8255 芯片复位,所有锁存器都处于输入方式,所以方式寄存器 ctrreg=9bh、C 口位控制字寄存器 bcttreg_v 和选择标志寄存器均为 0。＊/

```
            pa_latch <= "00000000";
            pb_latch <= "00000000";
            pc_latch <= "00000000";
            internal_bus_out <= "00000000";
            ctrreg <= "10000010";
            --A 组 B 组工作方式均为 0,最好设置单输入,其他为输出
            bctrreg_v := "0000";
            ctrregf := '0';
        End If;
        If(cs='0' And wr='0')Then
        --If()Then, 片选和写信号有效时
        ad <= a1 & a0;                  --ad 为端口的地址
        ctrregf := d(7);                --把 D 端口的第八位的值赋予变量 ctrregf
        internal_bus_out <= d;          --把数据线 D 上的数据送数据总线
        --End If;
        If(ctrregf ='1' And ad ="11" And cs = '0')Then
            ctrreg <= internal_bus_out;
        --把数据总线上的数据写入方式控制寄存器 ctrreg
        Elsif(ctrreg(7) = '1' And ad ="00" And cs = '0')Then
            pa_latch <=internal_bus_out;
        --把数据总线上的数据写入 pa 输出锁存器
        Elsif(ctrreg(7) = '1' And ad ="01" And cs = '0')Then
            pb_latch <= internal_bus_out;
        --把数据总线上的数据写入 pb 输出锁存器
        Elsif(ctrreg(7) = '1' And ad = "10" And cs = '0')Then
            pc_latch <= internal_bus_out;
        --把数据总线上的数据写入 pc 输出锁存器
        --Elsif(ctrreg(7) ='1' And ad = "11" And cs = '0')Then
        End If;
        If(ctrregf= '0' And ad="11" And cs='0') Then
            bctrreg_v := internal_bus_out(3 Downto 0);
            --把数据总线上第四位的数据送 C 口位控制字寄存器
            Case bctrreg_v Is
            --用 Case 语句来实现 C 口的位控功能
                When "0000"=>   pc_latch(0)<='0';
                When "0010"=>   pc_latch(1)<='0';
                When "0100"=>   pc_latch(2)<='0';
```

```
                    When "0110"=>   pc_latch(3)<='0';
                    When "1000"=>   pc_latch(4)<='0';
                    When "1010"=>   pc_latch(5)<='0';
                    When "1100"=>   pc_latch(6)<='0';
                    When "1110"=>   pc_latch(7)<='0';

                    When "0001"=>   pc_latch(0)<='1';
                    When "0011"=>   pc_latch(1)<='1';
                    When "0101"=>   pc_latch(2)<='1';
                    When "0111"=>   pc_latch(3)<='1';
                    When "1001"=>   pc_latch(4)<='1';
                    When "1011"=>   pc_latch(5)<='1';
                    When "1101"=>   pc_latch(6)<='1';
                    When "1111"=>   pc_latch(7)<='1';
                    When Others=>flag<="11";
                End Case;
            End If;
        End If;
    End Process p2;
---------------------------------------------------------
    --p3: --Process(pa_latch) Is              --pa、pb、pc口三态输出进程
    --Begin
        --If ctrreg(4)='0' Then
            --paout<=pa_latch;                --pa口为输出口,把锁存器的数据给pa
        --Else
            --paout<="ZZZZZZZZ";              --若pa不是输出口,则为高阻态输出
        --End If;
    --End Process p3;
------------------------------------------------------------
    p4: Process(pb_latch) Is                 --pb口三态输出进程
    Begin
        If(ctrreg(1)='0')Then
            pbout<=pb_latch;
        Else
            pbout<="ZZZZZZZZ";
        End If;
    End Process p4;
--------------------------------------------------------------
    p5: Process(pc_latch) Is                 --pc口低4位三态输出进程
```

```
    Begin
        If(ctrreg(0)='0')Then
            pcl<=pc_Latch(3 Downto 0);
        Else
            pcl<="ZZZZ";
        End If;
    End Process p5;
--------------------------------------------------------------
    p6: Process(pc_latch) Is                    --pc 口高 4 位三态输出进程
    Begin
        If(ctrreg(3)='0')Then
            pch<=pc_latch(7 Downto 4);
        Else
            pch<="ZZZZ";
        End If;
    End Process p6;
End Architecture a8255_arch;
```

程序说明:Quartus Ⅱ对 VHDL 总线设计描述中含三态功能的双端口的综合不是很理想,稳妥的办法是调用 Altera 设计原语中的双端口来实现设计。

习 题

9-1 通过 RS232 传送一组数据,在 LCD 显示出来,调通 RS232 收发自环测试。

9-2 设计开机界面,显示自己的学号,和可调整时钟等功能。

9-3 设计 1 个加减乘除计算器,键盘输入数据后,结果在 LCD 上显示出来,可以通过 RS232 超级终端看到结果,可以通过实验箱传输到开发板上看到结果。

9-4 设计 SPI 总线,并应用。(推荐参考:本书参考文献 [8])

9-5 仿真和实现 8255 功能,并应用。

推荐参考:范延滨,等. 微型计算机系统原理、接口与 EDA 设计技术[M]. 北京:北京邮电大学出版社,2006 年 9 月.

扩展学习与总结

参考 Cyclone 器件手册,对比 Open 的 8255 IP 核,扩展示例 9-4 的功能并给出具体的应用。

10 嵌入 51 单片机的设计型实验

本章的主要内容与方法：

(1)介绍对 MC8051 CPU 核的模块、部分模块的仿真以及应用。

(2)搭建 8051 系统，通过 Keil C51 程序实现单片机的 UART 通信、LED 流水灯以及定时器中断功能，并硬件调试。

(3)本章用到的 8051 开发软件是 Keil μVision2，所用的 8051 软核是 Oregano System 给出的开源免费 IP 核，结构与 51 单片机结构相似，使用 VHDL 语言实现。

本章希望达到的目的：

(1)学习具有哈佛结构(Harvard)的 8051 单片机结构和原理。

(2)初步掌握 SoC 设计 CPU 的步骤和验证方法。

(3)研究对 MC8051 核的功能模块扩展。

10.1 概　述

随着科技水平的发展，集成智能电子产品的需求不断地增加，在如今众多的电子产品当中，CPU 作为核心，已经成为了基本应用。CPU 设计的合理与否和性能的优劣甚至决定着电子产品市场推广的成败，决定着整个行业的良好发展。鉴于此，本章节为读者介绍在 FPGA 中嵌入 8051 核，在此基础上，将 PC 上编译完成的 C 语言程序，经过编译器的转换，下载到 CPU，实现让 CPU 真正地“跑”起来。

10.2 CPU 简述及应用

CPU(Central Processing Unit，中央处理器单元)是计算机的核心部件之一，它将用户的信息处理成数据和程序，并存储到计算机内存中，通过在计算机内存寻址(汇编中地址跳转)的方式，取出地址中的数据，同时取出的数据交由 CPU 进行译码，最后，CPU 将执行操作，并作用于数据。这样用户就可以看到程序执行的结果。以上是简单地阐述了程序在 CPU 中被执行的过程。可以看出，CPU 要完成以上基本功能，应该具有几个部件：算数逻辑运算单元(ALU)、定时器部件、控制部件、累加器、程序计数器(PC)和寄存器组(Register Files)。

在开始应用 CPU 8051 Core 之前，需要知道其功能、运行效率、性能以及功耗等，进而应用可集成的具有哈弗结构的 8051 CPU 核，通过典型实验对其时序仿真、功能仿真和实际硬件测试。

10.3 8051 核结构

本节将简单地介绍一下 8051 核的基本硬件结构和设计应用。设计采用的 8051 IP 核是在 Oregano Systems 公司提供的免费 8051 IP 核的基础上定制的，代码用 VHDL 硬件描述语言编写。

10.3.1　8051 核功能特点

8051 核功能与特点如下：

(1)采用完全同步设计。

(2)指令集和标准 8051 微控制器完全兼容。

(3)指令执行时间为 1～4 个时钟周期，执行性能优于标准 8051 微控制器 8 倍左右。

(4)用户可选择定时器/计数器、串行接口单元的数量。

(5)新增了特殊功能寄存器用于选择不同的定时器/计数器、串行接口单元。

(6)可选择是否使用乘法器(乘法指令 MUL)。

(7)可选择是否使用除法器(除法指令 DIV)。

(8)可选择是否使用十进制调整功能(十进制调整指令 DA)。

(9)I/O 端口不复用。

(10)内部带 256 B RAM。

(11)最多可扩展至 64 KB 的 ROM 和 64 KB 的 RAM。

10.3.2　8051 软核设计应用

MC8051 软核最基本的应用设计如图 10-1 所示。ROM、RAM 和 RAMX 模块并不包括于 8051 核心内，处于设计的顶层，以便于不同的应用设计及仿真。

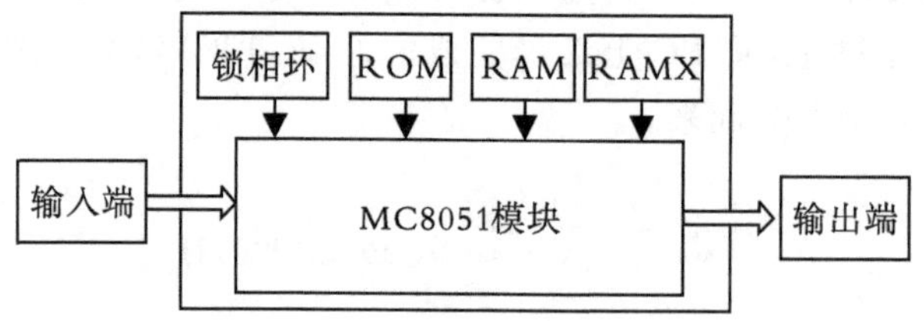

图 10-1　MC8051 软核应用设计框图

10.3.3　8051 设计层级

8051 核的层次结构及对应的 VHDL 文件如图 10-2 所示。

8051 核由定时器/计数器、ALU、串行接口和控制单元各模块组成。定时器/计数器和串行接口单元对应于图中 10-2 的 mc8051_tmrctr 和 mc8051_siu 模块，数量是可选择的。

8051 核是同步的设计，单一的时钟信号可以控制每一个存储模块的时钟输入，这里不使用时钟门控。时钟信号不能输入到任何组合逻辑单元。多个中断输入同步到全局时钟信号，通过使用标准二级同步模式，因为他们可能是由外部电路另一个时钟驱动。并行端口输入信号不能用这种方式同步。如果有需要同步这些信号，那也很容易添加进来；其次，由于优化结构导致的问题，8051 核与 RAM、ROM 数据之间的数据通信，是不能够被寄存的(Not Registered)。

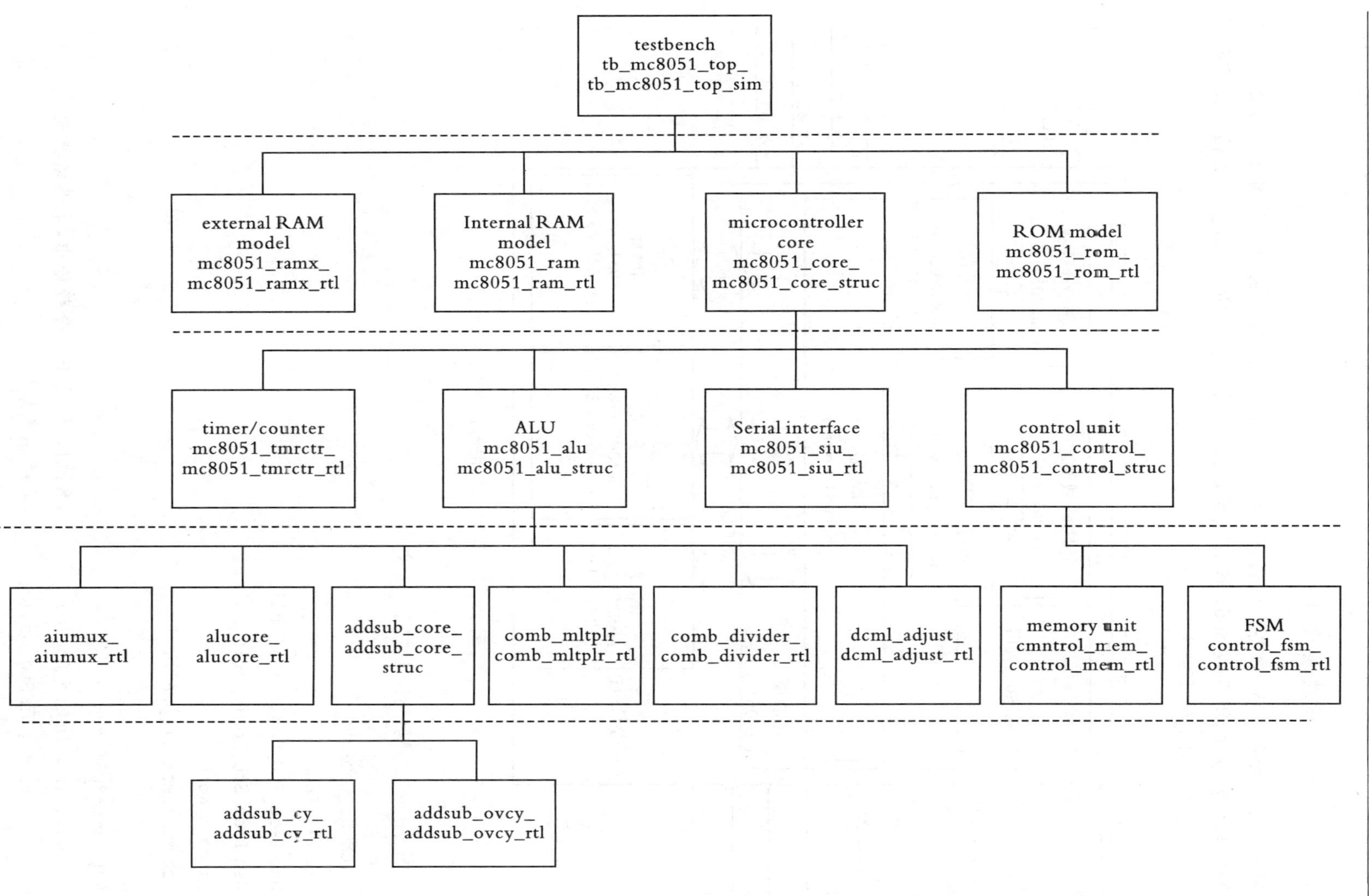

图10-2 8051 Core 顶层设计层级

10.3.4 8051 核顶层设计

根据以上介绍的 8051 核各个关键部件可知，对 8051 核的功能和逻辑组成可以将其按照逻辑顺序和组合顺序，组合成 8051 核的顶层设计，如图 10－3 所示的是 8051 核的顶层设计图，相应的源代码为示例 10－1。

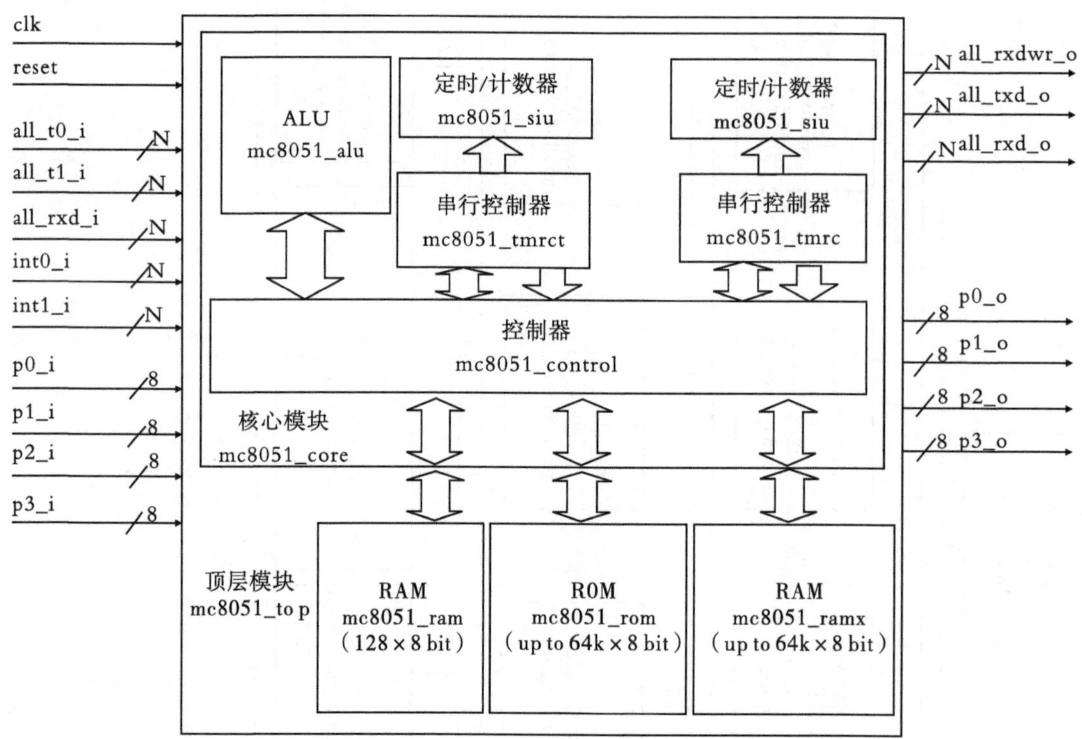

图 10－3 8051 核应用的顶层设计框图

示例 10－1 MC8051 核顶层文件

```
--mc8051_top.vhd
Library Ieee;
Use Ieee.Std_Logic_1164.All;
Use Ieee.Std_Logic_Arith.All;
Library Work;
Use Work.mc8051_p.All;

Entity mc8051_top Is
  Port (clk : In Std_Logic;          --系统时钟，由 PLL 分频给出，上升沿有效
        reset : In Std_Logic;        --系统异步复位
        int0_i: In Std_Logic_Vector(C_Impl_N_Ext - 1 Downto 0);      --外部中断 0
```

```
  int1_i: In Std_Logic_Vector(C_Impl_N_Ext - 1 Downto 0);          --外部中断 1
  --定时器或计数器输入 0
  all_t0_i: In Std_Logic_Vector(C_Impl_N_Tmr - 1 Downto 0);
  --定时器或计数器输入 1
  all_t1_i: In Std_Logic_Vector(C_Impl_N_Tmr - 1 Downto 0);
  --串口接受数据输入
  all_rxd_i : In Std_Logic_Vector(C_Impl_N_Siu - 1 Downto 0);
  --4 个 8 位并行输入
  p0_i: In Std_Logic_Vector(7 Downto 0);                 --p0 口输入
  p1_i: In Std_Logic_Vector(7 Downto 0);
  p2_i: In Std_Logic_Vector(7 Downto 0);
  p3_i: In Std_Logic_Vector(7 Downto 0);
  --4 个 8 位并行输出
  p0_o: Out Std_Logic_Vector(7 Downto 0);                --p0 口输出
  p1_o: Out Std_Logic_Vector(7 Downto 0);
  p2_o: Out Std_Logic_Vector(7 Downto 0);
  p3_o: Out Std_Logic_Vector(7 Downto 0);
  --模式 0 串口输出
  all_rxd_o: Out Std_Logic_Vector(C_IMPL_N_SIU - 1 Downto 0);
  --串口输出
  all_txd_o: Out Std_Logic_Vector(C_IMPL_N_SIU - 1 Downto 0);
  --rxd 方向设置(高电平输出)
  all_rxdwr_o : Out Std_Logic_Vector(C_IMPL_N_SIU - 1 Downto 0));
End Entity mc8051_top;

Architecture struc Of mc8051_top Is
Signal s_rom_adr: Std_Logic_Vector(15 Downto 0);         --ROM 地址
Signal s_rom_data: Std_Logic_Vector(7 Downto 0);         --存储 ROM 数据
Signal s_ram_data_out: Std_Logic_Vector(7 Downto 0);     --写 RAM 数据
Signal s_ram_data_in: Std_Logic_Vector(7 Downto 0);      --读 RAM 数据
Signal s_ram_adr:Std_Logic_Vector(6 Downto 0);           --内部 RAM 地址
Signal s_ram_wr:Std_Logic;                               --读写内部 RAM
Signal s_ram_en:Std_Logic;                               --使能内部 RAM
Signal s_ramx_data_out: Std_Logic_Vector(7 Downto 0);    --写外部 RAM 数据
Signal s_ramx_data_in:Std_Logic_Vector(7 Downto 0);      --读外部 RAM 数据
Signal s_ramx_adr:Std_Logic_Vector(15 Downto 0);         --外部 RAM 地址
Signal s_ramx_wr:Std_Logic;                              --读写外部 RAM
Signal s_rom_adr_sml:Std_Logic_Vector(12 Downto 0);
                                                         --宏模块调用 ROM,地址设置
```

```
Signal s_ramx_adr_sml:  Std_Logic_Vector(11 Downto 0);
                                                --宏模块调用外部 RAM,地址设置
Begin
  s_rom_adr_sml <= Std_Logic_Vector(s_rom_adr(12 Downto 0));
  s_ramx_adr_sml <= Std_Logic_Vector(s_ramx_adr(11 Downto 0));

  i_mc8051_core : mc8051_core                   --mc8051_core 元件例化
  Port Map(clk => clk, reset   => reset,
          rom_data_i => s_rom_data,  ram_data_i => s_ram_data_out,
          int0_i     => int0_i,      int1_i     => int1_i,
          all_t0_i   => all_t0_i,    all_t1_i   => all_t1_i,
          all_rxd_i  => all_rxd_i,
          p0_i => p0_i,    p1_i => p1_i,
          p2_i => p2_i,    p3_i => p3_i,
          p0_o => p0_o,    p1_o => p1_o,
          p2_o => p2_o,    p3_o => p3_o,
          ll_rxd_o     => all_rxd_o,    all_txd_o    => all_txd_o,
          all_rxdwr_o  => all_rxdwr_o, rom_adr_o    => s_rom_adr,
          ram_data_o   => s_ram_data_in, ram_adr_o   => s_ram_adr,
          ram_wr_o     => s_ram_wr,     ram_en_o     => s_ram_en,
          datax_i      => s_ramx_data_in, datax_o    => s_ramx_data_out,
          adrx_o       => s_ramx_adr,   wrx_o        => s_ramx_wr);
                                       --片上 RAM:最多支持 128×8 位同步片上 RAM

  i_mc8051_ram : mc8051_ram                     --mc8051 RAM 元件例化
  Port Map (clock   => clk,
           address => s_ram_adr, data => s_ram_data_in,
           wren    => s_ram_wr,   clken => s_ram_en,
           q       => s_ram_data_out);
                                                --片上 RAM 必须保留
  --片上 ROM:最多支持 64 k×8 位同步片上 ROM
  i_mc8051_rom : mc8051_rom                     --mc8051 ROM 元件例化
  Port Map (clock    => clk,
           address  => s_rom_adr_sml,
           q        => s_rom_data);
            --尽管 ROM 存储空间不大,但必须保留
  --外部 RAM:最多支持 64 K×8 位同步外部 RAM

  i_mc8051_ramx : mc8051_ramx                   --mc8051 RAMX 元件例化
```

```
    Port Map (clock    => clk,
              address  => s_ramx_adr_sml,
              data     => s_ramx_data_out,
              wren     => s_ramx_wr,
              q        => s_ramx_data_in);
  --这个RAM不一定保留,容量自定义
End Architecture struc;
```

以上就是应用 8051 核的顶层设计文件,其中可以看到 4 个部件:8051 内核、RAM、RAMX 和 ROM 基础部件,这些部件当中的内存器模块均是通过 Altera 宏模块调用的方式实现的。

10.4 8051 核设计研究

本节着重介绍 8051 核的内部部件和组成,根据标准 51 单片机的内部原理,可以看出 8051 核包含以下 4 个基本部件:

(1)ALU 算数运算逻辑单元;

(2)Timer 定时器控制器;

(3)Serial 串口控制器;

(4)控制单元(Memory 存储控制器)。

各个部件均由外部 PLL 时钟分频提供时钟信号,各个基础部件的协调工作由 8051 核内部的控制器来协调。各基础部件的具体结构和逻辑时序图下面将逐一进行介绍。

10.4.1 ALU 算数运算逻辑单元

ALU 运算单元包括的子模块有:加法(减法)模块 addsub_core、逻辑运算模块 alu_core、十进制码调整模块 dcml_adjust、除法模块 comb_divider、乘法模块 comb_mltplr。这几个模块各自都有一套独立的功能。

alucore 模块完成"与"运算、"或"运算、"异或"运算、左移运算、右移运算、循环左移、循环右移、带进位左移运算、带进位右移运算、取反运算、自加运算和比较运算。当模块每次采取运算时的命令常数分别为 cmd_i=land=3,cmd_i=lxor=6,cmd_i=rl=7,cmd_i=rlc=8,cmd_i=rr=9,cmd_i=rrc=10,cmd_i=inv=12,cmd_i=comp=11,cmd_i-lor=5 时,ALU 执行相应运算。

示例 10-2 ALU 算数运算逻辑单元

```
Library Ieee;
Use Ieee.Std_Logic_1164.All;
Use Ieee.Std_Logic_Arith.All;
Library Work;
Use Work.mc8051_P.All;
----------------------------ENTITY DECLARATION------------------------
```

```
Entity alucore Is
  Generic (DWIDTH : Integer := 8);                    --Data width of the ALU
  Port (op_a_i,op_b_i   : In Std_Logic_Vector(DWIDTH - 1 Downto 0);
        alu_cmd_i : In Std_Logic_Vector(3 Downto 0);
        cy_i       : In Std_Logic_Vector((DWIDTH - 1)/4 Downto 0);
        cy_o       : Out Std_Logic_Vector((DWIDTH - 1)/4 Downto 0);
        result_o   : Out Std_Logic_Vector(DWIDTH - 1 Downto 0));
End Entity alucore;

--op_a_i....... operand A,操作数 A
--op_b_i....... operand B,操作数 B
--alu_cmd_i.... command for the ALU core
--cy_i......... carry flags (MSB is CY, rest is AC),进位标志
--cy_o......... resulting carry out (MSB is CY, rest is AC)
--result_o..... result,结果

Architecture rtl Of alucore Is

Constant LAND : Std_Logic_Vector(3 Downto 0) := "0011";
Constant LOR  : Std_Logic_Vector(3 Downto 0) := "0101";
Constant LXOR : Std_Logic_Vector(3 Downto 0) := "0110";
Constant RL   : Std_Logic_Vector(3 Downto 0) := "0111";
Constant RLC  : Std_Logic_Vector(3 Downto 0) := "1000";
Constant RR   : Std_Logic_Vector(3 Downto 0) := "1001";
Constant RRC  : Std_Logic_Vector(3 Downto 0) := "1010";
Constant COMP : Std_Logic_Vector(3 Downto 0) := "1011";
Constant INV  : Std_Logic_Vector(3 Downto 0) := "1100";

Begin                                                  --architecture structural

  p_alu: Process (alu_cmd_i, op_a_i, op_b_i, cy_i) Is
  Begin
    Case alu_cmd_i Is
      When LAND =>                                     --op_a_i And op_b_i
        result_o <= op_a_i and op_b_i;
        cy_o <= cy_i;
      When LOR =>                                      --op_a_i Or op_b_i
        result_o <= op_a_i Or op_b_i;
        cy_o <= cy_i;
```

```
When LXOR =>                                      --op_a_i Xor op_b_i
  result_o <= op_a_i Xor op_b_i;
  cy_o <= cy_i;
When RL =>                                        --rotate left op_a_i
If DWIDTH > 1 Then
  result_o(DWIDTH - 1 Downto 1) <= op_a_i(DWIDTH - 2 Downto 0);
  result_o(0) <= op_a_i(DWIDTH - 1);
Else
  result_o <= op_a_i;
End If;
cy_o <= cy_i;
When RLC =>                                       --rotate left op_a_i with CY
If DWIDTH > 1 Then
  result_o(DWIDTH - 1 Downto 1) <= op_a_i(DWIDTH - 2 Downto 0);
  result_o(0) <= cy_i((DWIDTH - 1)/4);
Else
  result_o(0) <= cy_i((DWIDTH - 1)/4);
End If;
cy_o <= cy_i;
cy_o((DWIDTH - 1)/4) <= op_a_i(DWIDTH - 1);
When RR =>                                        --rotate right op_a_i
If DWIDTH > 1 Then
  result_o(DWIDTH - 2 Downto 0) <= op_a_i(DWIDTH - 1 Downto 1);
  result_o(DWIDTH - 1) <= op_a_i(0);
Else
  result_o <= op_a_i;
End If;
cy_o <= cy_i;
When RRC =>                                       --rotate right op_a_i with CY
If DWIDTH > 1 Then
  result_o(DWIDTH - 2 Downto 0) <= op_a_i(DWIDTH - 1 Downto 1);
  result_o(DWIDTH - 1) <= cy_i((DWIDTH - 1)/4);
Else
  result_o(0) <= cy_i((DWIDTH - 1)/4);
End If;
cy_o <= cy_i;
cy_o((DWIDTH - 1)/4) <= op_a_i(0);
When COMP =>                                      --Compare op_a_i with op_b_i
If op_a_i = op_b_i Then
```

```
        result_o <= (Others => '0');
      Else
        result_o <= (Others => '1');
      End If;
      cy_o <= cy_i;
      If Op_a_i < op_b_I Then
        cy_o((DWIDTH - 1)/4) <= '1';
      Else
        cy_o((DWIDTH - 1)/4) <= '0';
      End If;
      When INV =>                                --invert op_a_i
        result_o <= not(op_a_i);
        cy_o <= cy_i;
      When Others =>                             --turn unit off
        result_o <= (Others => '0');
        cy_o <= (Others => '0');
    End Case;
  End Process p_alu;
End Architecture rtl;
```

10.4.2 Timer 定时器控制器

串口通信的波特率等都要由 Timer 定时器模块控制的，而且在 RTOS 操作系统的移植中，遇到的中断问题同样要涉及定时器模块。定时器模块是应用程序编写定时和速率控制得以顺利进行的一个关键模块，图 10－4 是定时器模块 Symbol 图。

int0_i，int1_i 是外部中断源，但在此不是用做外部中断源而是负责启动定时器模块，它和定时器控制寄存器中的 TR 位共同控制定时器的运行。t0_i，t1_i 是外部时钟信号，这两个引脚在定时器的运行过程中充当着时钟信号的角色，但这种情况不是经常都发生的，只有在定时器控制寄存器中的 C/T 位等于 1 的时候才起作用，其他情况下它与定时器的运行没有关系。

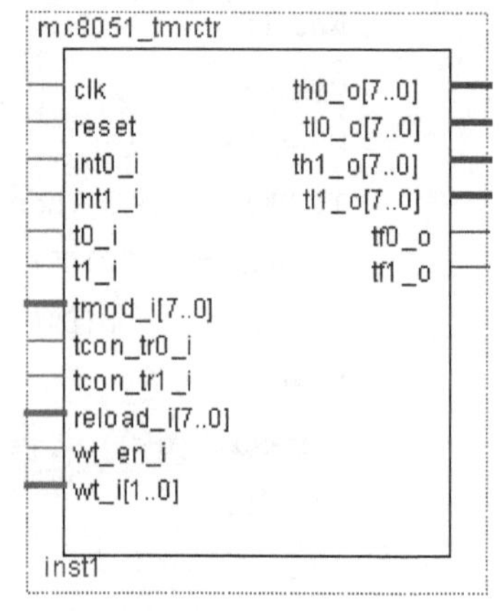

图 10－4　8051 核 Timer 模块 Symbol 图

tcon_tr0_i，tcon_tr1_i 是定时器的一个控制寄存器的一位，和 int0_i，int1_i 信号共同控制定时器的启动。wt_en_i 信号则是定时器数据位加载控制端的使能信号。tmod_i[7...0] 是定时器运行模式控制信号，直接控制定时器的运行模式。reload_i[7...0] 是数据位加载信号。wt_i[1...0] 是定时器数据位加载控制端，控制定时器到底该向哪个寄存器加载数据。tf_0，tf_1 是定时器溢出端，但还有控制串口通信的波特率的功能。th0_o[7...0]，tl0_o[7...0]，th1_o[7...0]，tl1_o[7...0] 是定时器计数寄存器。

图 10 - 5 是 Timer0 定时器模式 0 下 tmod_i=00 的仿真波形图。

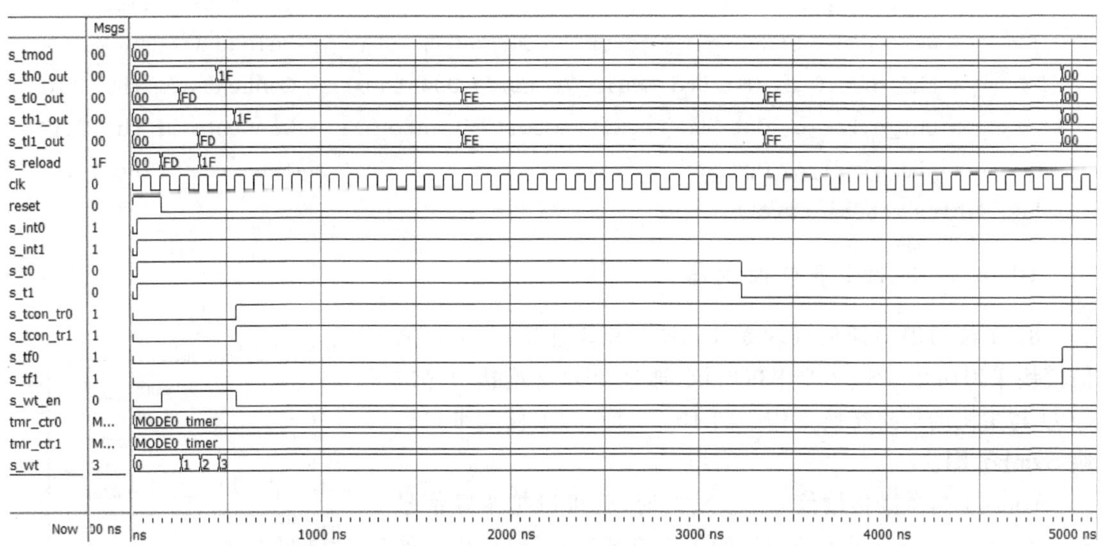

图 10 - 5　8051 核 Timer0 模式 0 下时序仿真

采用 MC8051 IP Core 所提供的 testbench 程序包进行仿真。

在方式 0 下，IP CORE 是 13 位的定时器，在 tmod_i=00 模式下定时器是定时方式，tmod_i 中的两位控制模式的位 M0，M1 均为 0，且定时器不受外部中断源和外部时钟信号控制。

从图 10 - 5 看出，定时器工作在定时方式，用来自内部的单片机时钟，这时启动 TR 就可以启动定时器。当 wt_i 分别等于 0，1，2，3 时，定时器把数据分别加载到 T0 和 T1 中，定时器计数从 FDh 到 FEh 经过了 1500 ns。定时器的定时计算为在单片机的频率为 10 MHz，定时器的定时周期约 1 μs，同步输出前后可能有半个周期的误差，所以证明单片机的定时器方式 0 能正确工作。

示例 10 - 3　Timer 定时器控制器

```
Library Ieee;
Use Ieee.Std_Logic_1164.All;
Use Ieee.Std_Logic_Arith.All;
-----------------------------ENTITY DECLARATION------------------------
Entity mc8051_Tmrctr Is
  Port (clk,reset : In  Std_Logic;                         --System Clock,Reset
        int0_i,int1_i : In  Std_Logic;                     --Interrupt 0,1
        t0_i,t1_i : In  Std_Logic;
                                  --External Clock For Timer/Counter0,Timer/Counter1
        tmod_i : In Std_Logic_Vector(7 Downto 0);          --From SFR Register
```

```
        tcon_tr0_i,tcon_tr1_i : In  Std_Logic;                   --Timer Run 0,Run 1
        reload_i: In  Std_Logic_Vector(7 Downto 0);              --To Load Counter
        wt_en_i : In  Std_Logic;                                 --Indicates Reload
        wt_i : In  Std_Logic_Vector(1 Downto 0);                 --Reload Which Reg.
        th0_o, tl0_o: Out Std_Logic_Vector(7 Downto 0);          --Contents Of Th0, Tl0
        th1_o,tl1_o: Out Std_Logic_Vector(7 Downto 0);           --Contents Of Th1,Tl1
        tf0_o,tf1_o: Out Std_Logic);                             --Interrupt Flag 0,1
End Entity mc8051_tmrctr;
```

10.4.3　Serial 串口控制器

8051 核的串口通信模块是 UART 串口通信，本串口通信模块采用简化 RS-232 串口协议，通过 DB9 标准接口完成 CPU 与主机的信息交换工作。图 10-6 是 8051 核的串口模块 Symbol 图。

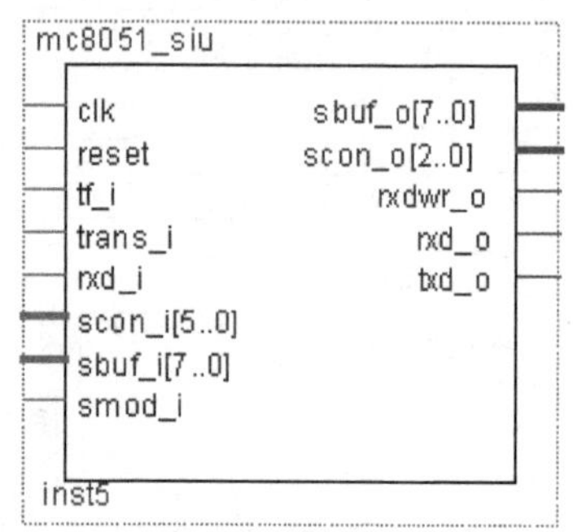

图 10-6　8051 核串口模块 Symbol 图

trans_i 是激励传输信号，它为 1 时，就可以把激励信号输入到传输端。smod_i 是串口的工作方式选择端口，串口的工作方式有 4 种，本仿真中采取的是方式 1。tf_i 是定时器 1 的溢出标志位。rxd_i 是串口输入，这里这个输入和 sbuf_i[7..0] 不同的是，前者是一个字节位，每次只能存放一个字节，是真正的串口输入，后者相当于串口输入的缓冲器。输出端口 scon_o[2..0]和输入端口 scon_i[5..0]放在一起介绍，原因是这两个端口在硬件中是一个寄存器，而在软核中则是一个输入端，一个输出端。

scon_i 和 s_mode 这两个信号之间的连接关系如下：

s_mode(1)<=scon_i(4),s_mode(0)<=scon_i(3),s_ren<=scon_i(1),

s_sm2<=scon_i(2),s_tb8<=scon_i(0),s_ri<=scon_i(5),

sbuf_o<=s_recv_buf,scon_o(0)<=s_recv_done,

scon_o(1)<=s_trans_done,scon_o(2)<=s_rb8。

sbuf_o[7..0]是输出缓冲器；txd_o 是串口输出，rxd_o 是串口在模式 0 情况下的输出；rxdwr_o 是 rxd_o 的控制位，控制其数据的流向。仿真分析串口模块时，给这个输入端加入激励，观测信号是否能加载到串口软核模块的输入端。

仿真的 Testbench 程序将示例 10-4 作为 2 个互为收发的模块，对串口通信的 4 个模式给出了输入激励，设定了接收到正确数据给出报告，如图 10-7 所示的是在模式 0 的情况下，仿真程序运行到 14550ns 正确接收数据 AAh 时的波形及报告。串行口的模式 1-3 工作方式的仿真，可以继续运行程序查看，在这里就不列举了。

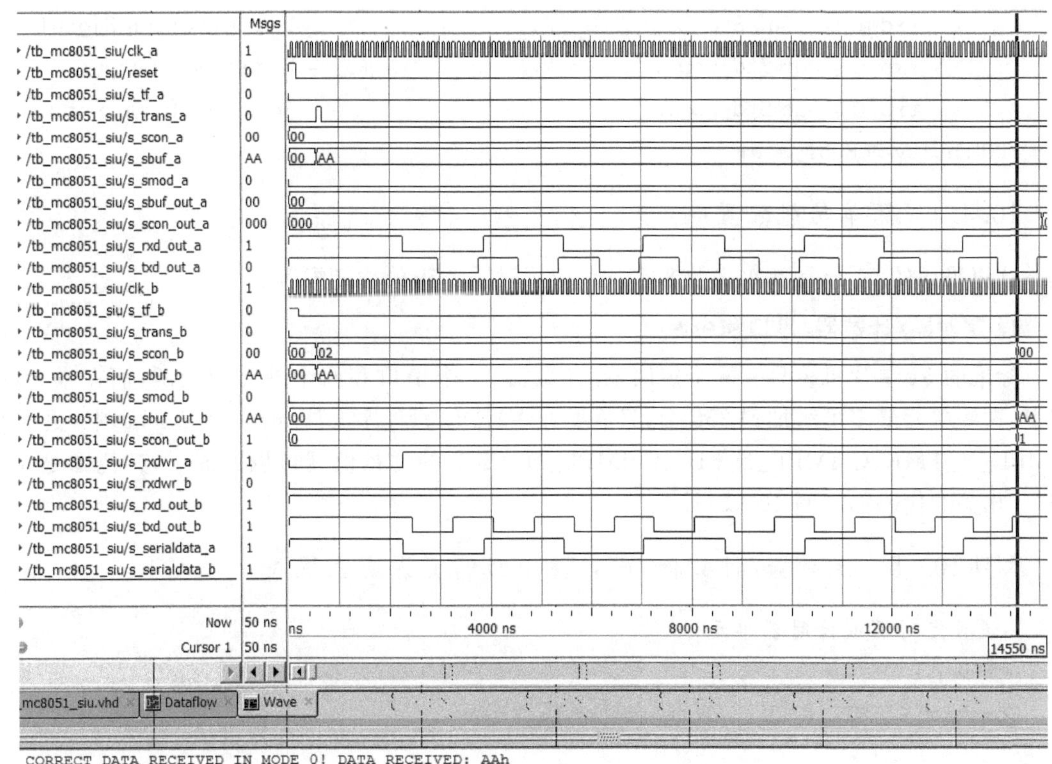

图 10-7 8051 核串口模块时序仿真

示例 10-4 Serial 串口控制器

```
Library Ieee;
Use Ieee.Std_Logic_1164.All;
Use Ieee.Std_Logic_Arith.All;

Entity mc8051_siu Is                                        --Entity Declaration
  Port (clk ,reset : In Std_Logic;                          --System Clock,Reset
        tf_i: In Std_Logic;                                 --Timer1 Overflow Flag
        trans_i: In Std_Logic;                              --1 Activates Transm.
        rxd_i: In Std_Logic;                                --Serial Data Input
        scon_i: In Std_Logic_Vector(5 Downto 0);            --From Sfr Register
                                                            --Bits 7 To 3

        sbuf_i: In Std_Logic_Vector(7 Downto 0);            --Data For Transm.
        smod_i: In Std_Logic;                               --Low(0)/High Baudrate
        sbuf_o: Out Std_Logic_Vector(7 Downto 0);           --Received Data
        scon_o: Out Std_Logic_Vector(2 Downto 0);           --To Sfr Register
                                                            --Bits 0 To 2
```

```
        rxdwr_0 : Out Std_Logic;                         --Rxd Direction Signal
        rxd_0 : Out Std_Logic;                           --Mode0 Data Output
        txd_0 : Out Std_Logic);                          --Serial Data Output
End Entity mc8051_siu;
```

10.4.4 简单功能配置

这里采用的 51 核是参数化的 8051 核,应用时需要简单功能配置。

1. 定时器/计数器、串口和中断

标准的 51 单片机核只有两个定时器/计数器、一个串口和两个外部中断源,而在本 8051 核中,这些单元最多可增加到 256 个,只需要在 VHDL 源程序文件 mc8051_p.vhd 中,更改 C_IMPL_N_TMR、C_IMPL_N_SIU、C_IMPL_N_EXT 的常量值就可以完成了,其范围是 1～256。相关的代码如示例 10-5 所示。

示例 10-5　定时器/计数器、串口及中断的资源配置程序

```
--设置定时器或计时器单元
--默认:1
Constant C_IMPL_N_TMR : Integer := 1;

--设置通信串口数量
--默认: C_IMPL_N_TMR
Constant C_IMPL_N_SIU : Integer := C_IMPL_N_TMR;

--设置外部中断输入数量
--默认: C_IMPL_N_TMR
Constant C_IMPL_N_EXT : Integer := C_IMPL_N_TMR;
```

C_IMPL_N_TMR、C_IMPL_N_SIU、C_IMPL_N_EXT 这三个常量是不能独立修改数值的,也就是说只能同时增减。

C_IMPL_N_TMR 加一,就意味 8051 核中同时添加了两个定时器或者计数器、一个串口单元和两个外部中断源。

为了控制这些新增的控制单元,CPU 在特殊寄存器内存空间增加了两个寄存器,分别是 TSEL(定时器/计数器选择寄存器,地址为 0X8E)和 SSEL(串口选择寄存器,地址为 0X9A),如果没有对这两个寄存器赋值,其默认值为 1。

如果一个中断发生在这个设备没有被 TSEL 寄存器选中时,相应的中断标志将保持到相应的中断服务程序执行。在这期间,其他的中断仅仅导致单一调用对中断服务程序。

2. 可选择的指令

在某些场合,有些指令用不到,可以通过禁用这些指令来节省片上(FPGA)资源。这些指令有 8 位乘法器(MUL)、8 位除法器(DIV)和 8 位十进制调整器(DA)。禁用时只需要在

VHDL 源程序文件 mc8051_p.vhd 中将 C_IMPL_MUL(乘法指令 MUL)、C_IMPL_DIV(除法指令 DIV)或 C_IMPL_DA(十进制调整指令 DA)的常量值设置为 0。相应的 VHDL 程序代码段如示例 10-6 所示。

示例 10-6 可选指令配置程序

```
--选择乘法指令
Constant C_IMPL_MUL : Integer := 1;            --默认: 1

--选择除法指令
Constant C_IMPL_DIV : Integer := 1;            --默认: 1

--选择十进制调整指令
Constant C_IMPL_DA : Integer := 1;             --默认: 1
```

这三条可选择指令如果没被设置执行，FPGA 可节省将近 10%的资源。

10.4.5 并行 IO 端口

8051 核的 IO 端口包括 4 个 8 位输入输出口、串行接口、计数器输入端和扩展存储器接口，为了便于 IC 设计，如果 IO 端口要做为双向口应用，电路连接如图 10-8 所示。图中的两个 D 触发器起同步输入信号的作用。输入端口做同步处理时，不用加 2 个触发器。因为 IO 端口输出高电平，所以上拉电阻是必要的。

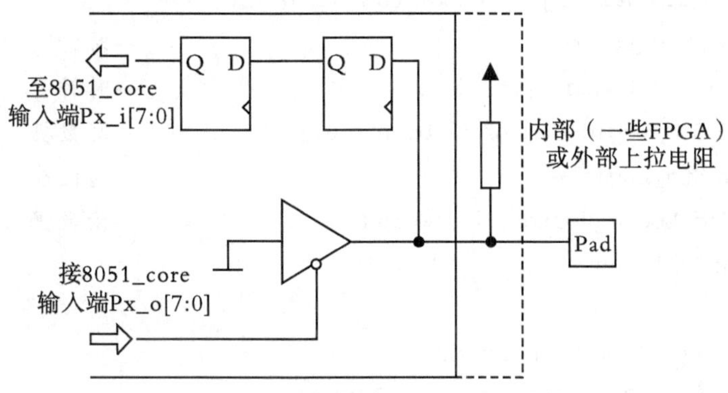

图 10-8 并行 IO 端口基本结构

Quartus Ⅱ 中上拉电阻的设置方法是 Assignments→Settings→Fitting Settings→More Settings→Name→Weak Pull-Up Resistor，Settings→On，Enable Bus Hold Circuitry→Off (On 为总线锁定)。上拉电阻的设置也可以在 Assignments→Assignment Editor 中设定。

10.4.6 杂项说明

(1) 8051 核的 Timer 和 Baudrate 的计算方式和普通的 51 单片机是一样的，即计数时钟是由系统时钟经 12 分频得到。

(2) 外部中断信号是经两级寄存器做同步处理后输入的。

(3) 8051 核的输入 IO 端口没有做同步处理,必要时可自行添加,如图 10-3 所示。

(4) 写执行程序的时候,IO 端口并没有做成双向口(如图 10-3 所示),而是输入和输出分开的。那么要特别注意,P0=~P0、P2^4=P2^4 这样的 IO 取反操作是无效的,原因是读回来的值不是 IO 寄存器的值,而仅仅是输入引脚的状态。

(5) 8051 核经 FPGA 设计软件综合编译后,需观看并分析时序分析报告,确定 Fmax(最高时钟频率),因为系统时钟不能超过时序报告的 Fmax。

10.4.7 内部数据存储器 RAM

8051 核内部的 RAM 和标准的 51 单片机一个重要的区别就是其中内部特殊寄存器和内部 RAM 两个独立分开。软核的内部 RAM 数据区是 128 个字节。这 128 字节和特殊功能寄存器共用地址总线。

示例 10-7 内部数据存储器 RAM

```
Library Ieee;
Use Ieee.Std_Logic_1164.All;
Library Altera_Mf;
Use Altera_Mf.All;

Entity mc8051_ram Is
    Port                                                  --内部数据存储器 RAM 端口
    ( address: In Std_Logic_Vector (6 Downto 0);
    clken: In Std_Logic   := '1';                         --时钟使能
    clock: In Std_Logic   := '1';                         --时钟
    data: In Std_Logic_Vector (7 Downto 0);               --写数据
    wren: In Std_Logic ;                                  --写使能
    q: Out Std_Logic_Vector (7 Downto 0));                --读数据
End Entity mc8051_ram;

Architecture syn Of mc8051_ram Is
Signal sub_wire0: Std_Logic_Vector (7 Downto 0);

    Component altsyncram Is                               --Altera Ram 元件声明
        Generic (address_aclr_a: String;
                 indata_aclr_a: String;
                 intended_device_family: String;
                 lpm_hint: String;
                 lpm_type: String;
                 numwords_a: Natural;
                 operation_mode: String;
```

```
                outdata_aclr_a: String;
                outdata_reg_a: String;
                power_up_uninitialized: String;
                widthad_a: Natural;
                width_a: Natural;
                width_byteena_a: Natural;
                wrcontrol_aclr_a: String);
        Port (address_a: In Std_Logic_Vector (6 Downto 0);
              clock0: In Std_Logic ;
              data_a: In Std_Logic_Vector (7 Downto 0);
              wren_a,clocken0: In Std_Logic ;
              q_a: Out Std_Logic_Vector (7 Downto 0));
    End Component altsyncram;

Begin
    q    <= sub_wire0(7 Downto 0);
    altsyncram_component : altsyncram                          --元件例化
    Generic Map (address_aclr_a => "none",
                indata_aclr_a => "none",
                intended_device_family => "cyclone",
                lpm_hint => "enable_runtime_mod=no",
                lpm_type => "altsyncram",
                numwords_a => 128,                             --128 字节
                operation_mode => "single_port",               --单端口
                outdata_aclr_a => "none",                      --无异步清零
                outdata_reg_a => "unregistered",               --无输出寄存器
                power_up_uninitialized => "false",
                widthad_a => 7,                                --地址宽度 7 位
                width_a => 8,                                  --数据宽度 8 位
                width_byteena a => 1,
                wrcontrol_aclr_a => "none")
    Port Map (address_a => address,
              clock0 => clock,
              data_a => data, wren_a => wren,
              clocken0 => clken,
              q_a => sub_wire0);
End Architecture syn;
```

10.4.8　内部数据存储器 RAMX

8051 核的内部 RAMX 寻址空间是 4K，这里端口和前面的 RAM 端口情况类似。内外数据存储器很多端口名一样，但在访问的时候由于使能信号不一样，所以软核可以区分内外数据存储器。

示例 10-8　内部数据存储器 RAMX

```
Library Ieee;
Use Ieee.Std_Logic_1164.All;
Library Altera_Mf;
Use Altera_Mf.All;

Entity mc8051_ramx Is
    Port
    (address: In Std_Logic_Vector (11 Downto 0);          --地址总线
     clock: In Std_Logic   := '1';                        --时钟
     data: In Std_Logic_Vector (7 Downto 0);              --数据输入
     wren: In Std_Logic ;                                 --写使能
     q: Out Std_Logic_Vector (7 Downto 0));               --输出端口
End Entity mc8051_ramx;

Architecture syn Of mc8051_ramx Is
Signal sub_wire0: Std_Logic_Vector (7 Downto 0);

    Component altsyncram Is                               --Altera Ram 元件声明
        Generic (address_aclr_a: String;
                 indata_aclr_a: String;
                 intended_device_family: String;
                 lpm_hint: String;
                 lpm_type: String;
                 numwords_a: Natural;
                 operation_mode: String;
                 outdata_aclr_a: String;
                 outdata_reg_a: String;
                 power_up_uninitialized: String;
                 widthad_a: Natural;
                 width_a: Natural;
                 width_byteena_a: NATURAL;
                 wrcontrol_aclr_a: String);
```

```
    Port ( address_a: In Std_Logic_Vector (11 Downto 0);
           clock0: In Std_Logic ;
           data_a: In Std_Logic_Vector (7 Downto 0);
           wren_a: In Std_Logic ;
           q_a: Out Std_Logic_Vector (7 Downto 0));
End Component altsyncram;

Begin
q <= sub_wire0(7 Downto 0);
altsyncram_component : altsyncram
Generic Map (
            address_aclr_a => "NONE",
            indata_aclr_a => "NONE",
            intended_device_family => "Cyclone",          --器件 Cyclone
            lpm_hint => "ENABLE_RUNTIME_MOD=NO",
            lpm_type => "altsyncram",
            numwords_a => 4096,                            --4K
            operation_mode => "SINGLE_PORT",               --单端口
            outdata_aclr_a => "NONE",                      --无异步清零
            outdata_reg_a => "UNREGISTERED",               --无输出寄存器
            power_up_uninitialized => "FALSE",             --无上电初始化
            widthad_a => 12,                               --地址宽度 12 位
            width_a => 8,                                  --数据宽度 8 位
            width_byteena_a => 1,
            wrcontrol_aclr_a => "NONE")
Port Map (
         address_a => address,
         clock0 => clock, data_a => data,
         wren_a => wren, q_a => sub_wire0);
End Architecture syn;
```

10.4.9　内部程序存储器 ROM

存储器的端口例化都映像为和存储器相关的名字，例如 ROM 的输出端口 q 映像为 rom_data_o。ROM 的寻址范围是 8K。

示例 10-9　内部程序存储器 ROM

```
Library Ieee;
Use Ieee.Std_Logic_1164.All;
Library Altera_Mf;
```

```
Use Altera_Mf.All;

Entity mc8051_rom Is
    Port(address: In Std_Logic_Vector (12 Downto 0);              --地址总线
         clock: In Std_Logic  := '1';                             --时钟
         q: Out Std_Logic_Vector (7 Downto 0));                   --数据输出
End Entity mc8051_rom;

Architecture syn Of mc8051_rom Is
Signal sub_wire0: Std_Logic_Vector (7 Downto 0);
    Component altsyncram Is
        Generic (address_aclr_a: String;
                 init_file: String;
                 intended_device_family: String;
                 lpm_hint: String;
                 lpm_type: String;
                 numwords_a: Natural;
                 operation_mode: String;
                 outdata_aclr_a: String;
                 outdata_reg_a: String;
                 widthad_a: Natural;
                 width_a: Natural;
                 width_byteena_a: Natural);
        Port ( address_a: In Std_Logic_Vector (12 Downto 0);
               clock0 : In Std_Logic ;
               q_a: Out Std_Logic_Vector (7 Downto 0));
    End Component altsyncram;
Begin
    q <= sub_wire0(7 Downto 0);
    altsyncram_component : altsyncram
    Generic Map (address_aclr_a => "NONE",
                 init_file => "mc8051test.hex",                   --ROM 初始文件
                 intended_device_family => "Cyclone",             --器件 Cyclone
                 lpm_hint => "Enable_Runtime_Mod=Yes,Instance_Name=rom",
                 lpm_type => "altsyncram",
                 numwords_a => 8192,                              --8 K
                 operation_mode => "ROM",
                 outdata_aclr_a => "None",                        --无异步清零
                 outdata_reg_a => "Unregistered",                 --无输出寄存器
```

```
            widthad_a => 13,                          --地址宽度13位
            width_a => 8,                             --数据宽度8位
            width_byteena_a => 1)
    Port Map (address_a => address,
            clock0 => clock, q_a => sub_wire0);
End Architecture syn;
```

10.5　Quartus Ⅱ建立8051核工程

本节介绍，利用Altera公司的Quartus Ⅱ v13.0软件开发8051核工程的步骤，调试硬件采用芯片为Cyclone系列EP1C12Q240C8的开发板。本节安排内容如下：

(1)建立8051核工程；

(2)8051核内部存储器模块(ROM、RAM)宏模块设计；

(3)8051核RTL级封装的生成；

(4)针对8051核设置外部时钟和初始外围电路；

(5)利用Quartus Ⅱ In－System Sources and Probes调测8051核。

10.5.1　建立8051核工程

开发项目应先建立工程，Quartus Ⅱ工程中存储着创建FPGA工程的各种管理文件等，便于开发人员的管理。

Quartus Ⅱ v13.0版本的软件不同于早期版本软件，需要在安装Quartus Ⅱ软件后，再根据开发需要安装相应的元件库，来支持所开发的器件。在确认以上信息设置正确后，开始新工程的建立。

第一步，打开Quartus Ⅱ v13.0软件界面，点击"File"→"New Project Wizard…"，即新建工程向导。

第二步，点击"Next"，并在指定的对话框中，填写新建工程路径、名称、顶层实体信息。注意在工程路径下不允许出现中文，否则编译不成功；这里顶层实体名设置为mc8051_top。

第三步，点击"Next"，这一步主要是将已经编写完成的VHDL或者Verilog文件加入到工程中，这里暂时不加入任何文件。

第四步，点击"Next"，配置器件(Device)型号。

第五步，工程信息汇总，显示工程配置的信息。

10.5.2　Mega Wizard创建8051核内部存储器

CPU一般需要有内部RAM和ROM，对8051核来说，也需要内部RAM和ROM，内部RAM固定为128字节(Bytes)，ROM最大可设置为64 K字节，还可选RAMX，RAMX容量最多亦支持64 KB。读者可以根据所使用的FPGA芯片资源，自行设置存储器容量大小，这里内部RAM设为128 B，ROM设为8 KB，RAMX设为4 KB。

8051核的RAM、RAMX、ROM的配置方法，与前面章节基本相同，只是名称和存储容量不一样。

1. 创建 RAM 模块

(1)选择 Memory Complier→ RAM:1 - PORT,指定器件系列为 Cyclone，以及 VHDL 和输出文件存放路径及名字为 mc8051_ram。

(2)配置 RAM 的数据宽度 8bit 和数据深度 128,其他默认即可。

(3)取消 RAM 的寄存器输出,其他选项默认。本 8051 核结构使用存储器不采用寄存器输出,否则,时钟会采样到不正确的数据。

(4)在 Page5 配置页,是关于 RAM 上电后初始化数据的加载,选择无数据加载项。因为 8051 核上电后,会从 ROM 中取数据存到 RAM,然后程序在 RAM 中执行。

(5)最后一页,在 . bsf 前面打勾,其他可不选，完成 RAM 配置。

2. 创建 ROM 模块

(1)选择 Memory Complier→ROM:1 - PORT 模块,命名为 mc8051_rom。

(2)设置数据宽度 8 bit,数据为 8192 个字;取消数据输出寄存器,取消时钟使能信号端;ROM 需要设置 Mem_Init 初始内容,将 51C 程序的下载文件 . hex 保存其中。

(3)选择允许在线下载升级或捕获 ROM 数据,设置 ID 号,这里的 ID 号为 ROM,其他的选项默认不变。

10.5.3 8051 核 RTL 级建立

实现 8051 核 RTL 级电路,通过 Quartus Ⅱ 软件综合来实现,首先要将已经编写好的 . vhd 文件加入到 8051 核工程中,还要对一些 8051 核的源文件进行更新修改,使之符合设计,也就是在 8051 核的顶层设计中,加入上一节配置的内部存储器模块,修改代码如下:

示例 10 - 10 mc8051_p. vhd 存储器模块添加

```
Component mc8051_ram Is
    Port (address: In Std_Logic_Vector (6 Downto 0);            --128
          clken,clock: In Std_Logic ;
          data: In Std_Logic_Vector (7 Downto 0);               --8 位
          wren: In Std_Logic ;
          q: Out Std_Logic_Vector (7 Downto 0));
End Component mc8051_ram;

Component  mc8051_ramx Is
    Port (address: In Std_Logic_Vector (11 Downto 0);           --4096
          clock: In Std_Logic ;
          data: In Std_Logic_Vector (7 Downto 0);               --8 位
          wren: In Std_Logic ;
          q: Out Std_Logic_Vector (7 Downto 0));
End Component mc8051_ramx;
```

```
Component  mc8051_rom Is
    Port (address: In Std_Logic_Vector (12 Downto 0);          --8192
          clock: In Std_Logic ;
          q: Out Std_Logic_Vector (7 Downto 0));               --8 位
End Component mc8051_rom;
```

打开 8051 核工程下的 mc8051_top_struc. vhd 文件,将原文件中 ROM、RAM 模块调用部分的代码修改成如下的代码。这样就完成了源文件更新修改。

示例 10-11 mc8051_top_struc. vhd 存储器模块添加

```
...
Signal s_rom_adr_sml: Std_Logic_Vector(12 Downto 0);          --新建 ROM
Signal s_ramx_adr_sml: Std_Logic_Vector(11 Downto 0);         --新建 RAMX
...
s_rom_adr_sml <= Std_Logic_Vector(s_rom_adr(12 Downto 0));
s_ramx_adr_sml <= Std_Logic_Vector(s_ramx_adr(11 Downto 0));
...

i_mc8051_ram : mc8051_ram  --注意和宏模块命名一样
    Port Map (  clock   => clk,
                address => s_ram_adr, data => s_ram_data_in,
                wren => s_ram_wr, clken => s_ram_en,
                q => s_ram_data_out);

i_mc8051_rom : mc8051_rom
    Port Map ( clock => clk,
               address => s_rom_adr_sml, q => s_rom_data);

i_mc8051_ramx : mc8051_ramx
    Port Map ( clock => clk,
               address => s_ramx_adr_sml, data => s_ramx_data_out,
               wren => s_ramx_wr,  q => s_ramx_data_in);
```

以上两个文件修改后,可以对 8051 核工程进行全编译,编译通过后,查看时序分析报告,确定 8051 核的最高工作频率 Fmax,然后新建 Block Diagram/Schematic File 文件,将 8051 核模块化,并在其外围配置相应的辅助元件。也可以在图形界面用绘图来实现 51 单片机与三个存储器的连接。

10.5.4 8051 核外围电路搭建

本节内容和 8051 单片机硬件和程序有一定的相关性,建议读者可以先看看关于 C51 程序开发、8051 单片机应用以及程序测试和测试流程的资料。

要对 8051 核工程增加一些外围设计，首先是调用 LPM 模块锁相环(PLL)。

单片机稳定地运行和工作，需要一个外部时钟信号，我们的 FPGA 开发板有 50 MHz 有源晶振，分频，直至给出 18 MHz 时钟信号。PLL 可输出精度高、信号稳定时钟可调频率，如倍频、分频、相位调整等。调用 LPM PLL 模块，配置模块参数的方法如下：

(1)利用 Mega Wizard Plug-In Manager 创建一个新的宏功能模块；

(2)选择 IO 选项下的 ALTPLL，选择器件类型，命名 PLL 宏模块为 mcu_pll；

(3)设置 inclk0，填写 FPGA 开发板上的有源晶振值 50 MHz；

(4)设置 PLL 分频输出选项，设置 c0 输出频率为 36 MHz，其实需要为 8051 核提供 18 MHz，然而 EP1C12Q240C8 FPGA 芯片上的 PLL 分频无法达到所需频率，先设 M=16，N=25 得到 36 MHz，再编写 2 分频模块分频达到要求，对于 EP2C35F672C8 芯片，设 M=9，N=25 即可得到所需频率；

(5)最后一页，这时除默认选项，选择 .bsf 选项，至此 PLL 配置完毕。

将 mc8051_top_. vhd 文件生成符号文件，方法：单击 Files→Creat a Symbol Files for Current File 选项，生成 mc8051_top. bsf 文件，用于顶层原理图设计。

打开顶层原理图文件，双击空白页面，在 Libraries 下打开 Project 文件夹，选择 mc8051_top 模块，拖进原理图界面，再将 PLL 时钟模块加入原理图中，配置 8051 核的外围电路，然后添加输入/输出引脚，在 Name 栏目中，输入 input 添加输入管脚，添加 output 输出管脚和非门 not。

10.5.5 In-System Sources and Probes 调测 8051 核

Quartus Ⅱ主要的三种在线硬件调试的工具是嵌入式逻辑分析仪 SignalTap Ⅱ、存储器内容系统编辑器 In-System Memory Content Editor 和系统信号与信号源编辑器 In-System Sources and Probes 。前两个工具为逻辑系统设计。测试和调试带来巨大的便利，但是它们仍有不足之处，比如 SignalTap Ⅱ，数据只能单向调试，而且数据空间占用量较大，信号观测有限；In-System Memory Content Editor，只能在片上 ROM 存储器内进行数据的双向读写，受限于数据存储区；相比较前两个调试工具，In-System Sources and Probes 逻辑系统调试模块可以作为补充，对需要观测的信号进行双向数据操作。

调用 In-System Sources and Probes 模块调试 8051 核的详细步骤如下：

(1)通过 Mega Wizard Plug-In Manager，创建一个新的宏功能模块；

(2)在 JTAG 通信扩展 JTAG-accessible Extensions 中，选择 In-System Sources and Probes，选择器件类型 Cyclone 系列和 VHDL 方式。输入此模块文件名；

(3)设定该模块参数，按如图 10-9 所示，设置 JTAG 通信模块的 ID 号为 test，测试端口 probe 为 9bit，源端口 sources 为 1 bit，最后单击 Finish 项，并保存被勾选的文件。

将设置好的 JTAG 模块加入到 8051 核原理图中，并连接好相应的测试信号，如图 10-10 所示。

图中的连接为 Cyclone Ⅱ器件。JTAG 通信模块的检测信号端 probe[8..0]信号总线，连接 8051 核的 p0[7..0]，检测 p0 总线信号的输出情况；probe[8]连接 8051 核 UART 的发送端 txd 信号，检测 UART 通信情况；JTAG 通信模块的控制信号端 source[0]信号线，连接 8051 核的 reset 复位信号，通过控制 8051 核的复位信号，达到控制 8051 核的上电与复位。如果需

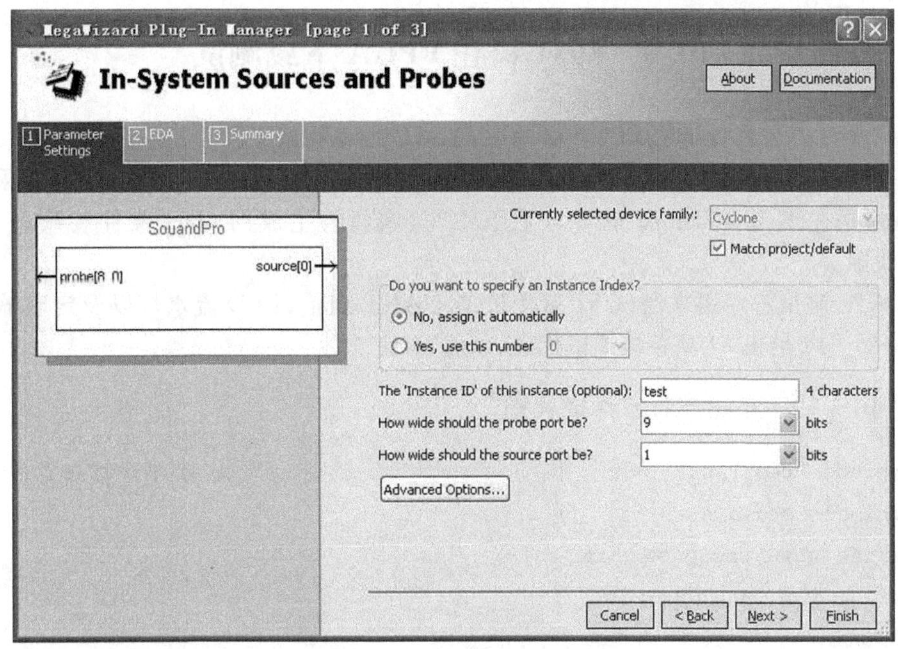

图 10-9　JTAG 通信模块参数设置

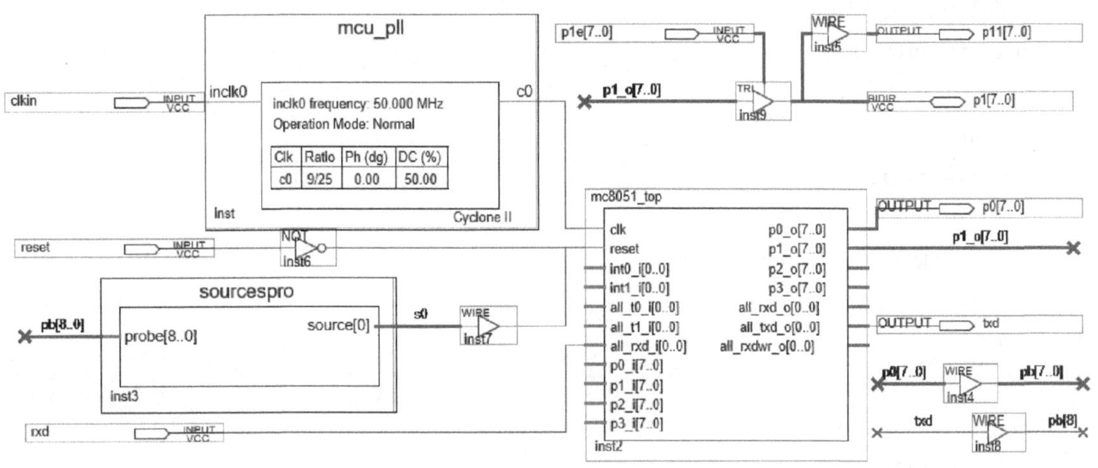

图 10-10　设计过程中添加 In-System Sources and Probes

要双向口，要增加双向口逻辑，设置电阻上拉。p1e 是双向口三态门控制信号，当执行 p1 口读入指令时，p1e 输出为高电平，外部数据通过双向口 p1 进入到 p11；当执行向 p1 口写入指令时，若 p1 口的输出口 p1_o 中的位为低电平时，控制信号 p1e 的对应位也为低电平，信号可从 p1 输出；当输出信号 p1_o 口中的位为高电平时，控制信号 p1e 的对应位也为高电平。

值得注意的是，与其他在线工具一样 In-System Sources and Probes 测试工具需要和开发板相连接，不能依靠 PC 模拟完成。

10.6 8051 核在 FPGA 下载测试

将 sof 文件下载到指定的 FPGA 芯片上，就可以准备测试了。

将一段 51C 单片机的程序运行在已装载 8051 核的 FPGA 上，并展现其运行效果。通常用到的程序是在 Keil uVision2 软件环境下编写，编译将生成的 .hex 文件装载到 8051 核 ROM 模块中。

该段 51C 测试程序用来测试 51 单片机的 UART 通信、LED 流水灯以及定时器中断功能。示例 10－12 给出 51 单片机测试代码，生成 .hex 文件。

示例 10－12 mc8051.c 测试代码

```
#include<reg51.h>                              //定义 51 单片机头文件
#include<stdio.h>
#define uchar unsigned char
#define uint  unsigned int
uchar num;
uchar led=0xff;
uchar counter;
//delay 函数声明
void delay(uint n)
{
  uint k;
  while(n--);
  {
    for(k=0;k<10000;k++)
    {;}
  }
}
//定时器 0 中断操作
void timer0(void) interrupt 1
{
  ET0=0;
  TR0=0;
  TH0=0x8a;                                    //重新装载定时器 0
  TL0=0xd0;
  TR0=1;
  if(++counter==50)                            //保证 LED 流水灯有足够延时
  {
    counter=0;
```

```
        if(led==0)
          led=0xff;
        else
          led<<=1;
    }
    P0=led;
    ET0=1;
}
//串口发送一个字符
char putchar(char c)
{
    SBUF=c;
    if(c=='\n')SBUF = 0x0D;
    while(! TI);
    TI=0;
    return (c);
}
//main()函数初始化、无限循环
main()
{
    SCON=0x50;
    PCON |=0x80;
    TMOD=0x21;
    TH0=0x8a;
    TL0=0xd0;
    TH1=0xf6;
    //注意此处主时钟为 18 MHz,65536 - time * (fsoc/12)=xxxxx=xxxxH
    TL1=0xf6;                                  //设置波特率为 9600 bps,主时钟 18 MHz
    TR0=1;
    TR1=1;
    ET0=1;
    EA=1;
    while(1)
    {
      printf("num = %02bX\n",num++);  //串口打印数据
      delay(1000);
}}
```

关于 Keil μVision2 的使用,读者可以查阅相关资料,或者登录 Keil 公司的官方网站获取更多软件信息。

利用 Keil 公司的 51 单片机开发工具 keil μVision2 建立工程，编写测试程序，编译工程，输出以 Intel HEX 文件格式保存下来的 hex 文件。关于 Intel HEX 文件是由一行行符合Intel HEX 文件格式的数据构成的 ASCII 码文本文件。Intel HEX 可由任意数量的十六进制记录组成。每个记录包含 5 个域，它们排列格式为":LLAAAATT[DD...]CC"，每一组字母对应一个不同的域，每一个字母对应一个十六进制编码的数字。每一个域至少由两个十六进制编码数字组成，它们构成一个字节，就像以下描述的那样(每个 Intel HEX 记录都由冒号开头)：

(1)LL 是数据长度域，它代表记录当中数据字节(DD)的数量。

(2)AAAA 是地址域，它代表记录当中数据的起始地址。

(3)TT 是代表 HEX 记录类型的域，它可能是以下数据当中的一个：

00：数据记录，01：文件结束记录，02：扩展段地址记录，04：扩展线性地址记录。

(4)DD 是数据域，它代表一个字节的数据。一个记录可以有许多数据字节，记录当中数据字节的数量必须和数据长度域(LL)中指定的数字相符。

(5)CC 是校验和域，它表示这个记录的校验和。校验和的计算是通过将记录当中所有十六进制编码数字对的值相加，以 256 为模进行补足。

根据上述提及的 .hex 文件数据格式和文件规则，下面来分析一下编译 mc8051.c 文件后生成 .hex 文件的内容，这里给文件的部分数据为

```
:10051B0058636F756E74203D2025303262580A0087
:03000300011DFFDD
:1004F700EF1FAC0670011E4C70F6900000AF82AE85
:1005070083E4FCFD7B407A9CF9F8D312036A40032D
...
:1003D800FAE6FB0808E6F925F0F618E6CA3AF62226
:1003E800D083D082F8E4937012740193700DA3A3A4
:1003F80093F8740193F5828883E4737402936860B8
:06040800EFA3A3A380DFB7
:00000001FF
```

选择 .hex 文件的第一行数据分析":10051B0058636F756E74203D2025303262580A0087"，此行数据意义代表如下：

(1)：冒号代表 HEX 文件数据起始；

(2)10 表示本行当中数据字节的数量；

(3)051B 表示下载有效数据到存储器地址；

(4)00 是记录类型，即数据记录；

(5)58636F756E74203D2025303262580A00 有效数据，总共 16 个十六进制字符；

(6)87 即本行数据校验和。

以上是对 .hex 文件的数据内容分析结果，读者可以自行分析后面的数据。细心的读者也许会发现将以上的十六进制数据转换成二进制数，就是程序的机器码，汇编指令对应的机器码和 8051 核定义的机器码完全相同。

接下来，需要将生成的 .hex 文件更新到 8051 核的 ROM 模块中，除了将 .hex 文件初始

化到 ROM 外，可以用在线工具 In-System Memory Content Editor 对 FPGA 芯片内部的 ROM 进行读写操作，这样可以观察到 ROM 内部情况，同时还可以快速地更新程序到 8051 核，避免了对 8051 核的再次编译，节约时间。

在 In-System Memory Content Editor 对话框，如图 10-11 所示，右上侧可以看到下载器信息及使用器件的相关信息，然后点击下载按钮下载 8051 核 .sof 文件到 FPGA 芯片；同时，左上侧可以看到 8051 核 ROM 的相关信息，右击 ROM 信息，会有下拉菜单出现，单击"Import Data from File…"选项，选择已编译好的 .hex 文件并导入。最后单击下载文件按钮，这时 .hex 文件便下载到 8051 上运行，整个过程类似 PC 对 51 单片机的程序烧写，但该过程读写数据更快。

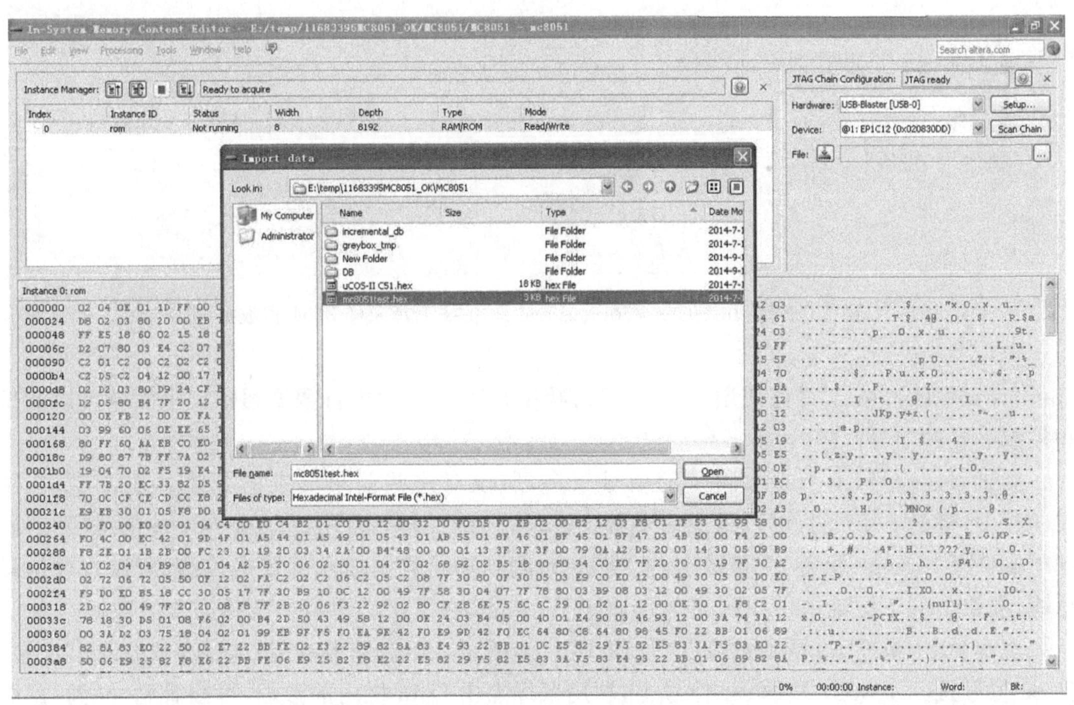

图 10-11　In-System Memory Content Editor 工具下载 .hex 文件

51C 程序下载到 8051 核上，程序在 FPGA 上运行起来，可以观察到 LED 灯会产生变化。LED 灯和 8051 的 P0 相连，为了更好地观察 P0 的状态，可用在线工具 In-System Sources and Probes，有效地控制 8051 核的启动与停止，动态地检测 P0 端口的变化。

可以看到整个 8051 核工程，如图 10-10 所示，probe[8..0]连接 8051 核的 TXD 信号和 P0 端口信号，source[0]接到 8051 核的 rst 信号。单击"Tools"菜单栏下的"In-System Sources and Probes"选项，打开工具 In-System Sources and Probes，弹出对话框，如图 10-12 所示。在对话框的右上侧，单击"Setup…"按钮，弹出"Hardware Setup"对话框默认"Hardware Settings"设置选项卡，选择下拉菜单中的"USB Blaster"并单击，然后单击"ADD Hardware…"按钮，最后单击"Close"按钮，关闭设置选项卡。这时，可以看到在对话框的右上侧"Hardware"栏出现 USB-Blaster 等字样，同时在下一栏显示开发板上的 FPGA 型号，这表明 In-System

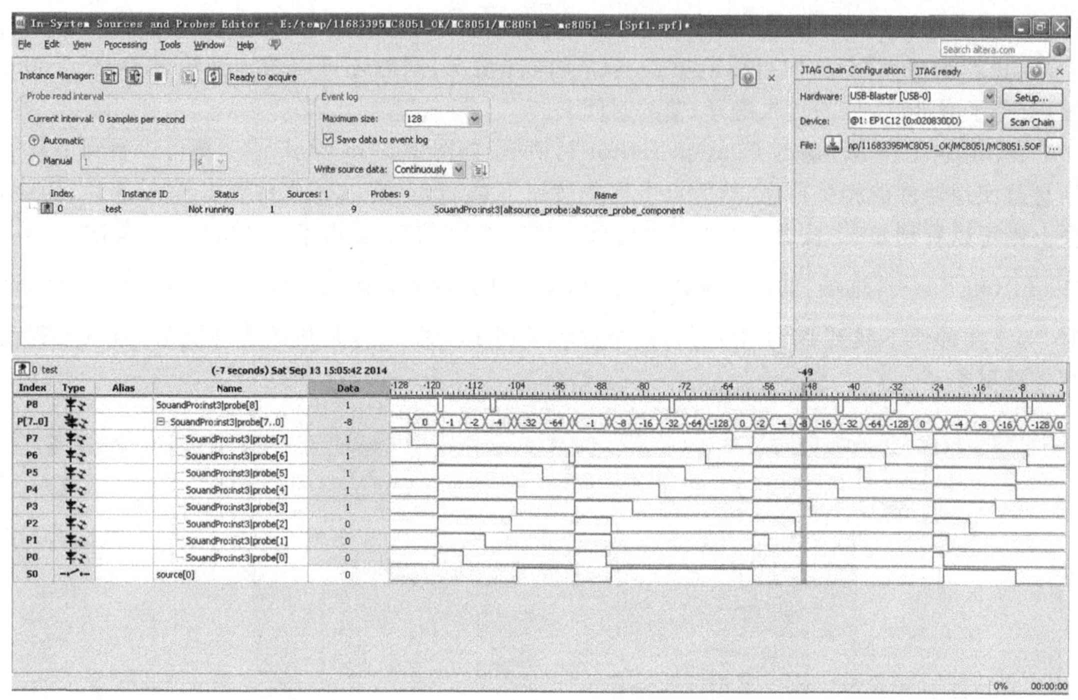

图 10－12　In-System Sources and Probes 工具采样 8051 核数据

Sources and Probes 工具可使用 JTAG 下载器和 FPGA 器件进行数据通信。下面就可以测试和控制设置的信号。

测试添加信号，由于本次测试的信号较多，所以对这些信号类别的归整和重命名，在图 10－12中，将 P0～P7 的信号，组成一个总线，即按住 Ctrl 键，逐一选择 P0～P7 的信号，然后右键单击选择“Group”标签，并更改名称。若想观测到不同数制表示的数据，读者可以右键信号总线，在弹出的下拉菜单中找到 Bus Display Format，在下级菜单栏中选择需要的数据类型。

观察 8051 核的 P0 端口的变化，如图 10－12 所示，S0 信号控制 source 信号的输出状态，且 S0 信号接到了 8051 核的 rst 信号上，利用鼠标点击 Data 栏内的数据，可以看到‘0’到‘1’或‘1’到‘0’的逻辑电平变换，也就是控制 8051 核的运行状态，即复位或者启动。这里通过控制 rst 信号的逻辑电平，观察 8051 核的 P0 端口的变化。

设置采样模式和采样宽度。在图 10－12 窗口栏中，采样模式可以选择一次采样或者连续采样，最大采样宽度可以设为 8、16、32 等采样点，这里为了观测方便设置为 128，data 数据栏的右侧标尺改为－128～0，至此，完成了 In-System Sources and Probes 工具的设置，可以采样测试 8051 核的数据。测试结果如图 10－12 所示。

综上所述，利用 In-System Sources and Probes 工具对 CPU 设计验证以及其他的 FPGA 系统设计测试具有非常高效和直观的特点。根据图 10－12 所示，将 In-System Sources and Probes 采样 8051 核的 P0 数据和实际开发板运行情况相比较，证实采样的数据和开发板上 LED 流水灯显示是一致的。开发板上实际运行效果图略。

当设计最终完成时，烧写程序就不要采用 *.sof 文件、JTAG 模式下载了，这种方式关电设计会消失。通常开发板还有另一个下载口，可能为 AS 模式，选择这个模式下载 *.pof，设计功能会在关电重新开启时仍然存在。

设计说明：

(1)8051核的设计给我们做了一个标准设计、平台设计的示范，简单的理解就是硬件的标准化换取了更多的软件空间。

(2)通过ROM存储器存储程序控制和数据，使得改变设计更加容易，相对于纯硬件连线的控制，改变设计的花费更小。在指令、接口等方面向标准的归化，可以利用现有的软件工具，如编译器、汇编器、仿真工具和调试器等。但是，比起纯硬件连线的控制，这种方法受到存储器读写时间较长的限制，操作速度较慢。使用标准的控制器，使得设计者的设计受限于控制器内置的功能。

习　题

10-1　设计实现既有8051逻辑，又有一般数字逻辑的电路。

10-2　学习8051设计，参阅资料，自行设计一个简易CPU。

10-3　尝试2个以上CPU的系统实现。

扩展学习与总结

设想一种应用，在FPGA开发版上，采用VHDL语言设计的逻辑、嵌入8051核/NIOS核、嵌入8255模块等，完成具有多种输入输出外设的可协同工作的系统，如扩展红外/手机遥控，控制颜色传感器TCS3200D、温湿度传感器DHT11、温度传感器DS1820、加速度传感器MMA7455L，有LCD/TFT显示，同时可通过RS232连接的上位机实时监测与控制，尝试嵌入操作系统如μC/OS-Ⅱ、算法应用和通信协议应用。

附录 A　FPGA 文档指南与规范说明

本规范的目的：

(1)规范整个设计，实现开发的合理性、一致性、高效性。

(2)形成风格良好和完整的文档。

(3)实现在 FPGA 不同厂家之间以及从 FPGA 到 ASIC 的顺利移植。

FPGA 器件以 Altera 的为主，工具组合为 Modelsim ＋ Quartus Ⅱ。

A.1　文件头

文件头有必须包含注释内容：正确的版权信息和声明。如果需要，它可以包括简短说明，组织的名字，设计工程师的名字，设计项目的名字，文件创建和最后修改的时间，文件或代码修改的历史，用到的工具及其版本，电子邮箱、版本以及对该文件进行更改的列表。

文件头最重要的问题是版权问题。源文件和脚本文件属于知识产权。

A.2　文件组织与目录结构

虽然有些 FPGA 编译器不要求模块名与文件名一致，但是通常仍坚持模块名与文件名相匹配。按照这一原则，可以很容易处理和管理项目。

FPGA 开发环境默认将所有文件放在当前工作目录中，通常要自己利用文件的工具将文件规划好分类按树形存放。采用合理、条理清晰的设计目录结构有助于提高设计的效率、可维护性。建议采用如图 A-1 和 A-2 所示的目录结构或者是统一采用本书 3、8 章中介绍的目录结构。

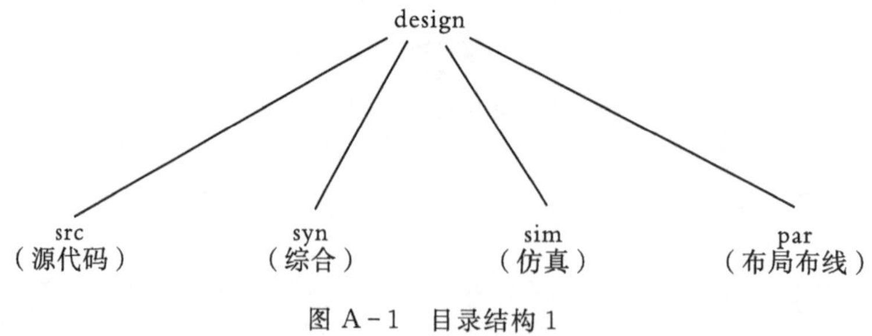

图 A-1　目录结构 1

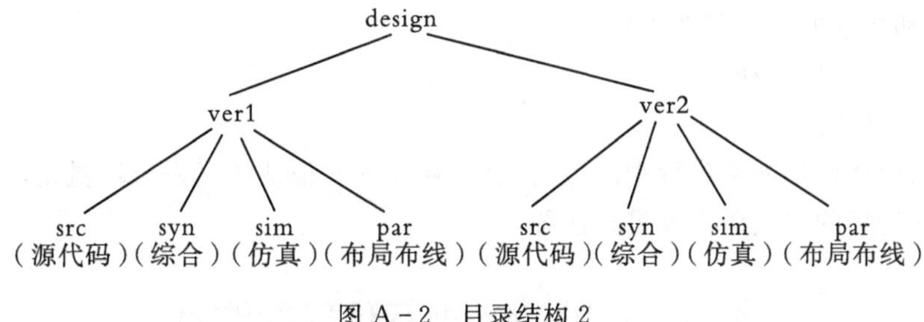

图 A-2　目录结构 2

A.3　文件名和目录名

虽然 Windows 和 Linux 操作系统允许文件和目录名中包含空格和其他特殊字符，但使用空格会导致一些工具软件和脚本语言出现混乱。

Linux 的命令行和 shell 用空格分割字段值，而不是作为文件和目录名的一部分。

文件名中包含空格和特殊字符还可能在文件要通过多个 subshell 和 pipe 时不工作。

文件名和目录名命名规则如下：

(1)文件和目录名中不包含空格和其他特殊字符。

(2)文件和目录名中应只使用字母、数字和下划线字符。

(3)应尽可能给出有助于描述文件内容并且是唯一的文件和目录名。

A.4　大写和小写

Verilog 和 System Verilog 语言中的网表、变量、模块、实例和其他名称是区分大小写的。

VHDL 是不区分大小写的，综合工具默认执行这种规则，一些工具提供一些灵活性。如 XST 提供- case 选项，它决定写入网表的名称是否使用大写或者小写，或者维持源代码的大小写不变。

建议命名参数、宏、常量、类属和枚举类型值使用大写字母；模块、文件、函数、实例和任务的名称可以是小写字母。

A.5　注释

为了提高代码的可读性，并帮助了解设计的意图，鼓励使用有意义的标识，注释的一致性是一个重要的要求。

加入详细、清晰的注释行可以增强代码的可读性和可移植性，要对并非显而易见的内容进行注释。一些开发单位对注释量有明确规定，如注释内容占代码篇幅不应少于 30%。

详细注释具体要求如下：

(1)体现关键功能/性能的程序段。

(2)使用了技巧的程序段。

(3)有独到之处和创新的程序段。

(4)信号和更新的信号要做注释。

(5)端口信号要做注释。

(6)模块要做注释。

在设计课程代码学习的过程中的系列报告,还要求对语法语义进行增量型注释,即每出现前面报告没有的新的语法语义要进行注释。

A.6 使用 Tab 进行代码的缩进

用 Tab 进行代码的缩进。通常 Tab 相当于 4 个字符。

A.7 换行符

Windows 操作系统代码编辑器使用回车和移行 (CR/LF)字符表示换行,UNiX/Linux 使用移行 (LF)字符表示换行。有些编译器或工具提供 2 者的转换。

A.8 限制行宽

大部分终端、编辑器和打印机的标准行宽为 80 个字符。对于不同系统,行宽选择 80 个字符,以使代码更具可读性。

A.9 标识符

流行的 HDL 编码的风格如下:

(1)大小写混写,没有下划线。每一个字从大写字母开始,连续两个或两个以上的大写字母是不允许的。

(2)全部小写,字与字之间使用下划线。

这种风格更加普遍。大多数工具和 IP 核供应商都采用这种风格。如标识符采用即在单词之间以"_"分开,max_delay、data_size 等。对于设计者来说,重要的是在所有开发工具和不同项目都按照同一种风格来书写代码。

(3)采用有意义的、能反映对象特征、作用和性质的单词命名标识符,以增强程序的可读性。

(4)为避免标识符过于冗长,对较长的单词应当采用适当的缩写形式,如用'buff'代替"buffer","ena"代替"enable","addr"代替"address"等。

A.10 转义标识符

虽然 HDL 通过加反斜线"\"对空格或 tab 字符提供了转义标识符,但是 FPGA 设计者应该避免使用转义标识符,因为它会降低代码的可读性,并导致很难找到错误,如键入错误的尾随空格。

A.11　名称前缀或后缀

使用特定含义的名称前缀对信号、寄存器或其他标识符进行分类。例如所有存储器信号用“mem_”作为名称前缀。

使用后缀名用来为信号、寄存器或其他标识符提供相关的额外信息，见表 A-1。

表 A-1　后缀的例子

后缀	描述
_p,_n	表示正极性或负极性
_ff,_q	寄存器输出
_c	由组合门驱动的信号
_cur	现态
_next	次态
_tb	测试平台

1. 时钟名称

时钟的名称要能描述出对象的特征，包括频率和其他特征，如单端或差分时钟。作为惯例，将 clk 作为时钟名称的一部分。如 clk50，单端 50 MHz 时钟；clk_200_p，clk_200_n，200 MHz差分时钟；clk_en，时钟使能；clk_333_mem，333 MHz 存储器控制时钟。

2. 复位信号名称

复位信号名称可以包括复位极性(高或低有效)、同步或异步、全局或局部信息。

例如 reset，phy_reset，高电平有效复位；reset_n，reset_b，低电平有效；rst_async，异步复位。

3. 端口名称

例如：

add_i 输入，write_enable_o 输出，data_io 双向，datai 输入，configb 双向。

A.12　空行和空格

(1)适当地在代码的不同部分中插入空行，避免因程序拥挤不利阅读。

(2)在表达式中插入空格，避免代码拥挤。通常赋值符号两边要有空格；双目运算符两边要有空格；单目运算符和操作数之间可没有空格。具体示例如下：

```
a  <=  b;
c  <=  a  +  b;
If (a  =  b) Then ...
a  <=  Not a  And  c;
```

A.13 对齐和缩进

(1)不要使用连续的空格来进行语句的对齐。

(2)采用制表符 Tab 对语句对齐和缩进,Tab 键采用 4 个字符宽度,可在编辑器中设置。各种嵌套语句尤其是 If... Else 语句,必须严格的逐层缩进对齐。

A.14 参数化设计

为了源代码的可读性和可移植性起见,不要在程序中直接写特定数值,在 Verilog 语言中,尽可能采用 define 语句或 parameter 语句定义常数或参数;在 VHDL 语言中尽可能采用 Constant 语句或 Generic 语句定义常数或参数。

A.15 可综合设计

用 HDL 实现电路,设计人员对可综合风格的 RTL 描述的掌握不仅会影响到仿真和综合的一致性,而且它也是逻辑综合后电路可靠性和质量好坏最主要的因素,对此应当予以充分的重视。

A.16 使用预编译库

在进行功能仿真和后仿真时都需要某些行为仿真模型和门级仿真模型,为避免在不同的设计目录中多次编译这些模型,应当采用一次编译,多次使用的方法。

A.17 逻辑仿真

考虑到性能和易用性,首选的逻辑仿真器是 Mentor Graphics 的 Modelsim。

A.18 测试程序(test bench)

测试程序对于设计功能和时序的验证有着举足轻重的影响,测试激励的完备性和真实性是关键所在,有以下原则须遵循:

(1)测试激励输入和响应输出采集的时序应当兼顾功能仿真(无延时)和时序仿真(有延时)的情况。

(2)对于周期较多的测试,为提高效率,尽可能采用程序语句来判断响应与标准结果是否一致,给出成功或出错标志,而不是通过观察波形来判断。

(3)采用基于文件的测试是很好的办法,即由 Matlab 或 spw 等系统工具产生测试数据,测试程序将其读入产生激励,再把响应结果写入到文件,再交给上述工具进行处理或分析。

(4)仿真器支持几乎所有的 HDL 语法,而不仅仅是常用可综合的 RTL 的描述,应当利用这一点使测试程序尽可能简洁、清楚，篇幅长的要尽量采用 task 来描述。

A.19　逻辑综合的一些原则

(1)HDL 代码综合后电路质量的好坏主要取决于三个方面：RTL 实现是否合理、对厂家器件特点的理解和对综合器掌握的程度。

(2)当出现综合结果不能满足约束条件时，不要急于修改设计源文件，应当通过综合器提供的时序和面积分析命令找出关键所在，然后更改综合控制或修改代码。

A.20　大规模设计的综合

1. 分块综合

当设计规模很大时，综合也会耗费很多时间。当设计只更改某个模块时，可以分块综合。如有设计 top.vhd 包含 a.vhd 和 b.vhd 两个模块，当只修改 a.vhd 时，可以先单独综合 b.vhd，输出其网表 b.edf，编写一个 b 模块的黑盒子接口 b_syn.vhd，每次修改 a.vhd 后只综合 top.vhd，a.vhd，b_syn.vhd，将综合后的网表和 b.edf 送去布线，可以节约综合 b 模块的时间。

2. 采用脚本命令

当设计规模比较大时，综合控制也许会比较复杂，可以考虑采用脚本控制文件的方式进行综合控制，Modelsim 和 Quartus 都支持 TCL(Tool Command Language)语言，采用脚本控制可以提供比图形界面更灵活和更方便的控制手段。

A.21　必须重视工具产生的警告信息

综合工具对设计进行处理可能会产生各种警告信息，有些是可以忽略的，但设计者应该尽量去除，不去除必须确认每条警告的含义，避免因此使设计的实现产生隐患。这个原则对仿真和布局布线同样适用。

A.22　调用模块的黑盒子(Black box)方法

使用黑盒子方法的原因主要有以下两点：

一是，HDL 代码中调用了一些 FPGA 厂家提供的模块(如 Altera 的 LPM 模块)或第三方提供的 IP，这些模块不需要综合，而且有些综合器也不能综合(如 FPGA Compiler II/FPGA Express 可以综合包含 LPM 的代码而 LeonardoSpectrum 不能)。

二是，方便代码的移植，由于厂家提供的模块或第三方提供的 IP 通常都是与工艺有关的，直接在代码中调用的话将不利于修改，影响代码移植。

下面以调用 Altera 的 LPM 库中的乘法器为例来说明。调用这样一个模块需要这样一个文件：mult8x8.v(可由 QuartusII 的 MegaWizer Plug-in Manager 产生)，代码如下：

```
// mult8x8.v
module mult8x8 (dataa, datab, result);
input [7:0] dataa;
input [7:0] datab;
output [15:0] result;

// exemplar translate_off
// synopsys translate_off
lpm_mult    lpm_mult_component(
        .dataa(dataa),
        .datab(datab),
        .aclr(1'b0),
        .clock(1'b0),
        .clken (1'b0),
        .sum(1'b0),
        .result(result)
        );
    defparam
        lpm_mult_component.lpm_widtha          = 8,
        lpm_mult_component.lpm_widthb          = 8,
        lpm_mult_component.lpm_widths          = 16,
        lpm_mult_component.lpm_widthp          = 16,
        lpm_mult_component.lpm_representation = "SIGNED",
// exemplar translate_on
// synopsys translate_on

endmodule
```

注意上述的代码有两对编译指示：

// exemplar translate_off 和// exemplar translate_on (LeonardoSpectrum 支持)

// synopsys translate_off 和 // synopsys translate_on(LeonardoSpectrum 和 FPGA Compiler Ⅱ都支持)

对于相应的综合器，在这些编译指示中间的语句将会被忽略，可以看到在综合过程中模块 mult8x8 实际变成了一个只有 I/O 定义的空盒子(即 Black box)，所以该部分的代码没有连接，在 Quartus Ⅱ布局布线的时候，LPM 模块的代码才连接到整个设计，在仿真的时候，编译指示不影响模块的完整性。

附录 B VHDL 保留字

VHDL 1076－1987 版保留字：

ABS
Access
After
Alias
All
And
Architecture
Array
Assert
Attribute
Begin
Block
Body
Bus
Case
Component
Configuration
Constant
Disconnector
Downto
Else
Elsif
End
Entity
Exit
File
For
Function
Generate
Generic
Guarded
If
In
Inout
Is
Label
Library
Linkage
Loop
Map
Mod
Nand
New
Next
Nor
Not
Null
Of
On
Open
Or
Others
Out
Package
Port
Procedure
Process
Range
Record
Register
Rem
Report
Return
Select
Severity
Signal
Subtype
Then
To
Transport
Type
Units
Until
Use
Variable
Wait
When
While
With
Xor

VHDL 1076－1993 版增加的保留字：

Group，Inpure，Ineratial，Postponed，Pure，Reject，Rol，Ror，Shared，Sla，Sll，Sra，Srl，Unaffected，Xnor

附录C　VHDL 1993版与2008版的特点

C.1　VHDL 1993版特点

(1)语法结构更统一，更规范。举例如下：

```
Entity a Is                          Procedure  p(……) Is
End Entity a;                        End Procedure p;
Architecture  a_arch Of  a Is        Process (……) Is
End Architecture a;                  End Process ;
Component  c Is                      Function f(……) Is
End Component  c;                    End Function f;
Package pk Is                        Configuretion  a_conf Of  a Is
End Package pk ;                     End Configuretion a_conf;
```

在87版中，上述结构都可以直接结束，93版标识更加清楚，结构好似更加繁琐，但是这在复杂设计中却十分必要。

(2)扩展标号标注。在任何顺序语句前面都可以加标号进行标注，例如

```
l1: If a=b  Then
      c<=d;
End If l1;
```

(3)直接说明。Component语句、实体、构造体或配置的直接说明，例如

```
Architecture direct_Nand Of Nand_2 Is
Begin
  G:Entity Work Nand_2
  Generic Map(thLH=>13 ns,tpHL=>14ns);

  G:Configuration Nand_2_Final
  Genric Map(tpLH=>13 ns,tpHL =>15 ns)
  Port Map(i1=>a,I2=>b,o=>c);

End Architecture direct_Nand ;
```

(4)Generate语句可包含端口说明部分。例如

```
Label:If N Mod 2=1 Generate
  Instance: Component_Name Port Map(t1,t2)
```

```
End Generate;
```

(5)在端口映射中使用常量表达式。例如

```
M1:mux Port Map
  (sel=>To_MVL(code) , d0=>To_MVL(bus0),
    d0=>To_MVL(bus1), To_Bit(2)=>ctrl);
```

这里 To_MVL 函数返回的值是常量。

(6)定义了共享变量．共享变量可以使变量在进程、子程序间 Variable 传递。

```
Shared Variable 变量名:子类型名[:=初值];
Architecture sample Of test Is
Shared Variable notclk: Std_Logic;                --在结构体说明位置声明
Signal clk:Std_Logic;
Begin

  p1: Process(clk) Is
  Begin
    If (clk'Event And clk='1') Then
        notclk:='0';
    End If;
  End Process p1;

  p2: Process(clk) Is
  Begin
    If (clk'Event And clk='1') Then
        notclk:='1';
    End If;
  End Process p2;
End Architecture sample;
```

共享变量不能在进程和子程序说明域中使用。

(7)可对信号赋无效值，表明不改变当前驱动器的输出值。

```
A<= Null;
```

执行该语句，A 信号将不会发生变化。

(8)定义了文件操作。

(9)增加了标识、新的保留字、属性。增加了预定义属性，例如

```
'Ascending,'Image,'Value,'Driving_Value
'Imple_name,'Instance_name,'Path_name
```

(10)定义了新属性 Foreign。该语句用于构造体和子程序中连接非 VHDL 语句

①用于结构体。

```
Entity nand Is
Generic(N:Positive:=2);
```

```
  Port(input:In Bit_Vector(N Downto 1);
       Output:Out Bit);
End Entity nand;
Architecture Nonvhdl Of Nand Is
Attribute Foreigen Of Nonvhdl:Architecture Is
"Nonvhdl _nand(a,b,c)";
Begin
End Architecture Nonvhdl;
```

②用于子程序,用于过程。

```
Procedure print_line(a:String;) Is
Attribute Foreigen Of print_line : Procedure  Is
"print_line(a)";
End Procedure print_line;
```

③用于子程序,用于函数。

```
Package p Is
Function atoi(s:String) Is
Return Integer;
Attribute Foreign Of  atoi:Function Is "/bin/ sh atoi";
End Package p;
```

(11)函数可以设计成纯函数和非纯函数。

①纯函数,例如:

```
Pure Function “And”(L,R,X01Z)
Return
Begin
Return Table_And(L,R);
End Function "And";
```

一个纯函数指的是在所带参数值相同时,调用后应返回一个相同的值。

②非纯函数,例如:

```
Impure Function Random Return Real;
```

Random 是一个产生随机值的函数,不同层次调用该函数就会得到不同的返回值。该函数没有输入参数。

(12)定义了 Group(组)。一个组是一些以定义项目的集合。

Group 组名:组属性名(项目,项目,……);

项目可以是一个标号名,可以是信号名,可以是简单的值。属性名相同的组,组结构相同。

(13)延迟过程。一个过程可以表示为延时过程,该过程仅仅在某一时段结束时才执行,也就是在 1 个时段的所有 Δ 延时之后才执行,延时的标识是“Postponed”。例如

```
P1:Postponed nand_1 (a,b:In Bit; c: Out Bit);
```

(14)信号延时可指定脉冲宽度。在信号延时表达式中 Reject 用来限制脉冲宽度。

```
dout1<=a And b After 5 ns;
dout2<= Reject 3 ns Inertial a And b After 5 ns;
--脉冲宽度限制为 3 ns
```

(15)重新定义了文件。

……93 版的特点不再一一例举。

C.2　VHDL 2008 版的特点

(1)注释。在 VHDL 2008 版以前,注释是逐行的。现在可以是多行的。

例如,“--”表示注释,从“--”开始到行末结束,编译器不对注释编译。在 VHDL 2008 版以后,注释可以是多行的。

例如:

```
/*
    Descriptive comment goes here.
    It can span multiple lines.
*/
```

(2) If…Generate 语句。原来 If…generate 语句不能有 Else 语句,现在可以有 Else 语句。

```
<generate_label>:                  --生成语句标号
If <optional_label1>: <condition> Generate
    --declarations 声明语句
    Begin
        --Concurrent Statement(s)并行语句
    End <optional_label1>;
Elsif <optional_label2>: <condition> Generate
    --declarations
    Begin
        --Concurrent Statement(s)
    End <optional_label2>;
Else <optional_label3>: Generate
    --declarations
    Begin
        --Concurrent Statement(s)
    End <optional_label3>;
End Generate;
```

(3)增加了 Case…Generater 语句。具体格式:

```
/* 在 VHDL Case…Generate 语句表达式的所有选择必须是常数,也必须是唯一的、全覆盖的,或者它必须包含一个 Others 语句 */
<generate_label>:
    Case <expression> Generate
```

```
    When <optional_label1>: <constant_expression> =>
        --declarations
        Begin
            --Concurrent Statement(s)  并行语句
    End <optional_label1>;
    When <optional_label2>: <constant_expression> =>
        --declarations
        Begin
            --Concurrent Statement(s)
    End <optional_label2>;
    When <optional_label3>: Others =>
        --declarations
        Begin
            --Concurrent Statement(s)
    End <optional_label3>;
End Generate;
```

(4)支持更简洁的敏感表。

```
<optional_label>:
    Process(All) Is
        --Declaration(s)
    Begin
        --Sequential Statement(s)
    End Process;
```

可以看到敏感表可以是 All。

(5)非限制的数组元素(Unconstrained Array Elements)。

```
--Examples
--声明具有非限制元素的非限制数组类型
Type my_type1 Is Array(Natural Range <>) Of Std_Logic_Vector;
--3 层深度的无约束数组
Type my_type2 Is Array(Natural Range <>) Of my_type1;

--部分限制的类型/子类型
Type my_type3 Is Array(5 To 9) Of my_type1(8 Downto 4);
--Std_Logic_Vector 仍然没有限制
Type my_type4 Is Array(4 To 6) Of my_type1(Open)(8 Downto 4);
Subtype my_subtype1 Is my_type1(Open)(10 To 12);
Subtype my_subtype2 Is my_type2(3 Downto 0)(Open)(3 To 5);
--全限制的类型/子类型
Type fully_constrained1 Is Array(7 Downto 3) Of
```

```
    my_type3(Open)(Open)(5 To 9);
Type fully_constrained2 Is Array(0 To 3) Of
    my_type1(7To 9)(5 Downto 2);
Subtype fully_constrained3 Is my_subtype1(4 To 6);
```

(6)匹配算符 Matching Operators。

```
? =
--matching equality:        等于匹配
--defined for STD_ULOGIC and BIT, and their one dimensional
--arrays ( a single '-' is considered to be equal to every scalar
--value of type BIT or STD_ULOGIC)
? /=--matching inequality:不等于匹配
--defined for STD_ULOGIC and BIT, and their one dimensional
--arrays ( a single '-' is considered to be equal to every scalar
--value of type BIT or STD_ULOGIC)
```

(7)增加了增强型的位-串文字的说明。

```
8B"XX_01LH" --"00XX01LH"
```

在进制符号前加数字表示位数，VHDL 2008 的规则可以自动在前面添‘0’，对齐位数。

```
10D"56"--"0000110110"
5UB"0010X1"--"010X1"
```

5 位无符号二进制“0010X1”即是“010X1”，自动去前面‘0’，对齐位数。

```
5SX"FW" --"1WWWW"
5UX"17"--"10111"
```

5 位无符号二进制用十六进制表示为“17”。

这里不再一一例举，请自行查阅学习。

附录 D　Ieee 库类型转换函数表

函数名	功能
程序包：Std_Logic_1164	
To_StdLogicVector(Bit_Vector)	Bit_Vector→Std_Logic_Vector
To_StdUlogicVector(Bit_Vector)	Bit_Vector→Std_uLogic_Vector
To_BitVector(Std_Logic_Vector)	Std_Logic_Vector→Bit_Vector
To_BitVector(Std_Ulogic_Vector)	Std_uLogic_Vector→Bit_Vector
To_Bit(Std_Logic)	Std_Logic→Bit
To_Bit(Std_Ulogic)	Std_uLogic→Bit
To_Std_Logic(Bit)	Bit→Std_Logic
To_Std_Ulogic(Bit)	Bit→Std_uLogic
程序包：Numeric_Std	
To_Signed(Integer,位长)	Integer→有符号数 Signed
To_Unsigned(Integer,位长)	Integer→无符号数 Unsigned
To_Integer (Signed)	Signed→Integer
To_Integer (Unsigned)	Unsigned→Integer
程序包：Numeric_Bit	
To_Integer (Signed)	Signed→Integer
To_Integer (Unsigned)	Unsigned→Integer
To_Signed(Integer,位长)	Integer→有符号数 Signed
To_Unsigned(Integer,位长)	Integer→无符号数 Unsigned
Signed(Bit_Vector)	Bit_Vector→Signed
Bit_Vector(Signed)	Signed→Bit_Vector
Bit_Vector(Unsigned)	Unsigned→Bit_Vector
程序包：Std_Logic_Unsigned，Std_Logic_Signed	
Conv_Integer(To_Std_logic_Vector)	Std_Logic_Vector →Integer
程序包：Std_Logic_Arith	
Conv_Std_Logic_Vector(Integer，位长)	Integer→Std_Logic_Vector
Conv_Std_uLogic_Vector(Integer，位长)	Integer→Std_uLogic_Vector
Conv_Integer(Signed 或 Unsigned)	Signed 或 Unsigned→Integer
Conv_Signed(Unsigned，位长)	Unsigned→Signed
Conv_Unsigned(Signed，位长)	Signed→Unsigned
Conv_Signed(Std_uLogic，位长)	Signed→Std_Ulogic
Conv_Unsigned(Std_Logic，位长)	Signed→Std_Logic

注：Std_uLogic 类型有些综合器不支持。

参考文献

[1] 梁松海. 数字系统设计与 VHDL(英文版)[M]. 北京:电子工业出版社, 2010.

[2] [美] Kenneth L Short. VHDL 大学实用教程[M]. 乔庐峰,等,译. 北京:电子工业出版社, 2011.

[3] 吴继华. Altera FPGA/CPLD 设计高级篇[M]. 2 版. 北京:人民邮电出版社,2011.

[4] 侯伯亨,刘凯,顾新. VHDL 硬件描述语言与数字逻辑电路设计[M]. 3 版. 西安:西安电子科技大学出版社,2009.

[5] 潘松. EDA 技术实用教程[M]. 4 版. 北京:科学出版社, 2010.

[6] 褚振勇,翁木云,高楷娟. FPGA 设计及应用[M]. 3 版. 西安:西安电子科技大学出版社, 2011.

[7] 林明权. VHDL 数字控制系统设计范例[M]. 北京:电子工业出版社,2003.

[8] 陈金鹰. FPGA 技术及应用[M]. 北京:机械工业出版社,2015.

[9] [美]Evgeni Stavinov. FPGA 高手设计实战真经 100 则[M]. 朱江,等,译. 北京:电子工业出版社, 2013.

[10] www. Altera. com. cn.

[11] 华为公司. FPGA 设计流程指南[EB/OL]. http://download. csdn. net/download/yuankang/1954511.

[12] Altera 中国区授权代理——骏龙科技有限公司(技术支持部). 使用 Cyclone 器件中的 PLL(节选)[EB/OL]. http://www. cytech. com.

[13] [英]Mark Zwoliski. VHDL 数字系统设计[M]. 李仁发,等,译 . 北京:电子工业出版社,2004.

[14] Oregano Systems. MC8051 IP Core User Guide[Z]. Oregano Systems,2002,6.